CONTE

34TH

INTERNATIONAL PHYSICS OLYMPIAD

TAIPEI, TAIWAN

2003

Theoretical Question 1

A Swing with a Falling Weight

A rigid cylindrical rod of radius R is held horizontal above the ground. With a string of negligible mass and length L ($L > 2\pi R$), a pendulum bob of mass m is suspended from point A at the top of the rod as shown in Figure 1a. The bob is raised until it is level with A and then released from rest when the string is taut. Neglect any stretching of the string. Assume the pendulum bob may be treated as a mass point and swings only in a plane perpendicular to the axis of the rod. Accordingly, the pendulum bob is also referred to as the *particle*. The acceleration of gravity is $\vec{g}$.

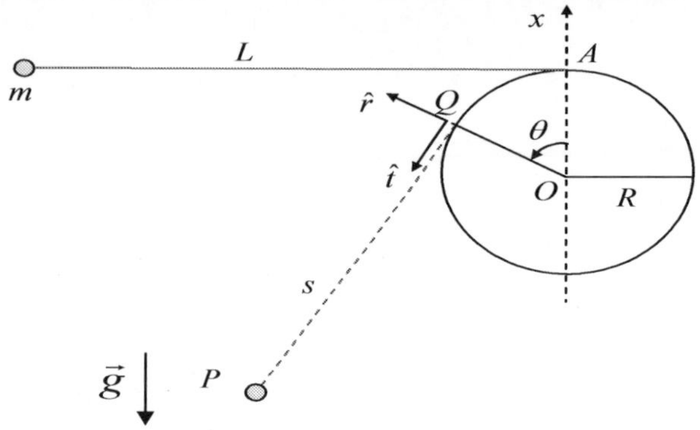

Figure 1a

Let O be the origin of the coordinate system. When the particle is at point P, the string is tangential to the cylindrical surface at Q. The length of the line segment QP is called s. The unit tangent vector and the unit radial vector at Q are given by $\hat{t}$ and $\hat{r}$, respectively. The angular displacement θ of the radius OQ, as measured counterclockwise from the vertical x-axis along OA, is taken to be positive.

When $\theta = 0$, the length s is equal to L and the gravitational potential energy U of the particle is zero. As the particle moves, the instantaneous time rates of change of θ and s are given by $\dot{\theta}$ and $\dot{s}$, respectively.

Unless otherwise stated, all the speeds and velocities are relative to the fixed point O.

Part A

In Part A, the string is taut as the particle moves. In terms of the quantities introduced above (i.e., s, $\theta, \dot{s}$, $\dot{\theta}$, R, L, g, $\hat{t}$ and $\hat{r}$), find:

(a) The relation between $\dot{\theta}$ and $\dot{s}$. [0.5 point]

(b) The velocity $\vec{v}_Q$ of the moving point Q relative to O. [0.5 point]

(c) The particle's velocity $\vec{v}\,'$ relative to the moving point Q when it is at P. [0.7 point]

(d) The particle's velocity $\vec{v}$ relative to O when it is at P. [0.7 point]

(e) The $\hat{t}$ -component of the particle's *acceleration* relative to O when it is at P. [0.7 point]

(f) The particle's gravitational potential energy U when it is at P. [0.5 point]

(g) The speed v_m of the particle at the lowest point of its trajectory. [0.7 point]

Part B

In Part B, the ratio L to R has the following value:

$$\frac{L}{R} = \frac{9\pi}{8} + \frac{2}{3}\cot\frac{\pi}{16} = 3.534 + 3.352 = 6.886$$

(h) What is the speed v_s of the particle when the string segment from Q to P is both straight and shortest in length? (in terms of g and R) [2.4 points]

(i) What is the speed v_H of the particle at its highest point H when it has swung to the other side of the rod? (in terms of g and R [1.9 points]

Part C

In Part C, instead of being suspended from A, the pendulum bob of mass m is connected by a string over the top of the rod to a heavier weight of mass M, as shown in Figure 1b. The weight can also be treated as a particle.

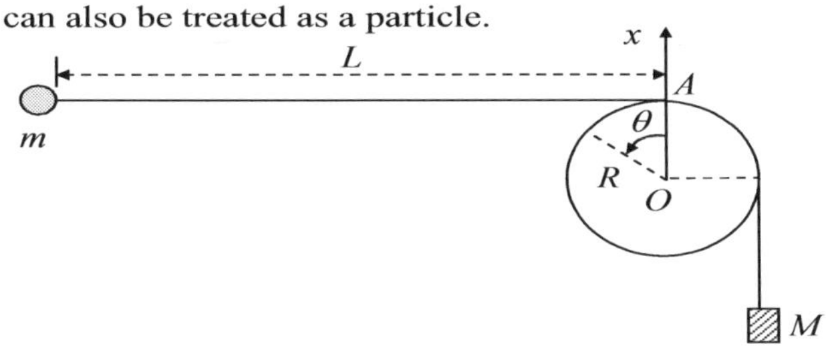

Figure 1b

Initially, the bob is held stationary at the same level as A so that, with the weight hanging below O, the string is taut with a horizontal section of length L. The bob is then released from rest and the weight starts falling. Assume that the bob remains in a vertical plane and can swing past the falling weight without any interruption.

The kinetic friction between the string and the rod surface is negligible. But the static friction is assumed to be large enough so that the weight will remain stationary once it has come to a stop (i.e. zero velocity).

(j) Assume that the weight indeed comes to a stop after falling a distance D and that $(L-D) >> R$. If the particle can then swing around the rod to $\theta = 2\pi$ while both segments of the string free from the rod remain straight, the ratio $\alpha = D/L$ must not be smaller than a critical value α_c. Neglecting terms of the order R/L or higher, obtain an estimate on α_c in terms of M/m. [3.4 points]

Theoretical Question 2

A Piezoelectric Crystal Resonator under an Alternating Voltage

Consider a uniform rod of unstressed length ℓ and cross-sectional area A (Figure 2a). Its length changes by $\Delta\ell$ when equal and opposite forces of magnitude F are applied to its ends faces normally. The *stress* T on the end faces is defined to be F/A. The *fractional change* in its length, i.e., $\Delta\ell/\ell$, is called the *strain* S of the rod. In terms of stress and strain, Hooke's law may be expressed as

$$T = Y\,S \quad \text{or} \quad \frac{F}{A} = Y\frac{\Delta\ell}{\ell} \tag{1}$$

where Y is called the *Young's modulus* of the rod material. Note that a *compressive* stress T corresponds to $F < 0$ and a decrease in length (i.e., $\Delta\ell < 0$). Such a stress is thus negative in value and is related to the pressure p by $T = -p$.

For a uniform rod of density ρ, the speed of propagation of longitudinal waves (i.e., sound speed) along the rod is given by

$$u = \sqrt{Y/\rho} \tag{2}$$

Figure 2a

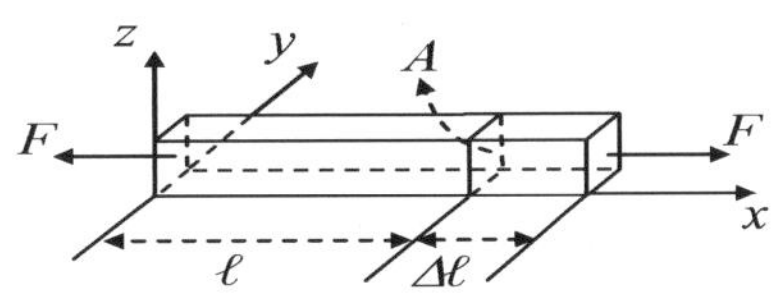

The effect of damping and dissipation can be ignored in answering the following questions.

Part A: mechanical properties

A uniform rod of semi-infinite length, extending from $x = 0$ to ∞ (see Figure 2b), has a density ρ. It is initially stationary and unstressed. A piston then steadily exerts a small pressure p on its left face at $x = 0$ for a very short time Δt, causing a pressure wave to propagate with *speed* u to the right.

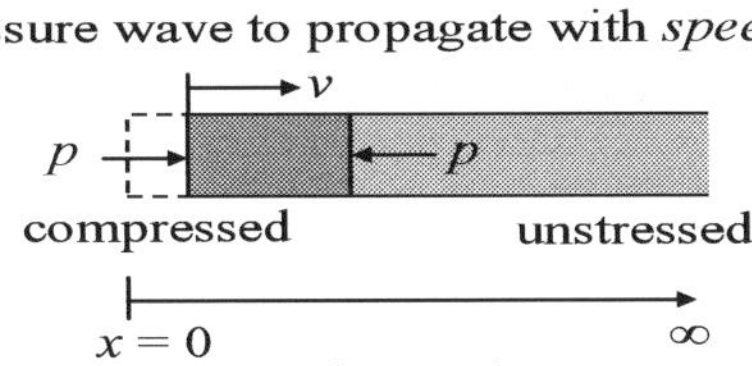

Figure 2b

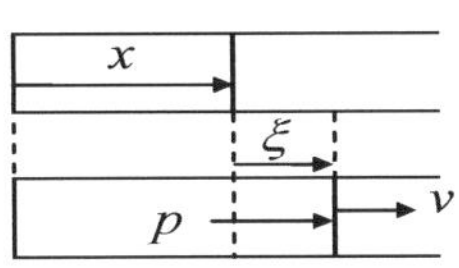

Figure 2c

(a) If the piston causes the rod's left face to move at a constant velocity v (Figure 2b), what are the strain S and pressure p at the left face during the time Δt? *Answers must be given in terms of ρ, u, and v only.* [1.6 points]

(b) Consider a longitudinal wave traveling along the x direction in the rod. For a cross section at x when the rod is unstressed (Figure 2c), let $\xi(x, t)$ be its displacement at time t and assume

$$\xi(x, t) = \xi_0 \sin k(x - ut) \tag{3}$$

where ξ_0 and k are constants. Determine the corresponding velocity $v(x, t)$, strain $S(x, t)$, and pressure $p(x, t)$ as a function of x and t. [2.4 points]

Part B: electromechanical properties (including piezoelectric effect)

Consider a quartz crystal slab of length b, thickness h, and width w (Figure 2d). Its length and thickness are along the x-axis and z-axis. Electrodes are formed by thin metallic coatings at its top and bottom surfaces. Electrical leads that also serve as mounting support (Figure 2e) are soldered to the electrode's centers, which may be assumed to be stationary for longitudinal oscillations along the x direction.

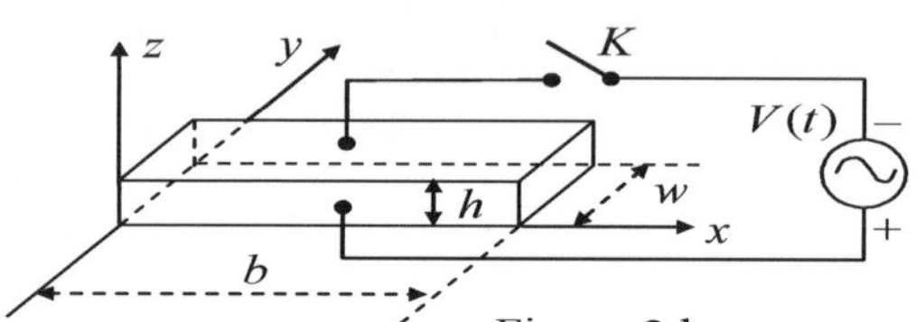

Figure 2d

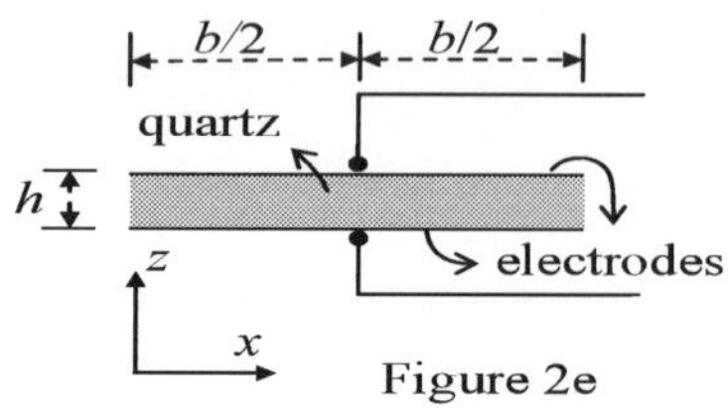

Figure 2e

The quartz crystal under consideration has a density ρ of 2.65×10^3 kg/m^3 and Young's modulus Y of 7.87×10^{10} N/m^2. The length b of the slab is 1.00 cm and the width w and height h of the slab are such that $h << w$ and $w << b$. With switch K left open, we assume only *longitudinal modes* of standing wave oscillation in the x direction are excited in the quartz slab.

For a standing wave of frequency $f = \omega / 2\pi$, the displacement $\xi(x, t)$ at time t of a cross section of the slab with equilibrium position x may be written as

$$\xi(x, t) = 2\xi_0\, g(x) \cos \omega t, \qquad (0 \le x \le b) \tag{4a}$$

where ξ_0 is a positive constant and the spatial function $g(x)$ is of the form

$$g(x) = B_1 \sin k(x - \frac{b}{2}) + B_2 \cos k(x - \frac{b}{2}). \tag{4b}$$

g(x) has the maximum value of one and $k = \omega / u$. Keep in mind that the centers of the electrodes are stationary and the left and right faces of the slab are free and must have zero stress (or pressure).

(c) Determine the values of B_1 and B_2 in Eq. (4b) for a longitudinal standing wave in the quartz slab. [1.2 point]

(d) What are the two lowest frequencies at which longitudinal standing waves may be excited in the quartz slab? [1.2 point]

The *piezoelectric* effect is a special property of a *quartz* crystal. Compression or dilatation of the crystal generates an electric voltage across the crystal, and conversely, an external voltage applied across the crystal causes the crystal to expand or contract depending on the polarity of the voltage. Therefore, mechanical and electrical oscillations can be coupled and made to resonate through a quartz crystal.

To account for the piezoelectric effect, let the surface charge densities on the upper and lower electrodes be $-\sigma$ and $+\sigma$, respectively, when the quartz slab is under an electric field E in the z direction. Denote the slab's strain and stress in the x direction by S and T, respectively. Then the piezoelectric effect of the quartz crystal can be described by the following set of equations:

$$S = (1/Y)T + d_p E \qquad \text{(5a)}$$

$$\sigma = d_p T + \varepsilon_T E \qquad \text{(5b)}$$

where $1/Y = 1.27\times10^{-11}$ m^2/N is the *elastic compliance* (i.e., inverse of Young's modulus) at constant electric field and $\varepsilon_T = 4.06\times10^{-11}$ F/m is the *permittivity* at constant stress, while $d_p = 2.25\times10^{-12}$ m/V is the *piezoelectric coefficient.*

Let switch K in Fig. 2d be closed. The alternating voltage $V(t) = V_m \cos \omega t$ now acts across the electrodes and a *uniform* electric field $E(t) = V(t)/h$ in the z direction appears in the quartz slab. When a steady state is reached, a longitudinal standing wave oscillation of angular frequency ω appears in the slab in the x direction.

With E being uniform, the wavelength λ and the frequency f of a longitudinal standing wave in the slab are still related by $\lambda = u/f$ with u given by Eq. (2). But, as Eq. (5a) shows, $T = YS$ is no longer valid, although the definitions of strain and stress remain unchanged and the end faces of the slab remain free with zero stress.

(e) Taking Eqs. (5a) and (5b) into account, the surface charge density σ on the lower electrode as a function of x and t is of the form,

$$\sigma(x,t) = \left[D_1 \cos k\left(x - \frac{b}{2}\right) + D_2 \right]\frac{V(t)}{h},$$

where $k = \omega/u$. Find the expressions for D_1 and D_2. [2.2 points]

(f) The total surface charge $Q(t)$ on the lower electrode is related to $V(t)$ by

$$Q(t) = [1 + \alpha^2 (\frac{2}{kb}\tan\frac{kb}{2} - 1)]C_0 V(t) \qquad \text{(6)}$$

Find the expression for C_0 and the expression and numerical value of α^2.

[1.4 points]

Theoretical Question 3

Part A

Neutrino Mass and Neutron Decay

A free neutron of mass m_n decays at rest in the laboratory frame of reference into three non-interacting particles: a proton, an electron, and an anti-neutrino. The rest mass of the proton is m_p, while the rest mass of the anti-neutrino m_ν is assumed to be nonzero and much smaller than the rest mass of the electron m_e. Denote the speed of light in vacuum by c. The measured values of mass are as follows:

$$m_n = 939.56563\ \text{MeV}/c^2,\ m_p = 938.27231\ \text{MeV}/c^2,\ m_e = 0.5109907\ \text{MeV}/c^2$$

In the following, all energies and velocities are referred to the laboratory frame. Let E be the total energy of the electron coming out of the decay.

(a) Find the maximum possible value $E_{\max}$ of E and the speed v_{m} of the anti-neutrino when $E = E_{\max}$. Both answers must be expressed in terms of the rest masses of the particles and the speed of light. Given that $m_\nu < 7.3\ \text{eV}/c^2$, compute $E_{\max}$ and the ratio v_{m}/c to 3 significant digits. [4.0 points]

Part B

Light Levitation

A transparent glass hemisphere with radius R and mass m has an index of refraction n. In the medium outside the hemisphere, the index of refraction is equal to one. A parallel beam of monochromatic laser light is incident uniformly and normally onto the central portion of its planar surface, as shown in Figure 3a. The acceleration of gravity $\vec{g}$ is vertically downwards. The radius δ of the circular cross-section of the laser beam is much smaller than R. Both the glass hemisphere and the laser beam are axially symmetric with respect to the z-axis.

The glass hemisphere does not absorb any laser light. Its surface has been coated with a thin layer of transparent material so that reflections are negligible when light enters and leaves the glass hemisphere. The optical path traversed by laser light passing through the non-reflecting surface layer is also negligible.

(b) Neglecting terms of the order $(\delta/R)^3$ or higher, find the laser power P needed to balance the weight of the glass hemisphere. [4.0 points]

Hint: $\cos\theta \approx 1-\theta^2/2$ when θ is much smaller than one.

Figure 3a

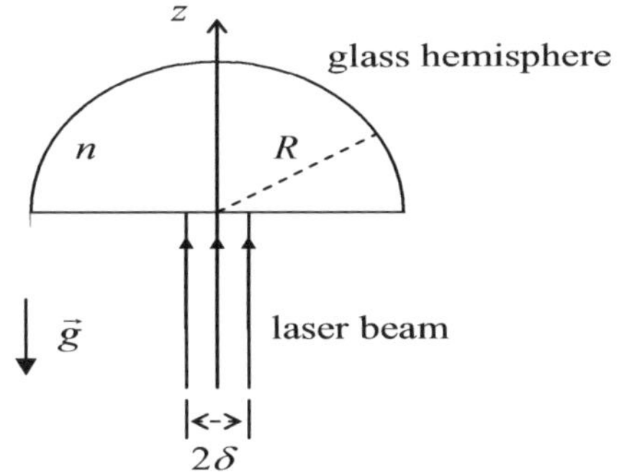

Solution- Theoretical Question 1

A Swing with a Falling Weight

Part A

(a) Since the length of the string $L = s + R\theta$ is constant, its rate of change must be zero. Hence we have

$$\dot{s} + R\dot{\theta} = 0 \qquad \text{(A1)*}[1]$$

(b) Relative to O, Q moves on a circle of radius R with angular velocity $\dot{\theta}$, so

$$\vec{v}_Q = R\dot{\theta}\hat{t} = -\dot{s}\hat{t} \qquad \text{(A2)*}$$

(c) Refer to Fig. A1. Relative to Q, the displacement of P in a time interval Δt is $\Delta\vec{r}\,' = (s\Delta\theta)(-\hat{r}) + (\Delta s)\hat{t} = [(s\dot{\theta})(-\hat{r}) + \dot{s}\hat{t}]\Delta t$. It follows

$$\vec{v}\,' = -s\dot{\theta}\hat{r} + \dot{s}\hat{t} \qquad \text{(A3)*}$$

Figure A1

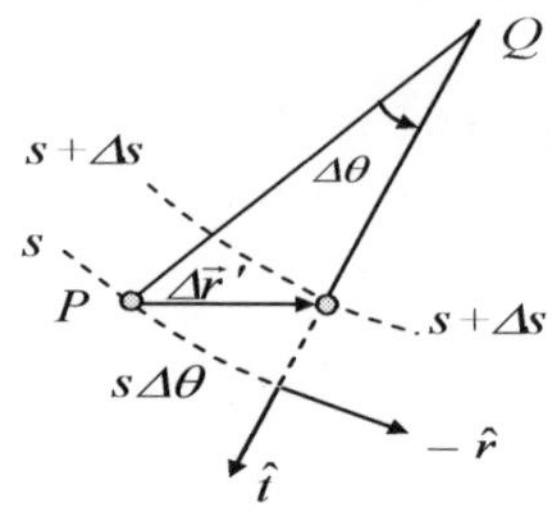

(d) The velocity of the particle relative to O is the sum of the two relative velocities given in Eqs. (A2) and (A3) so that

$$\vec{v} = \vec{v}\,' + \vec{v}_Q = (-s\dot{\theta}\hat{r} + \dot{s}\hat{t}) + R\dot{\theta}\hat{t} = -s\dot{\theta}\hat{r} \qquad \text{(A4)*}$$

(e) Refer to Fig. A2. The $(-\hat{t})$-component of the velocity change $\Delta\vec{v}$ is given by $(-\hat{t})\cdot\Delta\vec{v} = v\Delta\theta = v\dot{\theta}\,\Delta t$. Therefore, the $\hat{t}$-component of the acceleration $\vec{a} = \Delta\vec{v}/\Delta t$ is given by $\hat{t}\cdot\hat{a} = -v\dot{\theta}$. Since the speed v of the particle is $s\dot{\theta}$ according to Eq. (A4), we see that the $\hat{t}$-component of the particle's acceleration at P is given by

$$\vec{a}\cdot\hat{t} = -v\dot{\theta} = -(s\dot{\theta})\dot{\theta} = -s\dot{\theta}^2 \qquad \text{(A5)*}$$

Figure A2

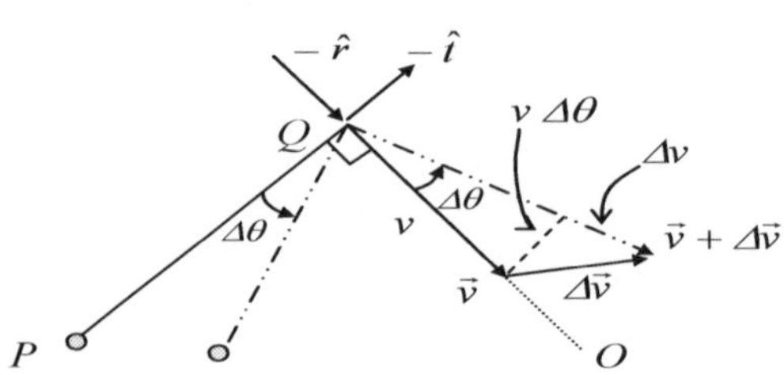

[1] An equation marked with an asterisk contains answer to the problem.

Note that, from Fig. A2, the radial component of the acceleration may also be obtained as $\vec{a}\cdot\hat{r} = -dv/dt = -d(s\dot{\theta})/dt$.

(f) Refer to Fig. A3. The gravitational potential energy of the particle is given by $U = -mgh$. It may be expressed in terms of s and θ as

$$U(\theta) = -mg[R(1-\cos\theta) + s\sin\theta] \quad \text{(A6)*}$$

Figure A3

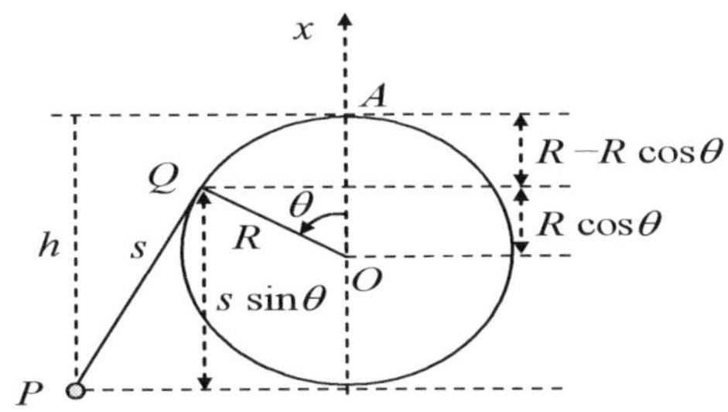

(g) At the lowest point of its trajectory, the particle's gravitational potential energy U must assume its minimum value U_m. If the particle's mechanical energy E were equal to U_m, its kinetic energy would be zero. The particle would then remain stationary and be in the static equilibrium state shown in Fig. A4. Thus, the potential energy reaches its minimum value when $\theta = \pi/2$ or $s = L - \pi R/2$.

Figure A4

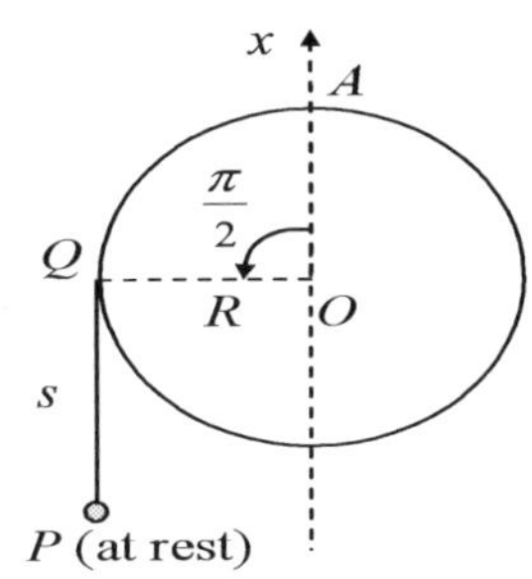

From Fig. A4 or Eq. (A6), the minimum potential energy is then

$$U_m = U(\frac{\pi}{2}) = -mg[R + L - (\pi R/2)] . \quad \text{(A7)}$$

Initially, the total mechanical energy E is 0. Since E is conserved, the speed v_m of the particle at the lowest point of its trajectory must satisfy

$$E = 0 = \frac{1}{2}mv_m^2 + U_m . \quad \text{(A8)}$$

From Eqs. (A7) and (A8), we obtain

$$v_m = \sqrt{-2U_m/m} = \sqrt{2g[R + (L - \pi R/2)]} . \quad \text{(A9)*}$$

Part B

(h) From Eq. (A6), the total mechanical energy of the particle may be written as

$$E = 0 = \frac{1}{2}mv^2 + U(\theta) = \frac{1}{2}mv^2 - mg[R(1-\cos\theta) + s\sin\theta] \qquad \text{(B1)}$$

From Eq. (A4), the speed v is equal to $s\dot{\theta}$. Therefore, Eq. (B1) implies

$$v^2 = (s\dot{\theta})^2 = 2g[R(1-\cos\theta) + s\sin\theta] \qquad \text{(B2)}$$

Let T be the tension in the string. Then, as Fig. B1 shows, the $\hat{t}$-component of the net force on the particle is $-T + mg\sin\theta$. From Eq. (A5), the tangential acceleration of the particle is $(-s\dot{\theta}^2)$. Thus, by Newton's second law, we have

$$m(-s\dot{\theta}^2) = -T + mg\sin\theta \qquad \text{(B3)}$$

Figure B1

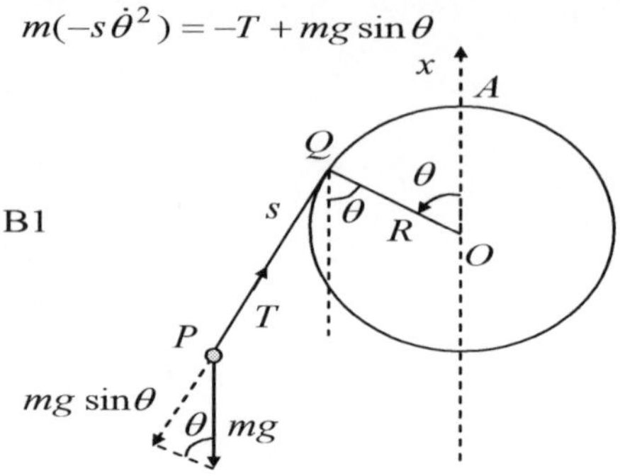

According to the last two equations, the tension may be expressed as

$$T = m(s\dot{\theta}^2 + g\sin\theta) = \frac{mg}{s}[2R(1-\cos\theta) + 3s\sin\theta]$$

$$= \frac{2mgR}{s}[\tan\frac{\theta}{2} - \frac{3}{2}(\theta - \frac{L}{R})](\sin\theta) \qquad \text{(B4)}$$

$$= \frac{2mgR}{s}(y_1 - y_2)(\sin\theta)$$

The functions $y_1 = \tan(\theta/2)$ and $y_2 = 3(\theta - L/R)/2$ are plotted in Fig B2.

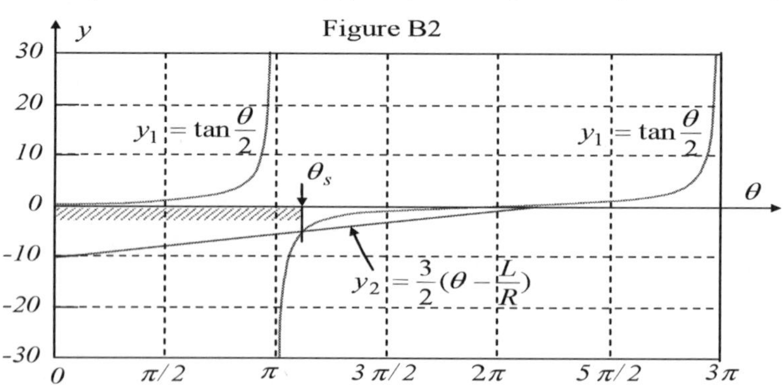

From Eq. (B4) and Fig. B2, we obtain the result shown in Table B1. The angle at which $y_2 = y_1$ is called $\theta_s(\pi < \theta_s < 2\pi)$ and is given by

$$\frac{3}{2}(\theta_s - \frac{L}{R}) = \tan\frac{\theta_s}{2} \qquad \text{(B5)}$$

or, equivalently, by

$$\frac{L}{R} = \theta_s - \frac{2}{3}\tan\frac{\theta_s}{2} \qquad \text{(B6)}$$

Since the ratio L/R is known to be given by

$$\frac{L}{R} = \frac{9\pi}{8} + \frac{2}{3}\cot\frac{\pi}{16} = (\pi + \frac{\pi}{8}) - \frac{2}{3}\tan\frac{1}{2}(\pi + \frac{\pi}{8}) \qquad \text{(B7)}$$

one can readily see from the last two equations that $\theta_s = 9\pi/8$.

Table B1

	$(y_1 - y_2)$	$\sin\theta$	tension T
$0 < \theta < \pi$	positive	positive	positive
$\theta = \pi$	$+\infty$	0	positive
$\pi < \theta < \theta_s$	negative	negative	positive
$\theta = \theta_s$	zero	negative	zero
$\theta_s < \theta < 2\pi$	positive	negative	negative

Table B1 shows that the tension T must be positive (or the string must be taut and straight) in the angular range $0<\theta < \theta_s$. Once θ reaches θ_s, the tension T becomes zero and the part of the string not in contact with the rod will not be straight afterwards. The shortest possible value $s_{\min}$ for the length s of the line segment QP therefore occurs at $\theta = \theta_s$ and is given by

$$s_{\min} = L - R\theta_s = R(\frac{9\pi}{8} + \frac{2}{3}\cot\frac{\pi}{16} - \frac{9\pi}{8}) = \frac{2R}{3}\cot\frac{\pi}{16} = 3.352R \qquad \text{(B8)}$$

When $\theta = \theta_s$, we have $T = 0$ and Eqs. (B2) and (B3) then leads to $v^2 = -gs\sin\theta$. Hence the speed v_s is

$$v_s = \sqrt{-gs_{\min}\sin\theta_s} = \sqrt{\frac{2gR}{3}\cot\frac{\pi}{16}\sin\frac{\pi}{8}} = \sqrt{\frac{4gR}{3}}\cos\frac{\pi}{16} \qquad \text{(B9)*}$$
$$= 1.133\sqrt{gR}$$

(i) When $\theta \geq \theta_s$, the particle moves like a projectile under gravity. As shown in Fig. B3, it is projected with an initial speed v_s from the position $P = (x_s, y_s)$ in a direction making an angle $\phi = (3\pi/2 - \theta_s)$ with the y-axis.

The speed v_H of the particle at the highest point of its parabolic trajectory is equal to the y-component of its initial velocity when projected. Thus,

$$v_H = v_s\sin(\theta_s - \pi) = \sqrt{\frac{4gR}{3}}\cos\frac{\pi}{16}\sin\frac{\pi}{8} = 0.4334\sqrt{gR} \qquad \text{(B10)*}$$

The horizontal distance H traveled by the particle from point P to the point of maximum height is

$$H = \frac{v_s^2\sin 2(\theta_s - \pi)}{2g} = \frac{v_s^2}{2g}\sin\frac{9\pi}{4} = 0.4535R \qquad \text{(B11)}$$

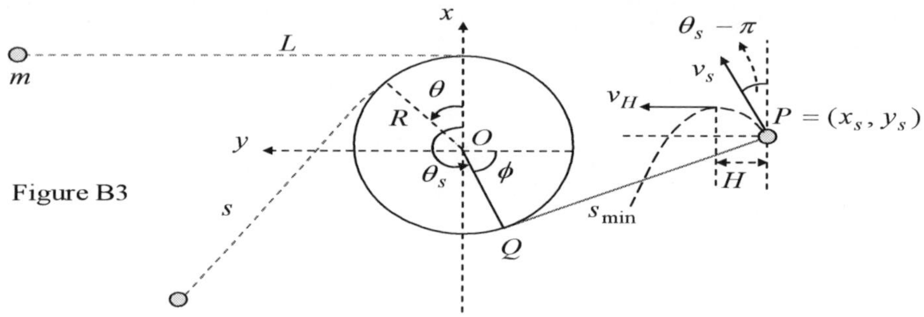

Figure B3

The coordinates of the particle when $\theta = \theta_s$ are given by

$$x_s = R\cos\theta_s - s_{\min}\sin\theta_s = -R\cos\frac{\pi}{8} + s_{\min}\sin\frac{\pi}{8} = 0.358R \qquad \text{(B12)}$$

$$y_s = R\sin\theta_s + s_{\min}\cos\theta_s = -R\sin\frac{\pi}{8} - s_{\min}\cos\frac{\pi}{8} = -3.478R \qquad \text{(B13)}$$

Evidently, we have $|y_s| > (R + H)$. Therefore the particle can indeed reach its maximum height without striking the surface of the rod.

Part C

(j) Assume the weight is initially lower than O by h as shown in Fig. C1.

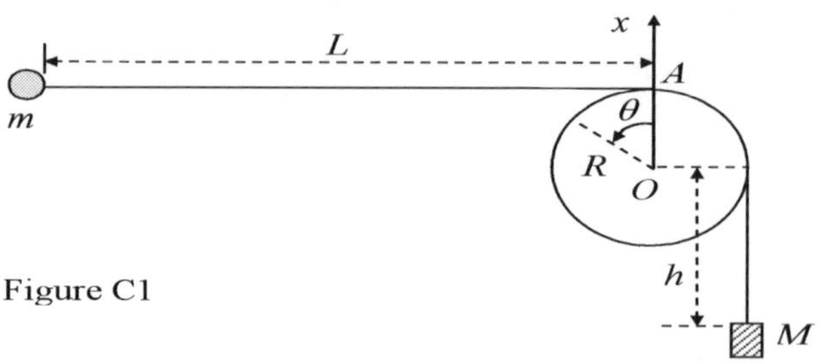

Figure C1

When the weight has fallen a distance D and stopped, the law of conservation of total mechanical energy as applied to the particle-weight pair as a system leads to

$$-Mgh = E' - Mg(h + D) \tag{C1}$$

where E' is the *total mechanical energy of the particle* when the weight has stopped. It follows

$$E' = MgD \tag{C2}$$

Let Λ be the total length of the string. Then, its value at $\theta = 0$ must be the same as at any other angular displacement θ. Thus we must have

$$\Lambda = L + \frac{\pi}{2}R + h = s + R(\theta + \frac{\pi}{2}) + (h + D) \tag{C3}$$

Noting that $D = \alpha L$ and introducing $\ell = L - D$, we may write

$$\ell = L - D = (1 - \alpha)L \tag{C4}$$

From the last two equations, we obtain

$$s = L - D - R\theta = \ell - R\theta \tag{C5}$$

After the weight has stopped, the total mechanical energy of the particle must be conserved. According to Eq. (C2), we now have, instead of Eq. (B1), the following equation:

$$E' = MgD = \frac{1}{2}mv^2 - mg[R(1 - \cos\theta) + s\sin\theta] \tag{C6}$$

The square of the particle's speed is accordingly given by

$$v^2 = (s\dot{\theta})^2 = \frac{2MgD}{m} + 2gR[(1 - \cos\theta) + \frac{s}{R}\sin\theta] \tag{C7}$$

Since Eq. (B3) stills applies, the tension T of the string is given by

$$-T + mg\sin\theta = m(-s\dot{\theta}^2) \tag{C8}$$

From the last two equations, it follows

$$\begin{aligned} T &= m(s\dot{\theta}^2 + g\sin\theta) \\ &= \frac{mg}{s}[\frac{2M}{m}D + 2R(1 - \cos\theta) + 3s\sin\theta] \\ &= \frac{2mgR}{s}[\frac{MD}{mR} + (1 - \cos\theta) + \frac{3}{2}(\frac{\ell}{R} - \theta)\sin\theta] \end{aligned} \tag{C9}$$

where Eq. (C5) has been used to obtain the last equality.

We now introduce the function

$$f(\theta) = 1 - \cos\theta + \frac{3}{2}(\frac{\ell}{R} - \theta)\sin\theta \tag{C10}$$

From the fact $\ell = (L - D) >> R$, we may write

$$f(\theta) \approx 1 + \frac{3}{2}\frac{\ell}{R}\sin\theta - \cos\theta = 1 + A\sin(\theta - \phi) \tag{C11}$$

where we have introduced

$$A = \sqrt{1 + (\frac{3}{2}\frac{\ell}{R})^2}\ , \qquad \phi = \tan^{-1}\frac{\frac{3\ell}{2R}}{\sqrt{1 + (\frac{3\ell}{2R})^2}} \tag{C12}$$

From Eq. (C11), the minimum value of $f(\theta)$ is seen to be given by

$$f_{\min} = 1 - A = 1 - \sqrt{1 + (\frac{3}{2}\frac{\ell}{R})^2} \tag{C13}$$

Since the tension T remains nonnegative as the particle swings around the rod, we have from Eq. (C9) the inequality

$$\frac{MD}{mR} + f_{\min} = \frac{M(L - \ell)}{mR} + 1 - \sqrt{1 + (\frac{3\ell}{2R})^2} \geq 0 \tag{C14}$$

or

$$(\frac{ML}{mR}) + 1 \ \geq \ (\frac{M\ell}{mR}) + \sqrt{1 + (\frac{3\ell}{2R})^2} \approx (\frac{M\ell}{mR}) + (\frac{3\ell}{2R}) \tag{C15}$$

From Eq. (C4), Eq. (C15) may be written as

$$(\frac{ML}{mR}) + 1 \ \geq \ [(\frac{ML}{mR}) + (\frac{3L}{2R})](1 - \alpha) \tag{C16}$$

Neglecting terms of the order (R/L) or higher, the last inequality leads to

$$\alpha \ \geq \ 1 - \frac{(\frac{ML}{mR}) + 1}{(\frac{ML}{mR}) + (\frac{3L}{2R})} = \frac{(\frac{3L}{2R}) - 1}{(\frac{ML}{mR}) + (\frac{3L}{2R})} = \frac{1 - \frac{2R}{3L}}{\frac{2M}{3m} + 1} \approx \frac{1}{1 + \frac{2M}{3m}} \tag{C17}$$

The critical value for the ratio D/L is therefore

$$\alpha_c = \frac{1}{(1 + \frac{2M}{3m})} \qquad \text{(C18)*}$$

Solution- Theoretical Question 2
A Piezoelectric Crystal Resonator under an Alternating Voltage

Part A

(a) Refer to Figure A1. The left face of the rod moves a distance $v\Delta t$ while the pressure wave travels a distance $u\Delta t$ with $u = \sqrt{Y/\rho}$. The strain at the left face is

$$S = \frac{\Delta \ell}{\ell} = \frac{-v\Delta t}{u\Delta t} = \frac{-v}{u} \qquad \text{(A1a)*}^{1}$$

From Hooke's law, the pressure at the left face is

$$p = -YS = Y\frac{v}{u} = \rho u v \qquad \text{(A1b)*}$$

Figure A1

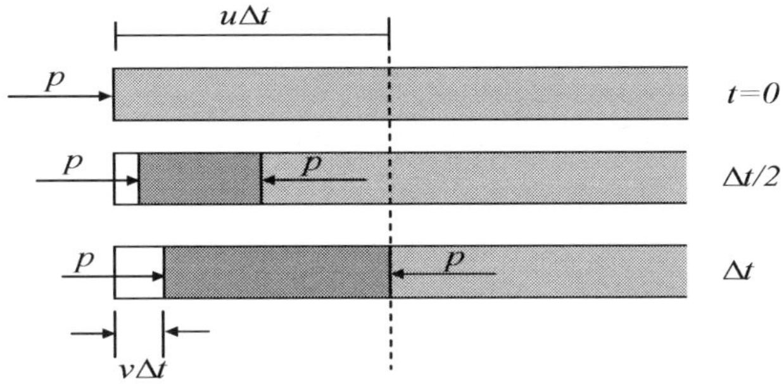

(b) The velocity v is related to the displacement ξ as in a simple harmonic motion (or a uniform circular motion, as shown in Figure A2) of angular frequency $\omega = ku$. Therefore, if $\xi(x,t) = \xi_0 \sin k(x-ut)$, then

$$v(x,t) = -ku\xi_0 \cos k(x-ut). \qquad \text{(A2)*}$$

The strain and pressure are related to velocity as in Problem (a). Hence,

$$S(x,t) = -v(x,t)/u = k\xi_0 \cos k(x-ut) \qquad \text{(A3)*}$$

$$p(x,t) = \rho u v(x,t) = -k\rho u^2 \xi_0 \cos k(x-ut) = -YS(x,t) = -kY\xi_0 \cos k(x-ut) \qquad \text{(A4)*}$$

Alternatively, the answers may be obtained by differentiations:

$$v(x,t) = \frac{\Delta\xi}{\Delta t} = -ku\xi_0 \cos k(x-ut),$$

$$S(x,t) = \frac{\Delta\xi}{\Delta x} = k\xi_0 \cos k(x-ut),$$

$$p(x,t) = -Y\frac{\Delta\xi}{\Delta x} = -kY\xi_0 \cos k(x-ut).$$

Figure A2

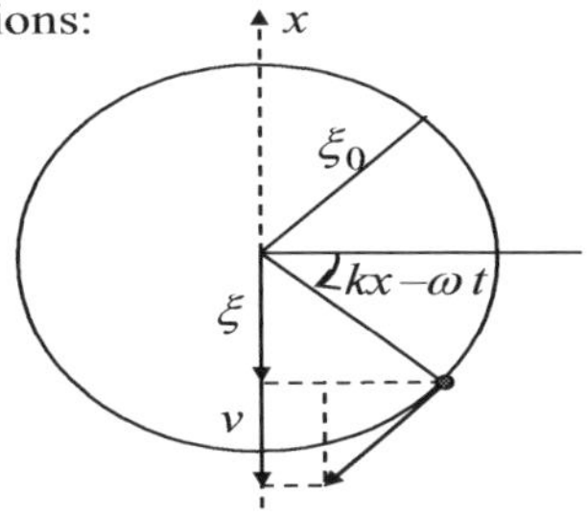

---

[1] An equations marked with an asterisk contains answer to the problem.

Part B

(c) Since the angular frequency ω and speed of propagation u are given, the wavelength is given by $\lambda = 2\pi / k$ with $k = \omega / u$. The spatial variation of the displacement ξ is therefore described by

$$g(x) = B_1 \sin k(x - \frac{b}{2}) + B_2 \cos k(x - \frac{b}{2}) \tag{B1}$$

Since the centers of the electrodes are assumed to be stationary, $g(b/2) = 0$. This leads to $B_2 = 0$. Given that the maximum of $g(x)$ is 1, we have $A = \pm 1$ and

$$g(x) = \pm \sin \frac{\omega}{u}(x - \frac{b}{2}) \tag{B2)*}$$

Thus, the displacement is

$$\xi(x,t) = \pm 2\xi_0 \sin \frac{\omega}{u}(x - \frac{b}{2}) \cos \omega t \tag{B3}$$

(d) Since the pressure p (or stress T) must vanish at the end faces of the quartz slab (i.e., $x = 0$ and $x = b$), the answer to this problem can be obtained, by analogy, from the resonant frequencies of sound waves in an open pipe of length b. However, given that the centers of the electrodes are stationary, all even harmonics of the fundamental tone must be excluded because they have antinodes, rather than nodes, of displacement at the bisection plane of the slab.

Since the fundamental tone has a wavelength $\lambda = 2b$, the fundamental frequency is given by $f_1 = u/(2b)$. The speed of propagation u is given by

$$u = \sqrt{\frac{Y}{\rho}} = \sqrt{\frac{7.87 \times 10^{10}}{2.65 \times 10^{3}}} = 5.45 \times 10^{3} \text{ m/s} \tag{B4}$$

and, given that $b = 1.00 \times 10^{-2}$ m, the two lowest standing wave frequencies are

$$f_1 = \frac{u}{2b} = 273 \text{ (kHz)}, \quad f_3 = 3f_1 = \frac{3u}{2b} = 818 \text{ (kHz)} \tag{B5)*}$$

[*Alternative solution to Problems (c) and (d)*]:

A longitudinal standing wave in the quartz slab has a displacement node at $x = b/2$. It may be regarded as consisting of two waves traveling in opposite directions. Thus, its displacement and velocity must have the following form

$$\begin{aligned} \xi(x,t) &= 2\xi_m \sin k(x - \frac{b}{2}) \cos \omega t \\ &= \xi_m [\sin k(x - \frac{b}{2} - ut) + \sin k(x - \frac{b}{2} + ut)] \end{aligned} \tag{B6}$$

$$\begin{aligned} v(x,t) &= -ku\xi_m [\cos k(x - \frac{b}{2} - ut) - \cos k(x - \frac{b}{2} + ut)] \\ &= -2\omega\xi_m \sin k(x - \frac{b}{2}) \sin \omega t \end{aligned} \tag{B7}$$

where $\omega = ku$ and the first and second factors in the square brackets represent waves traveling along the $+x$ and $-x$ directions, respectively. Note that Eq. (B6) is identical to Eq. (B3) if we set $\xi_m = \pm \xi_0$.

For a wave traveling along the $-x$ direction, the velocity v must be replaced by $-v$ in Eqs. (A1a) and (A1b) so that we have

$$S = \frac{-v}{u} \quad \text{and} \quad p = \rho u v \qquad \text{(waves traveling along } +x\text{)} \tag{B8}$$

$$S = \frac{v}{u} \quad \text{and} \quad p = -\rho u v \qquad \text{(waves traveling along } -x\text{)} \tag{B9}$$

As in Problem (b), the strain and pressure are therefore given by

$$\begin{aligned} S(x,t) &= -k\xi_m[-\cos k(x - \frac{b}{2} - ut) - \cos k(x - \frac{b}{2} + ut)] \\ &= 2k\xi_m \cos k(x - \frac{b}{2}) \cos \omega t \end{aligned} \tag{B10}$$

$$\begin{aligned} p(x,t) &= -\rho u \omega \xi_m [\cos k(x - \frac{b}{2} - ut) + \cos k(x - \frac{b}{2} + ut)] \\ &= -2\rho u \omega \xi_m \cos k(x - \frac{b}{2}) \cos \omega t \end{aligned} \tag{B11}$$

Note that v, S, and p may also be obtained by differentiating ξ as in Problem (b).

The stress T or pressure p must be zero at both ends ($x = 0$ and $x = b$) of the slab at all times because they are free. From Eq. (B11), this is possible only if $\cos(kb/2) = 0$ or

$$kb = \frac{\omega}{u} b = \frac{2\pi f}{\lambda f} b = n\pi, \qquad n = 1, 3, 5, \cdots \tag{B12}$$

In terms of wavelength λ, Eq. (B12) may be written as

$$\lambda = \frac{2b}{n}, \qquad n = 1, 3, 5, \cdots. \tag{B13}$$

The frequency is given by

$$f = \frac{u}{\lambda} = \frac{nu}{2b} = \frac{n}{2b}\sqrt{\frac{Y}{\rho}}, \qquad n = 1, 3, 5, \cdots. \tag{B14}$$

This is identical with the results given in Eqs. (B4) and (B5).

(e) From Eqs. (5a) and (5b) in the Question, the piezoelectric effect leads to the equations

$$T = Y(S - d_p E) \tag{B15}$$

$$\sigma = Y d_p S + \varepsilon_T (1 - Y \frac{d_p^2}{\varepsilon_T}) E \tag{B16}$$

Because $x = b/2$ must be a node of displacement for any longitudinal standing wave in the slab, the displacement ξ and strain S must have the form given in Eqs. (B6) and (B10), i.e., with $\omega = ku$,

$$\xi(x,t) = \xi_m \sin k(x - \frac{b}{2}) \cos(\omega t + \phi) \tag{B17}$$

$$S(x,t) = k\xi_m \cos k(x - \frac{b}{2}) \cos(\omega t + \phi) \tag{B18}$$

where a phase constant ϕ is now included in the time-dependent factors.

By assumption, the electric field E between the electrodes is uniform and depends only on time:

$$E(x,t)=\frac{V(t)}{h}=\frac{V_m\cos\omega t}{h}. \tag{B19}$$

Substituting Eqs. (B18) and (B19) into Eq. (B15), we have

$$T=Y[k\xi_m\cos k(x-\frac{b}{2})\cos(\omega t+\phi)-\frac{d_p}{h}V_m\cos\omega t] \tag{B20}$$

The stress T must be zero at both ends ($x=0$ and $x=b$) of the slab at all times because they are free. This is possible only if $\phi=0$ and

$$k\xi_m\cos\frac{kb}{2}=d_p\frac{V_m}{h} \tag{B21}$$

Since $\phi=0$, Eqs. (B16), (B18), and (B19) imply that the surface charge density must have the same dependence on time t and may be expressed as

$$\sigma(x,t)=\sigma(x)\cos\omega t \tag{B22}$$

with the dependence on x given by

$$\begin{aligned}\sigma(x)&=Yd_pk\xi_m\cos k(x-\frac{b}{2})+\varepsilon_T(1-Y\frac{d_p^2}{\varepsilon_T})\frac{V_m}{h}\\&=[Y\frac{d_p^2}{\cos\frac{kb}{2}}\cos k(x-\frac{b}{2})+\varepsilon_T(1-Y\frac{d_p^2}{\varepsilon_T})]\frac{V_m}{h}\end{aligned} \tag{B23)*}$$

(f) At time t, the total surface charge $Q(t)$ on the lower electrode is obtained by integrating $\sigma(x,t)$ in Eq. (B22) over the surface of the electrode. The result is

$$\begin{aligned}\frac{Q(t)}{V(t)}&=\frac{1}{V(t)}\int_0^b\sigma(x,t)w\,dx=\frac{1}{V_m}\int_0^b\sigma(x)w\,dx\\&=\frac{w}{h}\int_0^b[Y\frac{d_p^2}{\cos\frac{kb}{2}}\cos k(x-\frac{b}{2})+\varepsilon_T(1-Y\frac{d_p^2}{\varepsilon_T})]dx\\&=(\varepsilon_T\frac{bw}{h})[Y\frac{d_p^2}{\varepsilon_T}(\frac{2}{kb}\tan\frac{kb}{2})+(1-Y\frac{d_p^2}{\varepsilon_T})]\\&=C_0[\alpha^2(\frac{2}{kb}\tan\frac{kb}{2})+(1-\alpha^2)]\end{aligned} \tag{B24}$$

where

$$C_0=\varepsilon_T\frac{bw}{h},\qquad \alpha^2=Y\frac{d_p^2}{\varepsilon_T}=\frac{(2.25)^2\times10^{-2}}{1.27\times4.06}=9.82\times10^{-3} \tag{B25)*}$$

(The constant α is called the *electromechanical coupling coefficient*.)

Note: The result $C_0=\varepsilon_T bw/h$ can readily be seen by considering the static limit $k=0$ of Eq. (5) in the Question. Since $\tan x\approx x$ when $x<<1$, we have

$$\lim_{k\to0}Q(t)/V(t)\approx C_0[\alpha^2+(1-\alpha^2)]=C_0 \tag{B26}$$

Evidently, the constant C_0 is the capacitance of the parallel-plate capacitor formed by the electrodes (of area bw) with the quartz slab (of thickness h and permittivity

ε_T) serving as the dielectric medium. It is therefore given by $\varepsilon_T bw/h$.

(B47)*

Solution- Theoretical Question 3

Part A

Neutrino Mass and Neutron Decay

(a) Let $(c^2E_e, c\vec{q}_e)$, $(c^2E_p, c\vec{q}_p)$, and $(c^2E_\nu, c\vec{q}_\nu)$ be the energy-momentum 4-vectors of the electron, the proton, and the anti-neutrino, respectively, in the rest frame of the neutron. Notice that $E_e, E_p, E_\nu, \vec{q}_e, \vec{q}_p, \vec{q}_\nu$ are all in units of mass.

The proton and the anti-neutrino may be considered as forming a system of total rest mass M_c, total energy c^2E_c, and total momentum $c\vec{q}_c$. Thus, we have

$$E_c = E_p + E_\nu, \qquad \vec{q}_c = \vec{q}_p + \vec{q}_\nu, \qquad M_c^2 = E_c^2 - q_c^2 \tag{A1}$$

Note that the magnitude of the vector $\vec{q}_c$ is denoted as q_c. The same convention also applies to all other vectors.

Since energy and momentum are conserved in the neutron decay, we have

$$E_c + E_e = m_n \tag{A2}$$

$$\vec{q}_c = -\vec{q}_e \tag{A3}$$

When squared, the last equation leads to the following equality

$$q_c^2 = q_e^2 = E_e^2 - m_e^2 \tag{A4}$$

From Eq. (A4) and the third equality of Eq. (A1), we obtain

$$E_c^2 - M_c^2 = E_e^2 - m_e^2 \tag{A5}$$

With its second and third terms moved to the other side of the equality, Eq. (A5) may be divided by Eq. (A2) to give

$$E_c - E_e = \frac{1}{m_n}(M_c^2 - m_e^2) \tag{A6}$$

As a system of coupled linear equations, Eqs. (A2) and (A6) may be solved to give

$$E_c = \frac{1}{2m_n}(m_n^2 - m_e^2 + M_c^2) \tag{A7}$$

$$E_e = \frac{1}{2m_n}(m_n^2 + m_e^2 - M_c^2) \tag{A8}$$

Using Eq. (A8), the last equality in Eq. (A4) may be rewritten as

$$\begin{aligned} q_e &= \frac{1}{2m_n}\sqrt{(m_n^2 + m_e^2 - M_c^2)^2 - (2m_n m_e)^2} \\ &= \frac{1}{2m_n}\sqrt{(m_n + m_e + M_c)(m_n + m_e - M_c)(m_n - m_e + M_c)(m_n - m_e - M_c)} \end{aligned} \tag{A9}$$

Eq. (A8) shows that a maximum of E_e corresponds to a minimum of M_c^2. Now the rest mass M_c is the total energy of the proton and anti-neutrino pair in their center of mass (or momentum) frame so that it achieves the minimum

$$M = m_p + m_v \tag{A10}$$

when the proton and the anti-neutrino are both at rest in the center of mass frame. Hence, from Eqs. (A8) and (A10), the maximum energy of the electron $E = c^2 E_e$ is

$$E_{\max} = \frac{c^2}{2m_n}\left[m_n^2 + m_e^2 - (m_p + m_v)^2\right] \approx 1.292569 \text{ MeV} \approx 1.29\text{MeV} \quad \text{(A11)*}^1$$

When Eq. (A10) holds, the proton and the anti-neutrino move with the same velocity v_m of the center of mass and we have

$$\frac{v_m}{c} = (\frac{q_v}{E_v})|_{E=E_{\max}} = (\frac{q_p}{E_p})|_{E=E_{\max}} = (\frac{q_c}{E_c})|_{E=E_{\max}} = (\frac{q_e}{E_c})|_{M_c=m_p+m_v} \tag{A12}$$

where the last equality follows from Eq. (A3). By Eqs. (A7) and (A9), the last expression in Eq. (A12) may be used to obtain the speed of the anti-neutrino when $E = E_{\max}$. Thus, with $M = m_p + m_v$, we have

$$\frac{v_m}{c} = \frac{\sqrt{(m_n + m_e + M)(m_n + m_e - M)(m_n - m_e + M)(m_n - m_e - M)}}{m_n^2 - m_e^2 + M^2} \quad \text{(A13)*}$$
$$\approx 0.00126538 \approx 0.00127$$

[Alternative Solution]

Assume that, in the rest frame of the neutron, the electron comes out with momentum $c\vec{q}_e$ and energy c^2E_e, the proton with $c\vec{q}_p$ and $c^2 E_p$, and the anti-neutrino with $c\vec{q}_v$ and $c^2 E_v$. With the magnitude of vector $\vec{q}_\alpha$ denoted by the symbol q_α, we have

$$E_p^2 = m_p^2 + q_p^2, \qquad E_v^2 = m_v^2 + q_v^2, \qquad E_e^2 = m_e^2 + q_e^2 \tag{1A}$$

Conservation of energy and momentum in the neutron decay leads to

$$E_p + E_v = m_n - E_e \tag{2A}$$

$$\vec{q}_p + \vec{q}_v = -\vec{q}_e \tag{3A}$$

When squared, the last two equations lead to

$$E_p^2 + E_v^2 + 2E_pE_v = (m_n - E_e)^2 \tag{4A}$$

$$q_p^2 + q_v^2 + 2\vec{q}_p \cdot \vec{q}_v = q_e^2 = E_e^2 - m_e^2 \tag{5A}$$

Subtracting Eq. (5A) from Eq. (4A) and making use of Eq. (1A) then gives

$$m_p^2 + m_v^2 + 2(E_pE_v - \vec{q}_p \cdot \vec{q}_v) = m_n^2 + m_e^2 - 2m_nE_e \tag{6A}$$

or, equivalently,

$$2m_n E_e = m_n^2 + m_e^2 - m_p^2 - m_v^2 - 2(E_p E_v - \vec{q}_p \cdot \vec{q}_v) \qquad (7A)$$

If θ is the angle between $\vec{q}_p$ and $\vec{q}_v$, we have $\vec{q}_p \cdot \vec{q}_v = q_p q_v \cos\theta \le q_p q_v$ so that Eq. (7A) leads to the relation

$$2m_n E_e \le m_n^2 + m_e^2 - m_p^2 - m_v^2 - 2(E_p E_v - q_p q_v) \qquad (8A)$$

Note that the equality in Eq. (8A) holds only if $\theta = 0$, i.e., the energy of the electron $c^2 E_e$ takes on its maximum value only when the anti-neutrino and the proton *move in the same direction*.

Let the speeds of the proton and the anti-neutrino in the rest frame of the neutron be $c\beta_p$ and $c\beta_v$, respectively. We then have $q_p = \beta_p E_p$ and $q_v = \beta_v E_v$. As shown in Fig. A1, we introduce the angle ϕ_v ($0 \le \phi_v < \pi/2$) for the antineutrino by

$$q_v = m_v \tan\phi_v, \qquad E_v = \sqrt{m_v^2 + q_v^2} = m_v \sec\phi_v, \qquad \beta_v = q_v / E_v = \sin\phi_v \qquad (9A)$$

Figure A1

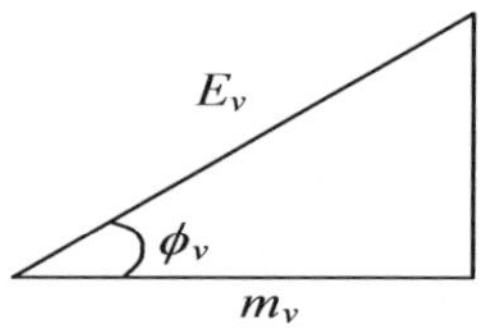

Similarly, for the proton, we write, with $0 \le \phi_p < \pi/2$,

$$q_p = m_p \tan\phi_p, \qquad E_p = \sqrt{m_p^2 + q_p^2} = m_p \sec\phi_p, \qquad \beta_p = q_p / E_p = \sin\phi_p \qquad (10A)$$

Eq. (8A) may then be expressed as

$$2m_n E_e \le m_n^2 + m_e^2 - m_p^2 - m_v^2 - 2m_p m_v \left(\frac{1 - \sin\phi_p \sin\phi_v}{\cos\phi_p \cos\phi_v}\right) \qquad (11A)$$

The factor in parentheses at the end of the last equation may be expressed as

$$\frac{1 - \sin\phi_p \sin\phi_v}{\cos\phi_p \cos\phi_v} = \frac{1 - \sin\phi_p \sin\phi_v - \cos\phi_p \cos\phi_v}{\cos\phi_p \cos\phi_v} + 1 = \frac{1 - \cos(\phi_p - \phi_v)}{\cos\phi_p \cos\phi_v} + 1 \ge 1 \qquad (12A)$$

and clearly assumes its minimum possible value of 1 when $\phi_p = \phi_v$, i.e., when the anti-neutrino and the proton *move with the same velocity* so that $\beta_p = \beta_v$. Thus, it follows from Eq. (11A) that the maximum value of E_e is

$$\begin{aligned}(E_e)_{\max} &= \frac{1}{2m_n}(m_n^2 + m_e^2 - m_p^2 - m_v^2 - 2m_p m_v) \\ &= \frac{1}{2m_n}[m_n^2 + m_e^2 - (m_p + m_v)^2]\end{aligned} \qquad (13A)^*$$

[1] An equation marked with an asterisk contains answer to the problem.

and the maximum energy of the electron $E = c^2 E_e$ is

$$E_{\max} = c^2 (E_e)_{\max} \approx 1.292569 \text{ MeV} \approx 1.29 \text{ MeV} \tag{14A)*}$$

When the anti-neutrino and the proton move with the same velocity, we have, from Eqs. (9A), (10A), (2A) ,(3A), and (1A), the result

$$\beta_\nu = \beta_p = \frac{q_p}{E_p} = \frac{q_\nu}{E_\nu} = \frac{q_p + q_\nu}{E_p + E_\nu} = \frac{q_e}{m_n - E_e} = \frac{\sqrt{E_e^2 - m_e^2}}{m_n - E_e} \tag{15A}$$

Substituting the result of Eq. (13A) into the last equation, the speed v_m of the anti-neutrino when the electron attains its maximum value $E_{\max}$ is, with $M = m_p + m_\nu$, given by

$$\frac{v_m}{c} = (\beta_\nu)_{\max E_e} = \frac{\sqrt{(E_e)_{\max}^2 - m_e^2}}{m_n - (E_e)_{\max}} = \frac{\sqrt{(m_n^2 + m_e^2 - M^2)^2 - 4m_n^2 m_e^2}}{2m_n^2 - (m_n^2 + m_e^2 - M^2)}$$

$$= \frac{\sqrt{(m_n + m_e + M)(m_n + m_e - M)(m_n - m_e + M)(m_n - m_e - M)}}{m_n^2 - m_e^2 + M^2} \tag{16A)*}$$

$$\approx 0.00126538 \approx 0.00127$$

Part B

Light Levitation

(b) Refer to Fig. B1. Refraction of light at the spherical surface obeys Snell's law and leads to

$$n \sin\theta_i = \sin\theta_t \tag{B1}$$

Neglecting terms of the order $(\delta/R)^3$ or higher in sine functions, Eq. (B1) becomes

$$n\theta_i \approx \theta_t \tag{B2}$$

For the triangle ΔFAC in Fig. B1, we have

$$\beta = \theta_t - \theta_i \approx n\theta_i - \theta_i = (n-1)\theta_i \tag{B3}$$

Let f_0 be the frequency of the incident light. If n_p is the number of photons incident on the plane surface per unit area per unit time, then the total number of photons incident on the plane surface per unit time is $n_p \pi\delta^2$. The total power P of photons incident on the plane surface is $(n_p \pi\delta^2)(hf_0)$, with h being Planck's constant. Hence,

$$n_p = \frac{P}{\pi\delta^2 h f_0} \tag{B4}$$

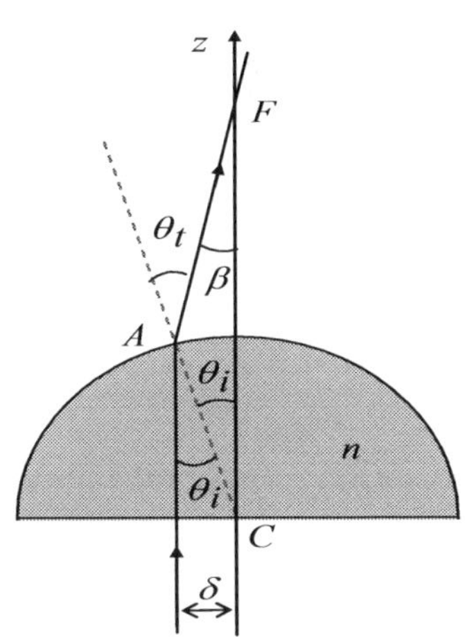

Fig. B1

The number of photons incident on an annular disk of inner radius r and outer radius $r + dr$ on the plane surface per unit time is $n_p(2\pi r dr)$, where

$r = R\tan\theta_i \approx R\theta_i$. Therefore,

$$n_p(2\pi r dr) \approx n_p(2\pi R^2)\theta_i d\theta_i \tag{B5}$$

The z-component of the momentum carried away per unit time by these photons when refracted at the spherical surface is

$$\begin{aligned} dF_z &= n_p \frac{hf_o}{c}(2\pi r dr)\cos\beta \approx n_p \frac{hf_0}{c}(2\pi R^2)(1-\frac{\beta^2}{2})\theta_i d\theta_i \\ &\approx n_p \frac{hf_0}{c}(2\pi R^2)[\theta_i - \frac{(n-1)^2}{2}\theta_i^3]d\theta_i \end{aligned} \tag{B6}$$

so that the z-component of the total momentum carried away per unit time is

$$\begin{aligned} F_z &= 2\pi R^2 n_p(\frac{hf_0}{c})\int_0^{\theta_{im}}[\theta_i - \frac{(n-1)^2}{2}\theta_i^3]d\theta_i \\ &= \pi R^2 n_p(\frac{hf_0}{c})\theta_{im}^2[1-\frac{(n-1)^2}{4}\theta_{im}^2] \end{aligned} \tag{B7}$$

where $\tan\theta_{im} = \frac{\delta}{R} \approx \theta_{im}$. Therefore, by the result of Eq. (B5), we have

$$F_z = \frac{\pi R^2 P}{\pi\delta^2 hf_0}(\frac{hf_0}{c})\frac{\delta^2}{R^2}[1-\frac{(n-1)^2\delta^2}{4R^2}] = \frac{P}{c}[1-\frac{(n-1)^2\delta^2}{4R^2}] \tag{B8}$$

The force of optical levitation is equal to the sum of the z-components of the forces exerted by the incident and refracted lights on the glass hemisphere and is given by

$$\frac{P}{c} + (-F_z) = \frac{P}{c} - \frac{P}{c}[1-\frac{(n-1)^2\delta^2}{4R^2}] = \frac{(n-1)^2\delta^2}{4R^2}\frac{P}{c} \tag{B9}$$

Equating this to the weight mg of the glass hemisphere, we obtain the minimum laser power required to levitate the hemisphere as

$$P = \frac{4mgcR^2}{(n-1)^2\delta^2} \qquad \text{(B10)*}$$

Part A: Optical Properties of Laser Diode

I. Introduction

1. Laser Diode

The light source in this experiment is a laser diode which emits laser light with wavelength 650 nm. When the current of the laser diode (LD) is greater than the threshold current, the laser diode can emit monochromatic, partially polarized and coherent light. When the current in the laser diode is less than the threshold, the emitted light intensity is very small. At above the threshold current, the light intensity increases dramatically with the current and keeps a linear relationship with the current. If the current increases further, then the increasing rate of the intensity with respect to the current becomes smaller because of the higher temperature of the laser diode. Therefore, the optimal operating current range for the laser diode is the region where the intensity is linear with the current. In general, the threshold current I_{th} is defined as the intersection point of the current axis with the extrapolation line of the linear region.

Caution: Do not look directly into the laser beam. You can damage your eyes!!

2. Photodetector

The photodetector used in this experiment consists of a photodiode and a current amplifier. When an external bias voltage is applied on the photodiode, the photocurrent is generated by the light incident upon the diode. Under the condition of a constant temperature and monochromatic incident light, the photocurrent is proportional to the light intensity. On the other hand, the current amplifier is utilized to transfer the photocurrent into an output voltage. There are two transfer ratios in our photodetector – high and low gains. In our experiment, only the low gain is used. However, because of the limitation of the photodiode itself, the output voltage would go into saturation at about 8 Volts if the light intensity is too high and the photodiode cannot operate properly any more. Hence the appropriate operating range of the photodetector is when the output voltage is indeed proportional to the light intensity. If the light intensity is too high so that the photodiode reaches the saturation, the reading of the photodetector can not correctly represent the incident light intensity.

II. Experiments and procedures

Characteristics of the laser diode & the photodetector

In order to make sure the experiments are done successfully, the optical alignment of light rays between different parts of an experimental setup is crucial. Also the light source and the detector should be operated at proper condition. Part A is related to these questions and the question of the degree of polarization.

1. Mount the laser diode and photodetector in a horizontal line on the optical bench, as shown in Fig. 5. Connect the variable resistor, battery set, ampere meter, voltage meter, laser diode and photodetector according to Fig. 6. Adjust the variable resistor so that the current passing through LD is around 25 mA and the laser diode emits laser light properly. Choose the low gain for the photodetector. Align the laser diode and the photodetector to make the laser light level at the small hole on the detector box and the reading of the photodetector reaches a maximum value.

Caution: Do not let the black and the red leads of the battery contact with each other to avoid short circuit.

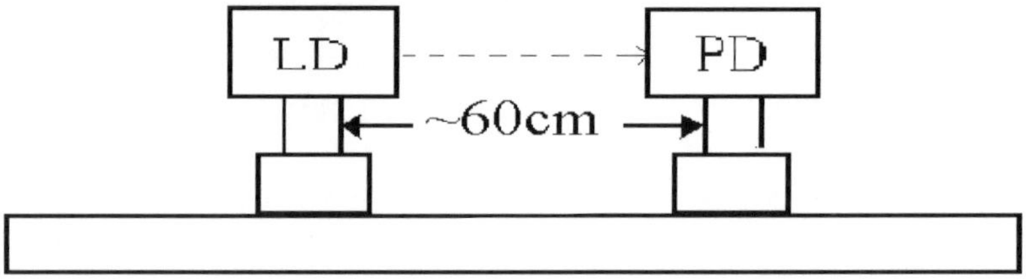

Fig. 5 Optical setup (LD : laser diode; PD : photodetector)

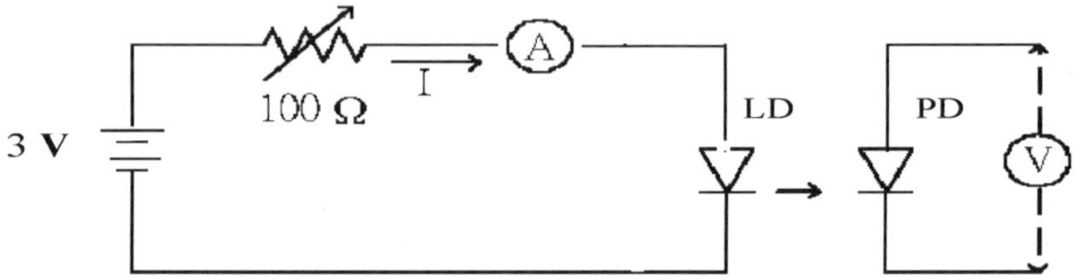

Fig. 6 Equivalent circuit for the connection of the laser diode

2. Use the output voltage of the photodetector to represent the laser light intensity $\mathcal{J}$. Adjust the variable resistor to make the current I of the laser diode varying from

zero to a maximum value and measure the $\mathcal{J}$ as I increases. Be sure to choose appropriate current increment in the measurement.

Question A-(1) (1.5 point) Measure, tabulate, and plot the $\mathcal{J}$ vs. I curve.

Question A-(2) (3.5 points)

Estimate the maximum current I_m with uncertainty in the linear region of the $\mathcal{J}$ vs. I curve. Mark the linear region on the $\mathcal{J}$ - I curve figure by using arrows (↓) and determine the threshold current I_{th} with uncertainty.

3. Choose the current of the laser diode as $I_{th} + 2(I_m - I_{th})/3$ to make sure the laser diode and photodetector are operated well.
4. **To prepare for the part B experiment:** Mount a polarizer on the optical bench close to the laser diode as shown in Fig. 7. Make sure the laser beam passing through the center portion of the polarizer. Adjust the polarizer so that the incident laser beam is perpendicular to the plane of the polarizer. (Hint: You can insert a piece of partially transparent paper as a test screen to check if the incident and reflected light spots coincide with each other.)

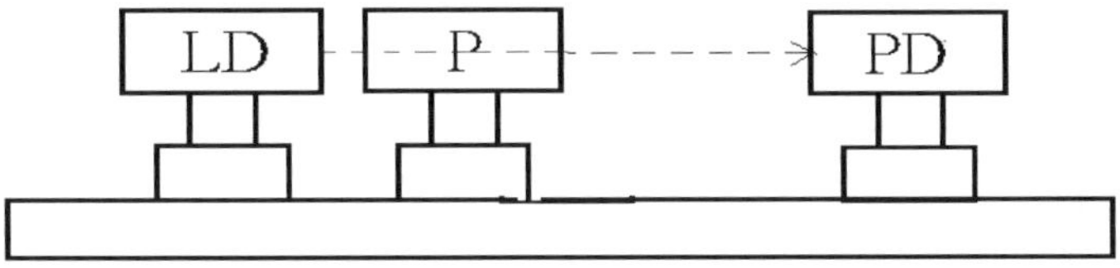

Fig. 7 Alignment of the polarizer (P : polarizer)

5. Keep the current of the laser diode unchanged, mount a second piece of polarizer on the optical bench and make sure proper alignment is accomplished, *i.e.*, set up the source, detector and polarizers in a straight line and make sure each polarizer plane is perpendicular to the light beam.

Part B Optical Properties of Nematic Liquid Crystal :

Electro-optical switching characteristic of 90° TN LC cell

I. Introduction

1. Liquid Crystal

Liquid crystal (LC) is a state of matter that is intermediate between the crystalline solid and the amorphous liquid. The nematic LCs are organic compounds consist of long-shaped needle-like molecules. The orientation of the molecules can be easily aligned and controlled by applying an electrical field. Uniform or well prescribed orientation of the LC molecules is required in most LC devices. The structure of the LC cell used in this experiment is shown in Fig 1. Rubbing the polyimide film can produce a well-aligned preferred orientation for LC molecules on substrate surfaces, thus due to the molecular interaction the whole slab of LC can achieve uniform molecular orientation. The local molecular orientation is called the director of LC at that point.

The LC cell exhibits the so-called double refraction phenomenon with two principal refractive indices. When light propagates along the direction of the director, all polarization components travel with the same speed $v_o = \frac{c}{n_o}$, where n_o is called the ordinary index of refraction. This propagation direction (direction of the director) is called the optic axis of the LC cell. When a light beam propagates in the direction perpendicular to the optic axis, in general, there are two speeds of propagation. The electric field of the light polarized perpendicular (or parallel) to the optic axis travels with the speed of $v_o = \frac{c}{n_o}$ (or $v_e = \frac{c}{n_e}$, where n_e is called the extraordinary index of refraction). The birefringence (optical anisotropy) is defined as the difference between the extraordinary and the ordinary indices of refraction $\Delta n \equiv n_e - n_o$.

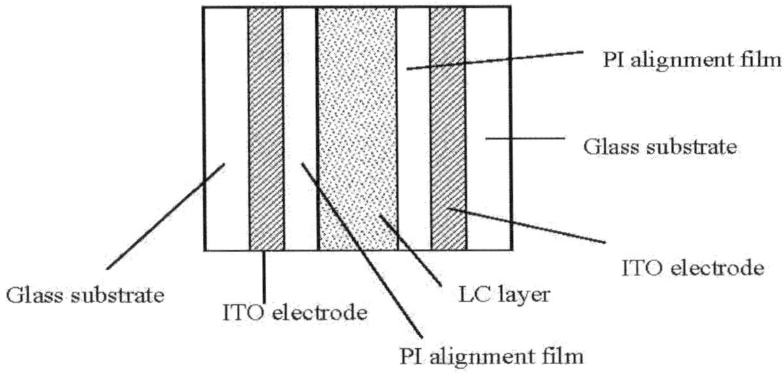

Fig. 1 LC cell structure

2. 90° Twisted Nematic LC Cell

In the 90° twisted nematic (TN) cell shown in Fig. 2, the LC director of the back surface is twisted 90° with respect to the front surface. The front local director is set parallel to the transmission axis of the polarizer. An incident unpolarized light is converted into a linearly polarized light by the front polarizer.

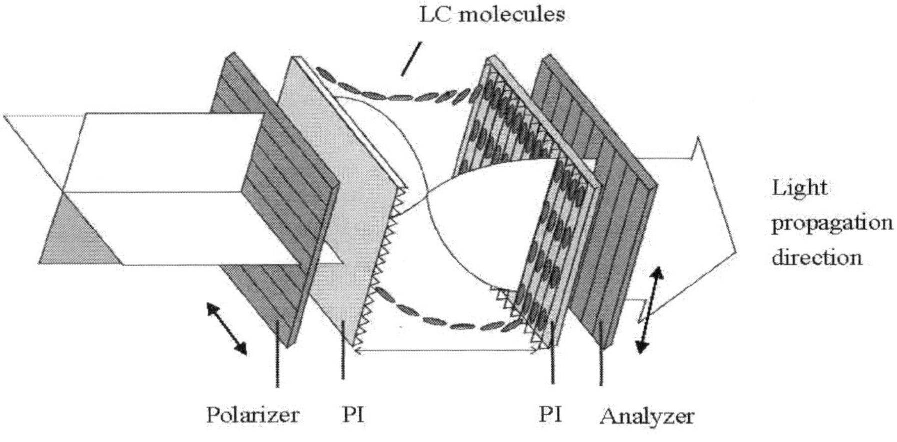

Fig. 2 90° TN LC cell

When a linearly polarized light traverses through a 90° TN cell, its polarization follows the twist of the LC directors (polarized light sees n_e only) so that the output beam remains linearly polarized except for that its polarization axis is rotated by 90° (it's called the polarizing rotary effect by n_e; similarly we can also find polarizing rotary effect by n_o). Thus, for a normally black (NB) mode using a 90° TN cell, the analyzer's (a second polarizer) transmission axis is set to be parallel to the polarizer's transmission axis, as shown in Fig. 3 (see next page). However, when the applied voltage V across the LC cell exceeds a critical value V_c, the director of LC molecules tends to align along the direction of applied external electrical field which is in the direction of the propagation of light. Hence, the polarization guiding effect of the LC cell is gradually diminishing and the light leaks through the analyzer. Its electro-optical switching slope γ is defined as $(V_{90}-V_{10})/V_{10}$, where V_{10} and V_{90} are the applied voltages enabling output light signal reaches up to 10% and 90% of its maximum light intensity, respectively.

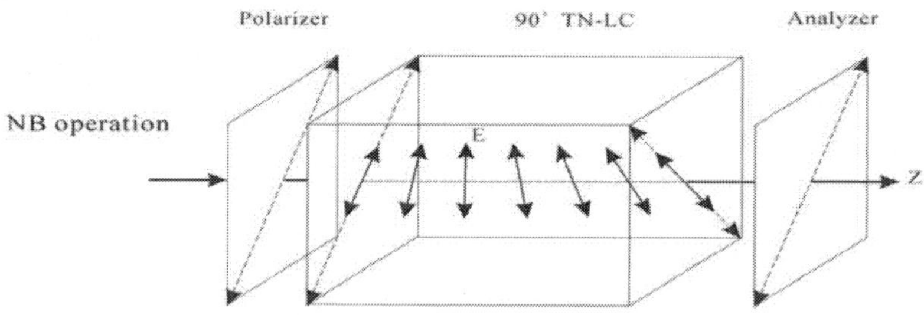

Fig. 3 NB mode operation of a 90° TN cell

II. Experiments and procedures

1. Setup a NB 90° TN LC mode between two polarizers with parallel transmission axes and apply 100 Hz square wave voltage using a function generator onto the ITO portions of two glass substrates and vary the applied voltage (V_{rms}) from 0 to 7.2 Volts.

   * In the crucial turning points, take more data if necessary.

Question B-(1) (5.0 points)

Measure, tabulate, and plot the electro-optical switching curve ($\mathcal{J}$ vs. V_{rms} curve) of the NB 90° TN LC, and find its switching slope γ, where γ is defined as $(V_{90}-V_{10})/V_{10}$.

Question B-(2) (2.5 points)

Determine the critical voltage V_c of this NB 90° TN LC cell. Show explicitly with graph how you determine the value V_c.

Hint:* When the external applied voltage exceeds the critical voltage, the light transmission increases rapidly and abruptly.

Part C Optical Properties of Nematic Liquid Crystal : Electro-optical switching characteristic of parallel aligned LC cell

I. Introduction

Homogeneous Parallel-aligned LC Cell

For a parallel-aligned LC cell, the directors in the front and back substrates are parallel with each other, as shown in Fig. 4. When a linearly polarized light impinges on a parallel-aligned cell with its polarization parallel to the LC director (rubbing direction), a pure phase modulation is achieved because the light behaves only as an extraordinary ray.

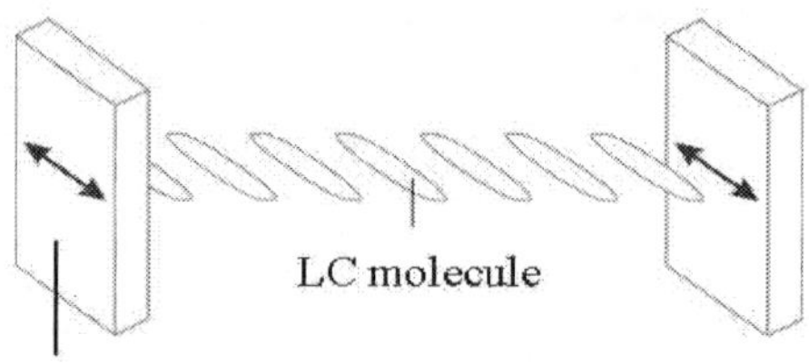

Fig. 4 Homogeneous parallel aligned LC cell

On the other hand, if a linearly polarized light is normally incident onto a parallel aligned cell but with its polarization making $\theta = 45°$ relative to the direction of the aligned LC directors (refer to Fig. 8 in part C), then phase retardation occurs due to the different propagating speed of the extraordinary and ordinary rays in the LC medium. In this $\theta = 45°$ configuration, when the two polarizers are parallel, the normalized transmission of a parallel aligned LC cell is given by

$$T_{\parallel} = \cos^2 \frac{\delta}{2}$$

The phase retardation δ is expressed as

$$\delta = 2\pi d \Delta n (V, \lambda) / \lambda$$

where d is the LC layer thickness, λ is the wavelength of light in air, V is the root mean square of applied AC voltage, and Δn, a function of λ and V, is the LC birefringence. It should be also noted that, at V = 0, $\Delta n (= n_e - n_o)$ has its maximum value, so does δ. Also Δn decreases as V increases.

In the general case, we have

$$T_{\parallel} = 1 - \sin^2 2\theta \sin^2 \frac{\delta}{2}$$

$$T_{\perp} = \sin^2 2\theta \sin^2 \frac{\delta}{2}$$

where $\parallel$ and $\perp$ represent that the transmission axis of analyzer is parallel and perpendicular to that of the polarizer, respectively.

II. Experiments and procedures

1. Replace NB 90° TN LC cell with parallel-aligned LC cell.
2. Set up θ= 45° configuration at V=0 as shown in Fig. 8. Let the analyzer's transmission axis perpendicular to that of the polarizer, then rotate the parallel-aligned LC cell until the intensity of the transmitted light reaches the maximum value ($T_{\perp}$). This procedure establishes the θ = 45° configuration. Take down $T_{\perp}$ value, then, measure the intensity of the transmitted light ($T_{\square}$) of the same LC cell at the analyzer's transmission axis parallel to that of the polarizer (also at V = 0).

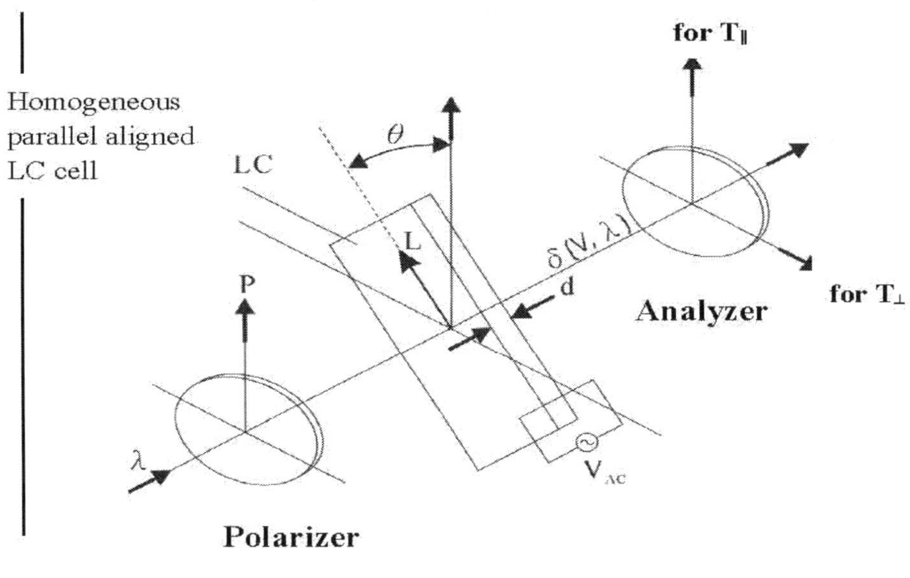

Fig. 8 Schematic diagram of experimental setup.

(The arrow L is the alignment direction)

Question C-(1) (2.5 points)
Assume that the wavelength of laser light 650 nm, LC layer thickness 7.7 μm, and approximate value of $\Delta n \approx 0.25$ are known. From the experimental data $T_{\perp}$ and $T_{\Box}$obtained above, calculate the accurate value of the phase retardation δ and accurate value of birefringence Δn of this LC cell at V=0.

3. Similar to the above experiment (1), in the $\theta = 45^{\circ}$ configuration, apply 100 Hz square wave voltage using a function generator onto the ITO portions of two glass substrates, vary the applied voltage (V_{rms}) from 0 to 7 Volts and measure the electro-optical switching curve ($T_{\parallel}$) at the analyzer's transmission axis parallel to the polarizer's transmission axis. (Hint: Measuring the $T_{\perp}$ switching curve is helpful to increase the data accuracy of the above $T_{\parallel}$ measurement; the data of $T_{\perp}$ are not needed in the following questions.)

* **In the crucial turning points, take more data if necessary (especially in the range of 0.5-4.0 Volts).**

Question_C-(2) (3.0 points)
Measure, tabulate, and plot the electro-optical switching curve for $T_{\parallel}$ of_this parallel aligned LC cell in the $\theta = 45^{\circ}$ configuration.

Question C-(3) (2.0 points)
From the electro-optical switching data, find the value of the external applied voltage V_{π}.

Hint: * V_{π} is the applied voltage which enables the phase retardation of this anisotropic LC cell become π (or 180°).

* Remember that Δn is a function of applied voltage, and Δn decreases as V increases.

* Interpolation is probably needed when you determine the accurate value of this V_{π}.

35TH

INTERNATIONAL PHYSICS OLYMPIAD

POHANG, SOUTH KOREA

2004

Theoretical Question 1:

"Ping-Pong" Resistor

A capacitor consists of two circular parallel plates both with radius R separated by distance d, where $d << R$, as shown in Fig. 1.1(a). The top plate is connected to a *constant* voltage source at a potential V while the bottom plate is grounded. Then a thin and small disk of mass m with radius r ($<< R, d$) and thickness t ($<< r$) is placed on the center of the bottom plate, as shown in Fig. 1.1(b).

Let us assume that the space between the plates is in vacuum with the dielectric constant ε_0; the plates and the disk are made of perfect conductors; and all the electrostatic edge effects may be neglected. The inductance of the whole circuit and the relativistic effects can be safely disregarded. The image charge effect can also be neglected.

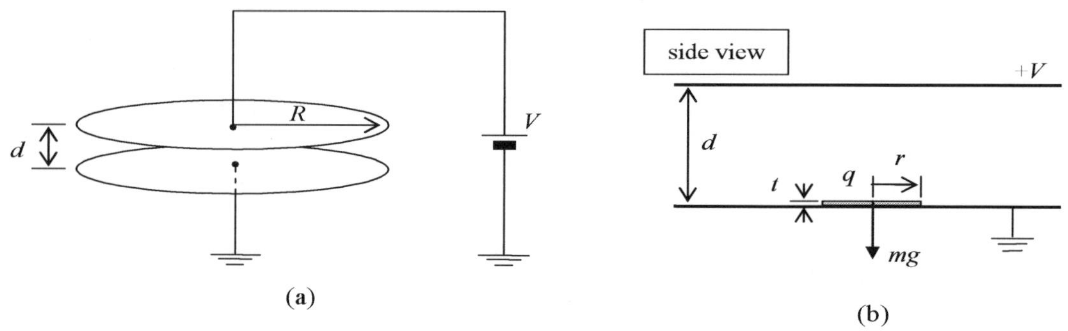

Figure 1.1 Schematic drawings of (a) a *parallel plate* capacitor connected to a constant voltage source and (b) a side view of the parallel *plates* with a small *disk* inserted inside the capacitor. (See text for details.)

(a) [1.2 points] Calculate the electrostatic force F_p between *the plates* separated by d before inserting the disk in-between as shown in Fig. 1.1(a).

(b) [0.8 points] When the disk is placed on the bottom plate, a charge q on *the disk* of Fig. 1.1(b) is related to the voltage V by $q = \chi V$. Find χ in terms of r, d, and ε_0.

(c) [0.5 points] The parallel plates lie perpendicular to a uniform gravitational field g. To lift up the disk at rest initially, we need to increase the applied voltage beyond a threshold voltage V_{th}. Obtain V_{th} in terms of m, g, d, and χ.

(d) [2.3 points] When $V > V_{\text{th}}$, the disk makes an up-and-down motion between the plates. (Assume that the disk moves only vertically *without any wobbling*.) The collisions between the disk and the plates are inelastic with the restitution coefficient $\eta \equiv (\text{v}_{\text{after}} / \text{v}_{\text{before}})$, where v_{before} and v_{after} are the speeds of the disk just before and after the collision respectively. The plates are stationarily fixed in position. The speed of the disk *just after* the collision at the bottom plate approaches a "steady-state speed" v_{s}, which depends on V as follows:

$$\text{v}_{\text{s}} = \sqrt{\alpha V^2 + \beta}\,. \tag{1.1}$$

Obtain the coefficients α and β in terms of m, g, χ, d, and η. Assume that the whole surface of the disk touches the plate evenly and simultaneously so that the complete charge exchange happens instantaneously at every collision.

(e) [2.2 points] After reaching its steady state, the time-averaged current I through the capacitor plates can be approximated by $I = \gamma V^2$ when $qV >> mgd$. Express the coefficient γ in terms of m, χ, d, and η.

(f) [3 points] When the applied voltage V is decreased (extremely slowly), there exists a critical voltage V_{c} below which the charge will cease to flow. Find V_{c} and the corresponding current I_{c} in terms of m, g, χ, d, and η. By comparing V_{c} with the lift-up threshold V_{th} discussed in (c), make a rough sketch of the $I-V$ characteristics when V is increased and decreased in the range from $V = 0$ to $3V_{\text{th}}$.

Theoretical Question 2

Rising Balloon

A rubber balloon filled with helium gas goes up high into the sky where the pressure and temperature decrease with height. In the following questions, assume that the shape of the balloon remains spherical regardless of the payload, and neglect the payload volume. Also assume that the temperature of the helium gas inside of the balloon is always the same as that of the ambient air, and treat all gases as ideal gases. The universal gas constant is R=8.31 J/mol·K and the molar masses of helium and air are $M_H = 4.00\times10^{-3}$ kg/mol and $M_A = 28.9\times10^{-3}$ kg/mol, respectively. The gravitational acceleration is $g = 9.8$ m/s^2.

[Part A]

(a) [1.5 points] Let the pressure of the ambient air be P and the temperature be T. The pressure inside of the balloon is higher than that of outside due to the surface tension of the balloon. The balloon contains n moles of helium gas and the pressure inside is $P+\Delta P$. Find the buoyant force $F_{\rm B}$ acting on the balloon as a function of P and ΔP.

(b) [2 points] On a particular summer day in Korea, the air temperature T at the height z from the sea level was found to be $T(z) = T_0(1 - z/z_0)$ in the range of $0 < z < 15$ km with z_0=49 km and T_0 =303 K. The pressure and density at the sea level were P_0 = 1.0 atm = 1.01×10^5 Pa and ρ_0 = 1.16 kg/m^3, respectively. For this height range, the pressure takes the form

$$P(z) = P_0(1 - z/z_0)^{\eta} \quad . \qquad (2.1)$$

Express η in terms of z_0, ρ_0, P_0, and g, and find its numerical value to the *two* significant digits. Treat the gravitational acceleration as a constant, independent of height.

[Part B]

When a rubber balloon of spherical shape with un-stretched radius r_0 is inflated to a sphere of radius r $(\geq r_0)$, the balloon surface contains extra elastic energy due to the stretching. In a simplistic theory, the elastic energy at constant temperature T can be expressed by

$$U = 4\pi r_0^2 \kappa RT(2\lambda^2 + \frac{1}{\lambda^4} - 3) \tag{2.2}$$

where $\lambda \equiv r/r_0$ (≥ 1) is the size-inflation ratio and κ is a constant in units of mol/m^2.

(c) [2 points] Express ΔP in terms of parameters given in Eq. (2.2), and sketch ΔP as a function of $\lambda = r/r_0$.

(d) [1.5 points] The constant κ can be determined from the amount of the gas needed to inflate the balloon. At T_0=303 K and P_0=1.0 atm = 1.01×10^5 Pa, an un-stretched balloon ($\lambda = 1$) contains n_0=12.5 moles of helium. It takes n=3.6n_0=45 moles in total to inflate the balloon to λ=1.5 at the same T_0 and P_0. Express the balloon parameter a, defined as $a = \kappa/\kappa_0$, in terms of n, n_0, and λ, where $\kappa_0 \equiv \dfrac{r_0 P_0}{4RT_0}$. Evaluate a to the two significant digits.

[Part C]

A balloon is prepared as in (d) at the sea level (inflated to $\lambda = 1.5$ with $n = 3.6n_0 = 45$ moles of helium gas at T_0=303 K and P_0=1 atm=1.01×10^5 Pa). The total mass including gas, balloon itself, and other payloads is $M_T = 1.12$ kg. Now let the balloon rise from the sea level.

(e) [3 points] Suppose that the balloon eventually stops at the height z_f where the buoyant force balances the total weight. Find z_f and the inflation ratio λ_f at that height. Give the answers in two significant digits. Assume there are no drift effect and no gas leakage during the upward flight.

Theoretical Question 3

Atomic Probe Microscope

Atomic probe microscopes (APMs) are powerful tools in the field of nano-science. The motion of a cantilever in APM can be detected by a photo-detector monitoring the reflected laser beam, as shown in Fig. 3.1. The cantilever can move only in the vertical direction and its displacement z as a function of time t can be described by the equation

$$m\frac{d^2z}{dt^2}+b\frac{dz}{dt}+kz=F, \qquad (3.1)$$

where m is the cantilever mass, $k=m\omega_0^2$ is the spring constant of the cantilever, b is a small damping coefficient satisfying $\omega_0 >> (b/m) > 0$, and finally F is an external driving force of the piezoelectric tube.

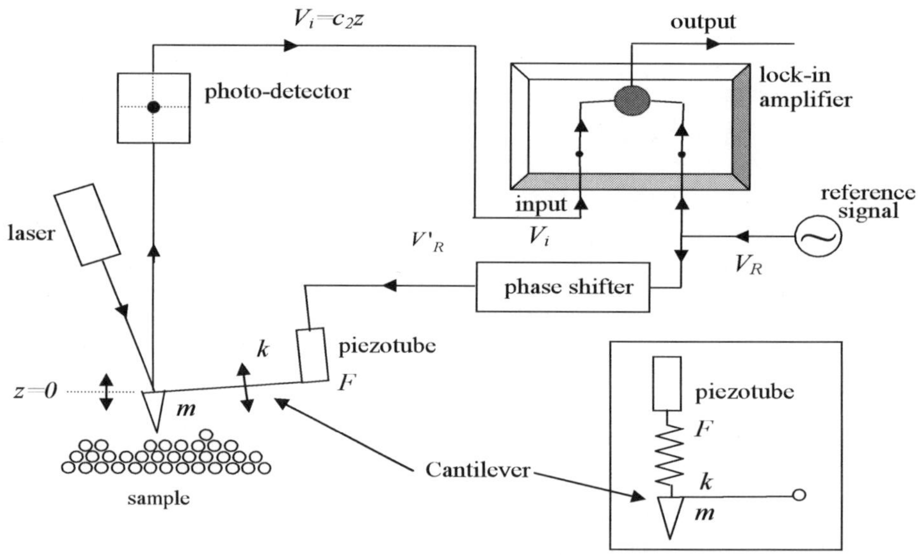

Figure 3.1 A schematic diagram for a scanning probe microscope (SPM). The inset in the lower right corner represents a simplified mechanical model to describe the coupling of the piezotube with the cantilever.

[Part A]

(a) [1.5 points] When $F=F_0\sin\omega t$, $z(t)$ satisfying Eq. (3.1) can be written as $z(t)=A\sin(\omega t-\phi)$, where $A>0$ and $0\le\phi\le\pi$. Find the expression of the amplitude A and $\tan\phi$ in terms of F_0, m, ω, ω_0, and b. Obtain A and the phase ϕ at the resonance frequency $\omega = \omega_0$.

(b) [1 point] A lock-in amplifier shown in Fig.3.1 multiplies an input signal by the lock-in reference signal, $V_R = V_{R0}\sin\omega t$, and then passes *only* the dc (direct current) component of the multiplied signal. Assume that the input signal is given by $V_i = V_{i0}\sin(\omega_i t - \phi_i)$. Here V_{R0}, V_{i0}, ω_i, and ϕ_i are all positive given constants. Find the condition on ω (>0) for a non-vanishing output signal. What is the expression for the magnitude of the non-vanishing *dc output signal* at this frequency?

(c) [1.5 points] Passing through the phase shifter, the lock-in reference voltage $V_R = V_{R0}\sin\omega t$ changes to $V'_R = V_{R0}\sin(\omega t + \pi/2)$. V'_R, applied to the piezoelectric tube, drives the cantilever with a force $F = c_1 V'_R$. Then, the photo-detector converts the displacement of the cantilever, z, into a voltage $V_i = c_2 z$. Here c_1 and c_2 are constants. Find the expression for the magnitude of the *dc output signal* at $\omega = \omega_0$.

(d) [2 points] The small change Δm of the cantilever mass shifts the resonance frequency by $\Delta\omega_0$. As a result, the phase ϕ at the original resonance frequency ω_0 shifts by $\Delta\phi$. Find the mass change Δm corresponding to the phase shift $\Delta\phi = \pi/1800$, which is a typical resolution in phase measurements. The physical parameters of the cantilever are given by $m = 1.0\times10^{-12}$ kg, $k = 1.0$ N/m, and $(b/m) = 1.0\times10^{3}$ s^{-1}. Use the approximations $(1+x)^a \approx 1 + ax$ and $\tan(\pi/2 + x) \approx -1/x$ when $|x| << 1$.

[Part B]

From now on let us consider the situation that some forces, besides the driving force discussed in Part A, act on the cantilever due to the sample as shown in Fig.3.1.

(e) [1.5 points] Assuming that the additional force $f(h)$ depends only on the distance h between the cantilever and the sample surface, one can find a new equilibrium position h_0. Near $h = h_0$, we can write $f(h) \approx f(h_0) + c_3(h - h_0)$, where c_3 is a constant in h. Find the new resonance frequency ω'_0 in terms of ω_0, m, and c_3.

(f) [2.5 points] While scanning the surface by moving the sample horizontally, the tip of the cantilever charged with $Q = 6e$ encounters an electron of charge $q = e$ trapped

(localized in space) at some distance below the surface. During the scanning around the electron, the maximum shift of the resonance frequency $\Delta\omega_0 (= \omega'_0 - \omega_0)$ is observed to be much smaller than ω_0. Express the distance d_0 from the cantilever to the trapped electron at the maximum shift in terms of m, q, Q, ω_0, $\Delta\omega_0$, and the Coulomb constant k_e. Evaluate d_0 in nm (1 nm = 1×10^{-9} m) for $\Delta\omega_0 = 20\ \text{s}^{-1}$.
The physical parameters of the cantilever are $m = 1.0\times10^{-12}$ kg and $k = 1.0$ N/m. Disregard any polarization effect in both the cantilever tip and the surface. Note that $k_e = 1/4\pi\varepsilon_0 = 9.0\times10^{9}\ \text{N}\cdot\text{m}^2/\text{C}^2$ and $e = -1.6\times10^{-19}$ C.

Theoretical Question 1: ***Ping-Pong Resistor***

1. Answers

(a) $F_R = -\frac{1}{2}\pi R^2 \varepsilon_0 \frac{V^2}{d^2}$

(b) $\chi = -\varepsilon_0 \frac{\pi r^2}{d}$

(c) $V_{th} = \sqrt{\frac{2mgd}{\chi}}$

(d) $v_s = \sqrt{\alpha V^2 + \beta}$

$$\alpha = \left(\frac{\eta^2}{1-\eta^2}\right)\left(\frac{2\chi}{m}\right), \quad \beta = \left(\frac{\eta^2}{1+\eta^2}\right)(2gd)$$

(e) $\gamma = \sqrt{\frac{1+\eta}{1-\eta}}\sqrt{\frac{\chi^3}{2md^2}}$

(f) $V_c = \sqrt{\frac{1-\eta^2}{1+\eta^2}}\sqrt{\frac{mgd}{\chi}}, \quad I_c = \frac{2\eta\sqrt{1-\eta^2}}{(1+\eta)(1+\eta^2)} g\sqrt{m\chi}$

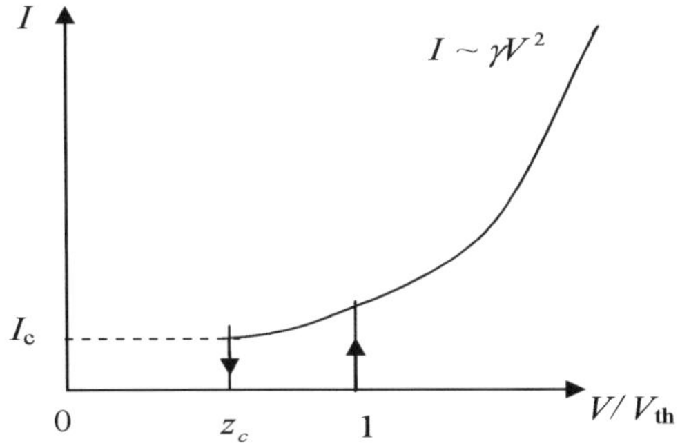

2. Solutions

(a) [1.2 points]

The charge Q induced by the external bias voltage V can be obtained by applying the Gauss law:

$$\varepsilon_0 \oint \vec{E} \cdot d\vec{s} = Q \qquad \text{(a1)}$$

$$Q = \varepsilon_0 E \cdot (\pi R^2) = \varepsilon_0 \left(\frac{V}{d}\right) \cdot (\pi R^2), \qquad \text{(a2)}$$

where $V = Ed$.

The energy stored in the capacitor:

$$U = \int_0^V Q(V')dV' = \int_0^V \varepsilon_0 \pi R^2 \left(\frac{V'}{d}\right) dV' = \frac{1}{2}\varepsilon_0 \pi R^2 \frac{V^2}{d}. \qquad \text{(a3)}$$

The force acting on the plate, when the bias voltage V is kept constant:

$$\therefore F_{\mathrm{R}} = +\frac{\partial U}{\partial d} = -\frac{1}{2}\varepsilon_0 \pi R^2 \frac{V^2}{d^2}. \qquad \text{(a4)}$$

[An alternative solution:]

Since the electric field E' acting on one plate should be generated by the other plate and its magnitude is

$$E' = \frac{1}{2}E = \frac{V}{2d}, \qquad \text{(a5)}$$

the force acting on the plate can be obtained by

$$F_{\mathrm{R}} = QE'. \qquad \text{(a6)}$$

(b) [0.8 points]

The charge q on the small disk can also be calculated by applying the Gauss law:

$$\varepsilon_0 \oint \vec{E} \cdot d\vec{s} = q. \qquad \text{(b1)}$$

Since one side of the small disk is in contact with the plate,

$$q = -\varepsilon_0 E \cdot (\pi r^2) = -\varepsilon_0 \frac{\pi r^2}{d} V = \chi V. \qquad \text{(b2)}$$

Alternatively, one may use the area ratio for $q = -\left(\frac{\pi r^2}{\pi R^2}\right)Q$.

$$\therefore \chi = -\varepsilon_0 \frac{\pi r^2}{d}. \tag{b3}$$

(c) [0.5 points]

The net force, F_{net}, acting on the small disk should be a sum of the gravitational and electrostatic forces:

$$F_{\text{net}} = F_{\text{g}} + F_{\text{e}}. \tag{c1}$$

The gravitational force: $F_{\text{g}} = -mg$.

The electrostatic force can be derived from the result of (a) above:

$$F_{\text{e}} = \frac{1}{2}\varepsilon_0 \frac{\pi r^2}{d^2}V^2 = \frac{\chi}{2d}V^2. \tag{c2}$$

In order for the disk to be lifted, one requires $F_{\text{net}} > 0$:

$$\frac{\chi}{2d}V^2 - mg > 0. \tag{c3}$$

$$\therefore V_{\text{th}} = \sqrt{\frac{2mgd}{\chi}}. \tag{c4}$$

(d) [2.3 points]

Let v_{s} be the steady velocity of the small disk just after its collision with the bottom plate. Then the *steady-state* kinetic energy K_{s} of the disk just above the bottom plate is given by

$$K_{\text{s}} = \frac{1}{2}m\text{v}_{\text{s}}^2. \tag{d1}$$

For each round trip, the disk gains electrostatic energy by

$$\Delta U = 2qV. \tag{d2}$$

For each inelastic collision, the disk lose its kinetic energy by

$$\Delta K_{\text{loss}} = K_{\text{before}} - K_{\text{after}} = (1-\eta^2)K_{\text{before}} = \left(\frac{1}{\eta^2} - 1\right)K_{\text{after}}. \tag{d3}$$

Since K_{s} is the energy after the collision at the bottom plate and $(K_{\text{s}} + qV - mgd)$ is the energy before the collision at the top plate, the total energy loss during the round trip can be written in terms of K_s:

$$\Delta K_{\text{tot}} = \left(\frac{1}{\eta^2} - 1\right) K_s + (1-\eta^2)(K_s + qV - mgd). \qquad \text{(d4)}$$

In its steady state, ΔU should be compensated by ΔK_{tot}.

$$2qV = \left(\frac{1}{\eta^2} - 1\right) K_s + (1-\eta^2)(K_s + qV - mgd). \qquad \text{(d5)}$$

Rearranging Eq. (d5), we have

$$\begin{aligned} K_s &= \frac{\eta^2}{1-\eta^4}\left[(1+\eta^2)qV + (1-\eta^2)mgd\right] \\ &= \left(\frac{\eta^2}{1-\eta^2}\right)qV + \left(\frac{\eta^2}{1+\eta^2}\right)mgd \\ &= \frac{1}{2}m\mathrm{v}_s^2. \end{aligned} \qquad \text{(d6)}$$

Therefore,

$$\mathrm{v}_s = \sqrt{\left(\frac{\eta^2}{1-\eta^2}\right)\left(\frac{2\chi V^2}{m}\right) + \left(\frac{\eta^2}{1+\eta^2}\right)(2gd)}. \qquad \text{(d7)}$$

Comparing with the form:

$$\mathrm{v}_s = \sqrt{\alpha V^2 + \beta}, \qquad \text{(d8)}$$

$$\alpha = \left(\frac{\eta^2}{1-\eta^2}\right)\left(\frac{2\chi}{m}\right), \quad \beta = \left(\frac{\eta^2}{1+\eta^2}\right)(2gd). \qquad \text{(d9)}$$

[An alternative solution:]

Let v_n be the velocity of the small disk just after n-th collision with the bottom plate. Then the kinetic energy of the disk just above the bottom plate is given by

$$K_n = \frac{1}{2}m\mathrm{v}_n^2. \qquad \text{(d10)}$$

When it reaches the top plate, the disk gains energy by the increase of potential energy:

$$\Delta U_{\text{up}} = qV - mgd. \qquad \text{(d11)}$$

Thus, the kinetic energy just before its collision with the top plate becomes

$$K_{n\text{-up}} = \frac{1}{2}m\mathrm{v}_{\text{up}}^2 = K_n + \Delta U_{\text{up}}. \qquad \text{(d12)}$$

Since $\eta = \mathrm{v}_{\text{after}} / \mathrm{v}_{\text{before}}$, the kinetic energy after the collision with the top plate becomes scaled down by a factor of η^2:

$$K'_{n\ \text{up}} = \eta^2 \cdot K_{n\ \text{up}}. \tag{d13}$$

Now the potential energy gain by the downward motion is:

$$\Delta U_{\text{down}} = qV + mgd \tag{d14}$$

so that the kinetic energy just before it collides with the bottom plate becomes:

$$K_{n\ \text{down}} = K'_{n\ \text{up}} + \Delta U_{\text{down}}. \tag{d15}$$

Again, due to the loss of energy by the collision with the bottom plate, the kinetic energy after its $(n+1)$-th collision can be obtained by

$$\begin{aligned} K_{n+1} &= \eta^2 \cdot K_{n\ \text{down}} \\ &= \eta^2 (K'_{n\ \text{up}} + \Delta U_{\text{down}}) \\ &= \eta^2 (\eta^2 (K_n + \Delta U_{\text{up}}) + \Delta U_{\text{down}}) \\ &= \eta^2 (\eta^2 (K_n + qV - mgd) + qV + mgd) \\ &= \eta^4 K_n + \eta^2 (1+\eta^2) qV + \eta^2 (1-\eta^2) mgd. \end{aligned} \tag{d16}$$

As $n \to \infty$, we expect the velocity $\mathrm{v}_n \to \mathrm{v}_s$, that is, $K_n \to K_s = \frac{1}{2} m\mathrm{v}_s^2$:

$$\begin{aligned} K_s &= \frac{1}{1-\eta^4} \left[\eta^2 (1+\eta^2) qV + \eta^2 (1-\eta^2) mgd \right] \\ &= \left(\frac{\eta^2}{1-\eta^2} \right) qV + \left(\frac{\eta^2}{1+\eta^2} \right) mgd \\ &= \frac{1}{2} m\mathrm{v}_s^2 \end{aligned} \tag{d17}$$

(e) [2.2 points]

The amount of charge carried by the disk during its round trip between the plates is $\Delta Q = 2q$, and the time interval $\Delta t = t_+ + t_-$, where $t_+ (t_-)$ is the time spent during the up- (down-) ward motion respectively.

Here $t_+ (t_-)$ can be determined by

$$\begin{aligned} \mathrm{v}_{0+} t_+ + \frac{1}{2} a_+ t_+^2 &= d \\ \mathrm{v}_{0-} t_- + \frac{1}{2} a_- t_-^2 &= d \end{aligned} \tag{e1}$$

where $\mathrm{v}_{0+} (\mathrm{v}_{0-})$ is the initial velocity at the bottom (top) plate and $a_+ (a_-)$ is the up-

(down-) ward acceleration respectively.

Since the force acting on the disk is given by

$$F = ma_{\pm} = qE \mp mg = \frac{qV}{d} \mp mg\,, \tag{e2}$$

in the limit of $mgd << qV$, $a_{\pm}$ can be approximated by

$$a_0 = a_+ = a_- \approx \frac{qV}{md}, \tag{e3}$$

which implies that the upward and down-ward motion should be symmetric. Thus, Eq.(e1) can be described by a single equation with $t_0 = t_+ = t_-$, $v_s = v_{0+} = v_{0-}$, and $a_0 = a_+ = a_-$. Moreover, since the speed of the disk just after the collision should be the same for the top- and bottom-plates, one can deduce the relation:

$$v_s = \eta(v_s + a_0 t_0), \tag{e4}$$

from which we obtain the time interval $\Delta t = 2t_0$,

$$\Delta t = 2t_0 = 2\left(\frac{1-\eta}{\eta}\right)\frac{v_s}{a_0}. \tag{e5}$$

From Eq. (d6), in the limit of $mgd << qV$, we have

$$K_s = \frac{1}{2}mv_s^2 \approx \left(\frac{\eta^2}{1-\eta^2}\right)qV. \tag{e6}$$

By substituting the results of Eqs. (e3) and (e6), we get

$$\Delta t = 2\left(\frac{1-\eta}{\eta}\right)\sqrt{\frac{2\eta^2}{1-\eta^2}}\sqrt{\frac{md^2}{qV}} = 2\sqrt{\frac{1-\eta}{1+\eta}}\sqrt{\frac{2md^2}{\chi V^2}}. \tag{e7}$$

Therefore, from $I = \frac{\Delta Q}{\Delta t} = \frac{2q}{\Delta t}$,

$$I = \frac{2q}{\Delta t} = \chi V\sqrt{\frac{1+\eta}{1-\eta}}\sqrt{\frac{\chi V^2}{2md^2}} = \sqrt{\frac{1+\eta}{1-\eta}}\sqrt{\frac{\chi^3}{2md^2}}V^2. \tag{e8}$$

$$\therefore \gamma = \sqrt{\frac{1+\eta}{1-\eta}}\sqrt{\frac{\chi^3}{2md^2}} \tag{e9}$$

[Alternative solution #1:]

Starting from Eq. (e3), we can solve the quadratic equation of Eq. (e1) so that

$$t_{\pm} = \frac{\mathrm{v}_{0\pm}}{a_0}\left(\sqrt{1+\frac{2da_0}{\mathrm{v}_{0\pm}^2}}-1\right). \tag{e10}$$

When it reaches the steady state, the initial velocities $\mathrm{v}_{0\pm}$ are given by

$$\mathrm{v}_{0+} = \mathrm{v}_{\mathrm{s}} \tag{e11}$$

$$\mathrm{v}_{0-} = \eta\cdot(\mathrm{v}_{\mathrm{s}} + a_0 t_+) = \eta\mathrm{v}_{\mathrm{s}}\sqrt{1+\frac{2da_0}{\mathrm{v}_{\mathrm{s}}^2}}, \tag{e12}$$

where v_{s} can be rewritten by using the result of Eq. (e6),

$$\mathrm{v}_{\mathrm{s}}^2 \approx \alpha V = \left(\frac{\eta^2}{1-\eta^2}\right)\frac{2qV}{m} = \left(\frac{\eta^2}{1-\eta^2}\right)2a_0 d. \tag{e13}$$

As a result, we get $\mathrm{v}_{0-} \cong \eta\mathrm{v}_{\mathrm{s}}\cdot\frac{1}{\eta} = \mathrm{v}_{\mathrm{s}}$ and consequently $t_{\pm} = \frac{\mathrm{v}_{\mathrm{s}}}{a_0}\left(\frac{1}{\eta}-1\right)$, which is equivalent to Eq. (e4).

[Alternative solution #2:]

The current I can be obtained from

$$I = \frac{2q}{\Delta t} = \frac{2q\bar{\mathrm{v}}}{d}, \tag{e14}$$

where $\bar{\mathrm{v}}$ is an average velocity. Since the up and down motions are symmetric with the same constant acceleration in the limit of $mgd << qV$,

$$\bar{\mathrm{v}} = \frac{1}{2}\left(\mathrm{v}_{\mathrm{s}} + \frac{\mathrm{v}_{\mathrm{s}}}{\eta}\right). \tag{e15}$$

Thus, we have

$$I = \frac{q}{2d}\left(1+\frac{1}{\eta}\right)\mathrm{v}_{\mathrm{s}}. \tag{e16}$$

Inserting the expression (Eq. (e15)) of v_{s} into Eq. (e16), one obtains an expression identical to Eq. (e8).

(f) [3 points]

The disk will lose its kinetic energy and eventually cease to move when the disk can not reach the top plate. In other words, the threshold voltage V_{c} can be determined from the condition that the velocity v_{0-} of the disk at the top plate is zero, i.e., $\mathrm{v}_{0-} = 0$.

In order for the disk to have $\mathrm{v}_{0-} = 0$ at the top plate, the kinetic energy $\bar{K}_{\mathrm{s}}$ at the top plate should satisfy the relation:

$$\overline{K}_s = K_s + qV_c - mgd = 0, \tag{f1}$$

where K_s is the *steady-state* kinetic energy at the bottom plate after the collision. Therefore, we have

$$\left(\frac{\eta^2}{1-\eta^2}\right)qV_c + \left(\frac{\eta^2}{1+\eta^2}\right)mgd + qV_c - mgd = 0, \tag{f2}$$

or equivalently,

$$(1+\eta^2)qV_c - (1-\eta^2)mgd = 0. \tag{f3}$$

$$\therefore \quad qV_c = \frac{1-\eta^2}{1+\eta^2}mgd \tag{f4}$$

From the relation $q = \chi V_c$,

$$\therefore V_c = \sqrt{\frac{1-\eta^2}{1+\eta^2}}\sqrt{\frac{mgd}{\chi}}. \tag{f5}$$

In comparison with the threshold voltage V_{th} of Eq. (c4), we can rewrite Eq. (f5) by

$$V_c = z_c V_{\text{th}} \tag{f6}$$

where z_c should be used in the plot of I vs. (V/V_{th}) and

$$z_c = \sqrt{\frac{1-\eta^2}{2(1+\eta^2)}}. \tag{f7}$$

[Note that an alternative derivation of Eq. (f1) is possible if one applies the energy compensation condition of Eq. (d5) or the recursion relation of Eq. (d17) at the top plate instead of the bottom plate.]

Now we can setup equations to determine the time interval $\Delta t = t_- + t_+$:

$$\mathrm{v}_{0-}t_- + \frac{1}{2}a_- t_-^2 = d \tag{f8}$$

$$\mathrm{v}_{0+}t_+ + \frac{1}{2}a_+ t_+^2 = d \tag{f9}$$

where the accelerations are given by

$$a_+ = \frac{qV_c}{md} - g = \left[\frac{1-\eta^2}{1+\eta^2} - 1\right]g = \left(\frac{-2\eta^2}{1+\eta^2}\right)g \tag{f10}$$

$$a_- = \frac{qV_c}{md} + g = \left[\frac{1-\eta^2}{1+\eta^2} + 1\right] g = \left(\frac{2}{1+\eta^2}\right) g \quad \text{(f11)}$$

$$\frac{a_+}{a_-} = -\eta^2 \quad \text{(f12)}$$

Since $v_{0-} = 0$, we have $v_{0+} = \eta(a_- t_-)$ and $t_-^2 = 2d/a_-$.

$$t_- = \sqrt{\frac{2d}{a_-}} = \sqrt{(1+\eta^2)\left(\frac{d}{g}\right)}, \quad \text{(f13)}$$

By using $v_{0+}^2 = \eta^2(2da_-) = -2da_+$, we can solve the quadratic equation of Eq. (f9):

$$t_+ = \frac{v_{0+}}{a_+}\left(\sqrt{1 + \frac{2da_+}{v_{0+}^2}} - 1\right) = -\frac{v_{0+}}{a_+} = \sqrt{\frac{2d}{|a_+|}} = \sqrt{\left(\frac{1+\eta^2}{\eta^2}\right)\left(\frac{d}{g}\right)} = \frac{t_-}{\eta}. \quad \text{(f14)}$$

$$\therefore \Delta t = t_- + t_+ = \left(1 + \frac{1}{\eta}\right)\sqrt{(1+\eta^2)\left(\frac{d}{g}\right)} \quad \text{(f15)}$$

$$I_c = \frac{\Delta Q_c}{\Delta t} = \frac{2q}{\Delta t} = \frac{2\chi V_c}{\Delta t} = \frac{2\eta\sqrt{1-\eta^2}}{(1+\eta)(1+\eta^2)} g\sqrt{m\chi}. \quad \text{(f16)}$$

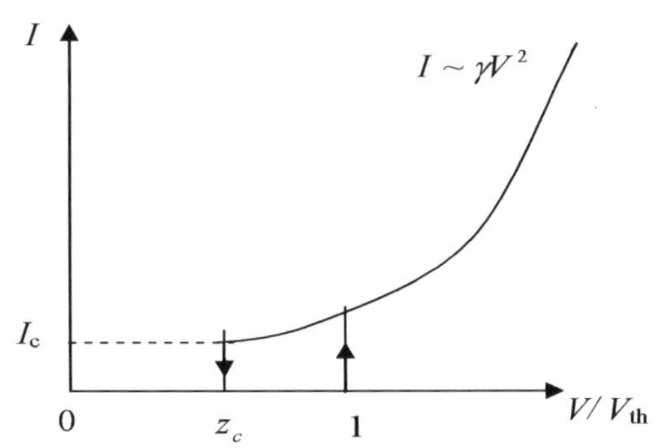

[A more elaborate Solution:]

One may find a general solution for an arbitrary value of V. By solving the quadratic equations of Eqs. (f8) and (f9), we have

$$t_{\pm} = \frac{\mathrm{v}_{0\pm}}{a_{\pm}}\left[-1+\sqrt{1+\frac{2da_{\pm}}{\mathrm{v}_{0\pm}^{2}}}\right]. \tag{f17}$$

(It is noted that one has to keep the smaller positive root.)

To simplify the notation, we introduce a few variables:

(i) $y = \dfrac{V}{V_{\mathrm{th}}}$ where $V_{\mathrm{th}} = \sqrt{\dfrac{2mgd}{\chi}}$,

(ii) $z_c = \sqrt{\dfrac{1-\eta^2}{2(1+\eta^2)}}$, which is defined in Eq. (f7),

(iii) $w_0 = 2\eta\sqrt{\dfrac{gd}{1-\eta^2}}$ and $w_1 = 2\sqrt{\dfrac{d}{(1-\eta^2)g}}$,

In terms of y, w, and z_c,

$$a_{+} = \frac{qV}{md} - g = g(2y^2-1) \tag{f18}$$

$$a_{-} = \frac{qV}{md} + g = g(2y^2+1) \tag{f19}$$

$$\mathrm{v}_{0+} = \mathrm{v}_s = w_0\sqrt{y^2+z_c^2} \tag{f20}$$

$$\mathrm{v}_{0-} = \eta(\mathrm{v}_s + a_+ t_+) = w_0\sqrt{y^2-z_c^2} \tag{f21}$$

$$t_{+} = w_1\frac{\sqrt{y^2-z_c^2}-\eta\sqrt{y^2+z_c^2}}{2y^2-1} \tag{f22}$$

$$t_{-} = w_1\frac{\sqrt{y^2+z_c^2}-\eta\sqrt{y^2-z_c^2}}{2y^2+1} \qquad \text{(f21)}$$

$$I = \frac{\Delta Q}{\Delta t} = \frac{2q}{t_+ + t_-} = (2\chi V_{\text{th}})\frac{y}{\Delta t} = \frac{\sqrt{8mgd\chi}}{w_1}F(y) \qquad \text{(f22)}$$

where

$$F(y) = y\left\{\frac{\sqrt{y^2 - z_c^2} - \eta\sqrt{y^2 + z_c^2}}{2y^2 - 1} + \frac{\sqrt{y^2 + z_c^2} - \eta\sqrt{y^2 - z_c^2}}{2y^2 + 1}\right\}^{-1} \qquad \text{(f23)}$$

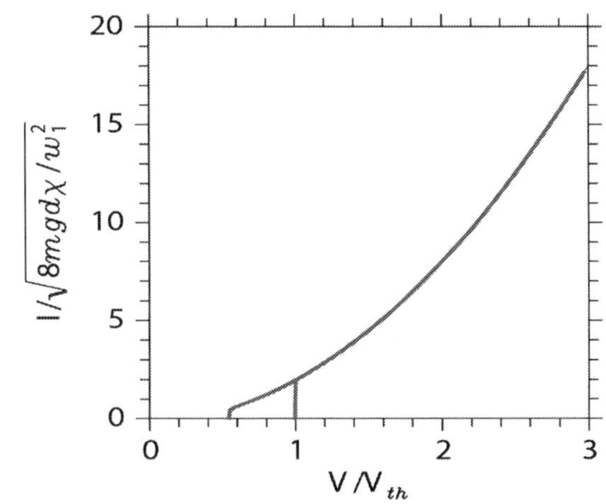

3. Mark Distribution

No.	Total Pt.	Partial Pt.	Contents	
(a)	1.2	0.3	Gauss law, or a formula for the capacitance of a parallel plate	
		0.5	Total energy of a capacitor at V	E'=electrical field by the other plate
		0.4	Force from the energy expression	$F = QE'$
(b)	0.8	0.3	Gauss law	Use of area ratio and result of (a)
		0.5	Correct answer	
(c)	0.5	0.1	Correct lift-up condition with force balance	
		0.2	Use of area ratio and result of (a)	
		0.2	Correct answer	
(d)	2.3	0.5	Energy conservation and the work done by the field	
		0.5	Loss of energy due to collisions	
		0.8	Condition for the steady state: energy balance equation (loss = gain)	Condition for the steady state: recursion relation
		0.5	Correct answer	
(e)	2.2	0.2	$\Delta Q = 2q$ per trip	
		0.5	Acceleration $a_\pm$ in the limit of $qV >> mgd$; $a_+ = a_-$ by symmetry	
		0.3	Kinetic equations for d , v , a , and t , solutions for t_+	By using the symmetry, derive the relation (e4)
		0.4	Expression of $v_{0\pm}$ and $t_\pm$ in its steady state	
		0.4	Solutions of $t_\pm$ in approximation	
		0.4	Correct answer	
(f)	3.0	0.5	Condition for V_c; $K_{\text{up}} = 0$ or $v_{s,\text{up}} = 0$	Using (d8), Recursion relations
		0.3	energy balance equation	
		0.3	Correct answer of V_c	
		0.7	Kinetic equations for Δt	
		0.3	Correct answer of I_c	
		0.9	Distinction between V_{th} and V_c, the asymptotic behavior $I = \gamma V^2$ in plots	
Total	10			

Theoretical Question 2: *Rising Balloon*

1. Answers

(a) $F_B = M_A ng \dfrac{P}{P + \Delta P}$

(b) $\gamma = \dfrac{\rho_0 z_0 g}{P_0} = 5.5$

(c) $\Delta P = \dfrac{4\kappa RT}{r_0}\left(\dfrac{1}{\lambda} - \dfrac{1}{\lambda^7}\right)$

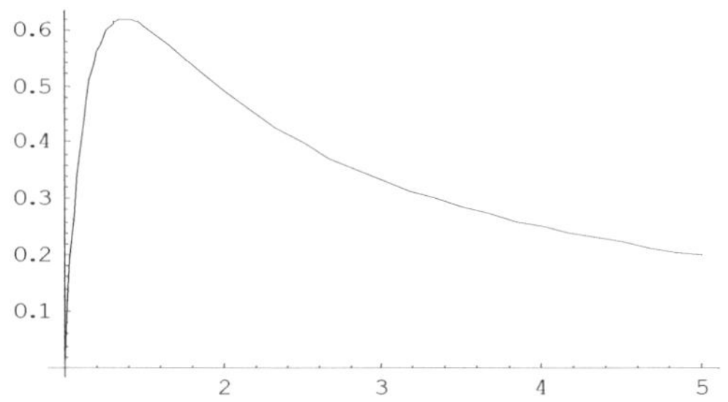

(d) a=0.110

(e) z_f=11 km, λ_f=2.1.

2. Solutions

[Part A]

(a) [1.5 points]

Using the ideal gas equation of state, the volume of the helium gas of n moles at pressure $P+\Delta P$ and temperature T is

$$V = nRT/(P+\Delta P) \qquad \text{(a1)}$$

while the volume of n' moles of air gas at pressure P and temperature T is

$$V = n'RT/P. \qquad \text{(a2)}$$

Thus the balloon displaces $n' = n\dfrac{P}{P+\Delta P}$ moles of air whose weight is $M_A n' g$.

This displaced air weight is the buoyant force, i.e.,

$$F_B = M_A n g \frac{P}{P+\Delta P}. \qquad \text{(a3)}$$

(Partial credits for subtracting the gas weight.)

(b) [2 points]

The pressure difference arising from a height difference of z is $-\rho g z$ when the air density ρ is a constant. When it varies as a function of the height, we have

$$\frac{dP}{dz} = -\rho g = -\frac{\rho_0 T_0}{P_0}\frac{P}{T} g \qquad \text{(b1)}$$

where the ideal gas law $\rho T/P$ = constant is used. Inserting Eq. (2.1) and $T/T_0 = 1 - z/z_0$ on both sides of Eq. (b1), and comparing the two, one gets

$$\gamma = \frac{\rho_0 z_0 g}{P_0} = \frac{1.16\times 4.9\times 10^4 \times 9.8}{1.01\times 10^5} = 5.52. \qquad \text{(b2)}$$

The required numerical value is 5.5.

[Part B]

(c) [2 points]

The work needed to increase the radius from r to $r+dr$ under the pressure difference ΔP is

$$dW = 4\pi r^2 \Delta P dr, \qquad \text{(c1)}$$

while the increase of the elastic energy for the same change of r is

$$dW = \left(\frac{dU}{dr}\right)dr = 4\pi\kappa RT(4r - 4\frac{r_0^6}{r^5})dr . \qquad \text{(c2)}$$

Equating the two expressions of dW, one gets

$$\Delta P = 4\kappa RT(\frac{1}{r} - \frac{r_0^6}{r^7}) = \frac{4\kappa RT}{r_0}\left(\frac{1}{\lambda} - \frac{1}{\lambda^7}\right). \qquad \text{(c3)}$$

This is the required answer.

The graph as a function of λ (>1) increases sharply initially, has a maximum at $\lambda = 7^{1/6}$ =1.38, and decreases as λ^{-1} for large λ. The plot of $\Delta P/(4\kappa RT/r_0)$ is given below.

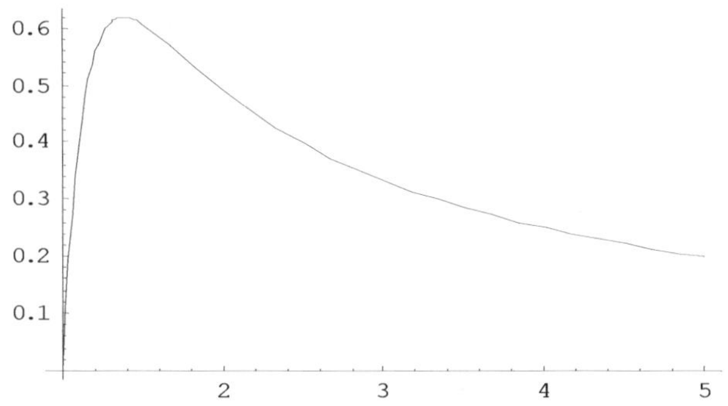

(d) [1.5 points]

From the ideal gas law,

$$P_0 V_0 = n_0 R T_0 \qquad \text{(d1)}$$

where V_0 is the unstretched volume.

At volume $V = \lambda^3 V_0$ containing n moles, the ideal gas law applied to the gas inside at $T = T_0$ gives the inside pressure P_{in} as

$$P_{\text{in}} = nRT_0 / V = \frac{n}{n_0 \lambda^3} P_0 . \qquad \text{(d2)}$$

On the other hand, the result of (c) at $T = T_0$ gives

$$P_{\text{in}} = P_0 + \Delta P = P_0 + \frac{4\kappa RT_0}{r_0}(\frac{1}{\lambda} - \frac{1}{\lambda^7}) = (1 + a(\frac{1}{\lambda} - \frac{1}{\lambda^7}))P_0 . \qquad \text{(d3)}$$

Equating (d2) and (d3) to solve for a,

$$a = \frac{n/(n_0\lambda^3) - 1}{\lambda^{-1} - \lambda^{-7}}. \tag{d5}$$

Inserting n/n_0=3.6 and λ=1.5 here, a=0.110.

[Part C]

(e) [3 points]

The buoyant force derived in problem (a) should balance the total mass of $M_{\rm T}$=1.12 kg. Thus, from Eq. (a3), at the weight balance,

$$\frac{P}{P+\Delta P} = \frac{M_{\rm T}}{M_A n}. \tag{e1}$$

On the other hand, applying again the ideal gas law to the helium gas inside of volume $V = \frac{4}{3}\pi r^3 = \lambda^3 \frac{4}{3}\pi r_0^{\;3} = \lambda^3 V_0$, for arbitrary ambient P and T, one has

$$(P+\Delta P)\lambda^3 = \frac{nRT}{V_0} = P_0 \frac{T}{T_0}\frac{n}{n_0} \tag{e2}$$

for n moles of helium. Eqs. (c3), (e1), and (e2) determine the three unknowns P, ΔP, and λ as a function of T and other parameters. Using Eq. (e2) in Eq. (e1), one has an alternative condition for the weight balance as

$$\frac{P}{P_0}\frac{T_0}{T}\lambda^3 = \frac{M_{\rm T}}{M_A n_0}. \tag{e3}$$

Next using (c3) for ΔP in (e2), one has

$$P\lambda^3 + \frac{4\kappa RT}{r_0}\lambda^2(1-\lambda^{-6}) = P_0\frac{T}{T_0}\frac{n}{n_0}$$

or, rearranging it,

$$\frac{P}{P_0}\frac{T_0}{T}\lambda^3 = \frac{n}{n_0} - a\lambda^2(1-\lambda^{-6}), \tag{e4}$$

where the definition of a has been used again.

Equating the right hand sides of Eqs. (e3) and (e4), one has the equation for λ as

$$\lambda^2(1-\lambda^{-6}) = \frac{1}{an_0}\left(n - \frac{M_{\rm T}}{M_A}\right) = 4.54. \tag{e5}$$

The solution for λ can be obtained by

$$\lambda^2 \approx 4.54/(1-4.54^{-3}) \approx 4.54: \;\; \lambda_f \cong 2.13. \tag{e6}$$

To find the height, replace $(P/P_0)/(T/T_0)$ on the left hand side of Eq. (e3) as a function of the height given in (b) as

$$\frac{P}{P_0}\frac{T_0}{T}\lambda^3 = (1 - z_f / z_0)^{\gamma-1}\lambda_f^3 = \frac{M_T}{M_A n_0} = 3.10 \,. \tag{e7}$$

Solution of Eq. (e7) for z_f with λ_f=2.13 and $\gamma-1$=4.5 is

$$z_f = 49 \times \left(1 - (3.10/2.13^3)^{1/4.5}\right) = 10.9 \text{ (km)}. \tag{e8}$$

The required answers are $\lambda_f = 2.1$, and $z_f = 11$ km.

3. Mark Distribution

No.	Total Pt.	Partial Pt.	Contents
(a)	1.5	0.5	Archimedes' principle
		0.5	Ideal gas law applied correctly
		0.5	Correct answer (partial credits 0.3 for subtracting He weight)
(b)	2.0	0.8	Relation of pressure difference to air density
		0.5	Application of ideal gas law to convert the density into pressure
		0.5	Correct formula for γ
		0.2	Correct number in answer
(c)	2.0	0.7	Relation of mechanical work to elastic energy change
		0.3	Relation of pressure to force
		0.5	Correct answer in formula
		0.5	Correct sketch of the curve
(d)	1.5	0.3	Use of ideal gas law for the increased pressure inside
		0.4	Expression of inside pressure in terms of a at the given conditions
		0.5	Formula or correct expression for a
		0.3	Correct answer
(e)	3.0	0.3	Use of force balance as one condition to determine unknowns
		0.3	Ideal gas law applied to the gas as an independent condition to determine unknowns
		0.5	The condition to determine λ_f numerically
		0.7	Correct answer for λ_f
		0.5	The relation of z_f versus λ_f
		0.7	Correct answer for z_f
Total	10		

Theoretical Question 3: *Scanning Probe Microscope*

1. Answers

(a) $A = \dfrac{F_0}{\sqrt{m^2(\omega_0^2 - \omega^2)^2 + b^2\omega^2}}$ and $\tan\phi = \dfrac{b\omega_0}{m(\omega_0^2 - \omega^2)}$. At $\omega = \omega_0$, $A = \dfrac{F_0}{b\omega_0}$

and $\phi = \dfrac{\pi}{2}$.

(b) A non-vanishing dc component exists only when $\omega = \omega_i$.

In this case the amplitude of the dc signal will be $\frac{1}{2}V_{i0}V_{R0}\cos\phi_i$.

(c) $\dfrac{c_1 c_2}{2}\dfrac{V_{R0}^2}{b\omega_0}$ at the resonance frequency ω_0.

(d) $\Delta m = 1.7\times10^{-18}$ kg.

(e) $\omega'_0 = \omega_0\left(1 - \dfrac{c_3}{m\omega_0^2}\right)^{1/2}$.

(f) $d_0 = \left(k_e \dfrac{qQ}{m\omega_0\Delta\omega_0}\right)^{1/3}$

$d_0 = 41$ nm.

2. Solutions

(a) [1.5 points]

Substituting $z(t) = A\sin(\omega t - \phi)$ in the equation $m\frac{d^2 z}{dt^2} + b\frac{dz}{dt} + m\omega_0^2 z = F_0 \sin\omega t$

yields,

$$-m\omega^2 \sin(\omega t - \phi) + b\omega\cos(\omega t - \phi) + m\omega_0{}^2 \sin(\omega t - \phi) = \frac{F_0}{A}\sin\omega t\ . \qquad \text{(a1)}$$

Collecting terms proportional to $\sin\omega t$ and $\cos\omega t$, one obtains

$$\left\{m(\omega_0^2 - \omega^2)\cos\phi + b\omega\sin\phi - \frac{F_0}{A}\right\}\sin\omega t + \left\{-m(\omega_0^2 - \omega^2)\sin\phi + b\omega\cos\phi\right\}\cos\omega t = 0 \quad \text{(a2)}$$

Zeroing the each curly square bracket produces

$$\tan\phi = \frac{b\omega}{m(\omega_0^2 - \omega^2)}, \qquad \text{(a3)}$$

$$A = \frac{F_0}{\sqrt{m^2(\omega_0^2 - \omega^2)^2 + b^2\omega^2}}\ . \qquad \text{(a4)}$$

At $\omega = \omega_0$,

$$A = \frac{F_0}{b\omega_0} \text{ and } \phi = \frac{\pi}{2}. \qquad \text{(a5)}$$

(b) [1 point]

The multiplied signal is

$$V_{i0}\sin(\omega_i t - \phi_i)V_{R0}\sin(\omega t) = \frac{1}{2}V_{i0}V_{R0}[\cos\{(\omega_i - \omega)t - \phi_i\} - \cos\{(\omega_i + \omega)t - \phi_i\}] \qquad \text{(b1)}$$

A non-vanishing dc component exists only when $\omega = \omega_i$. In this case the amplitude of the dc signal will be

$$\frac{1}{2}V_{i0}V_{R0}\cos\phi_i\ . \qquad \text{(b2)}$$

(c) [1.5 points]

Since the lock-in amplifier measures the ac signal of the same frequency with its reference signal, the frequency of the piezoelectric tube oscillation, the frequency of the cantilever, and the frequency of the photodiode detector should be same. The magnitude of the input signal at the resonance is

$$V_{i0} = c_2 \frac{F_0}{b\omega_0} = \frac{c_1 c_2 V_{R0}}{b\omega_0}. \qquad \text{(c1)}$$

Then, since the phase of the input signal is $-\frac{\pi}{2}+\frac{\pi}{2}=0$ at the resonance, $\phi_i = 0$ and the lock-in amplifier signal is

$$\frac{1}{2} V_{i0} V_{R0} \cos 0 = \frac{c_1 c_2}{2} \frac{V_{R0}^2}{b\omega_0}. \qquad \text{(c2)}$$

(d) [2 points]

The original resonance frequency $\omega_0 = \sqrt{\frac{k}{m}}$ is shifted to

$$\sqrt{\frac{k}{m+\Delta m}} = \sqrt{\frac{k}{m}}\left(1+\frac{\Delta m}{m}\right)^{-\frac{1}{2}} \cong \sqrt{\frac{k}{m}}\left(1-\frac{1}{2}\frac{\Delta m}{m}\right) = \omega_0\left(1-\frac{1}{2}\frac{\Delta m}{m}\right). \qquad \text{(d1)}$$

Thus

$$\Delta\omega_0 = -\frac{1}{2}\omega_0 \frac{\Delta m}{m}. \qquad \text{(d2)}$$

Near the resonance, by substituting $\phi \to \frac{\pi}{2} + \Delta\phi$ and $\omega_0 \to \omega_0 + \Delta\omega_0$ in Eq. (a3), the change of the phase due to the small change of ω_0 (not the change of ω) is

$$\tan\left(\frac{\pi}{2}+\Delta\phi\right) = -\frac{1}{\tan\Delta\phi} = \frac{b}{2m\Delta\omega_0}. \qquad \text{(d3)}$$

Therefore,

$$\Delta\phi \approx \tan\Delta\phi = -\frac{2m\Delta\omega_0}{b}. \qquad \text{(d4)}$$

From Eqs. (d2) and (d4),

$$\Delta m = \frac{b}{\omega_0}\Delta\phi = \frac{10^3 \cdot 10^{-12}}{10^6}\frac{\pi}{1800} = \frac{\pi}{1.8}10^{-18} = 1.7\times 10^{-18}\ \text{kg}. \qquad \text{(d5)}$$

(e) [1.5 points]

In the presence of interaction, the equation of motion near the new equilibrium position h_0 becomes

$$m\frac{d^2z}{dt^2}+b\frac{dz}{dt}+m\omega_0^2 z-c_3 z=F_0\sin\omega t \tag{e1}$$

where we used $f(h)\approx f(h_0)+c_3 z$ with $z=h-h_0$ being the displacement from the new equilibrium position h_0. Note that the constant term $f(h_0)$ is cancelled at the new equilibrium position.

Thus the original resonance frequency $\omega_0=\sqrt{\frac{k}{m}}$ will be shifted to

$$\omega'_0=\sqrt{\frac{k-c_3}{m}}=\sqrt{\frac{m\omega_0^2-c_3}{m}}=\omega_0\sqrt{1-\frac{c_3}{m\omega_0^2}}. \tag{e3}$$

Hence the resonance frequency shift is given by

$$\Delta\omega_0=\omega_0\left[\sqrt{1-\frac{c_3}{m\omega_0^2}}-1\right]. \tag{e4}$$

(f) [2.5 points]

The maximum shift occurs when the cantilever is on top of the charge, where the interacting force is given by

$$f(h)=k_e\frac{qQ}{h^2}. \tag{f1}$$

From this,

$$c_3=\left.\frac{df}{dh}\right|_{h=d_0}=-2k_e\frac{qQ}{d_0^3}. \tag{f2}$$

Since $\Delta\omega_0 << \omega_0$, we can approximate Eq. (e4) as

$$\Delta\omega_0\approx-\frac{c_3}{2m\omega_0}. \tag{f3}$$

From Eqs. (f2) and (f3), we have

$$\Delta\omega_0=-\frac{1}{2m\omega_0}\left(-2k_e\frac{qQ}{d_0^3}\right)=k_e\frac{qQ}{m\omega_0 d_0^3}. \tag{f4}$$

Here $q=e=-1.6\times10^{-19}$ Coulomb and $Q=6e=-9.6\times10^{-19}$ Coulomb. Using the values provided,

$$d_0 = \left(k_e \frac{qQ}{m\omega_0 \Delta\omega_0} \right)^{1/3} = 4.1 \times 10^{-8} \text{ m} = 41 \text{ nm}. \tag{f5}$$

Thus the trapped electron is 41 nm from the cantilever.

3. Mark Distribution

No.	Total Pt.	Partial Pt.	Contents
(a)	1.5	0.7	Equations for A and ϕ (substitution and manipulation)
		0.4	Correct answers for A and ϕ
		0.4	A and ϕ at ω_0
(b)	1.0	0.4	Equation for the multiplied signal
		0.3	Condition for the non-vanishing dc output
		0.3	Correct answer for the dc output
(c)	1.5	0.6	Relation between V_i and V_R
		0.4	Condition for the maximum dc output
		0.5	Correct answer for the magnitude of dc output
(d)	2.0	0.5	Relation between Δm and $\Delta\omega_0$
		1.0	Relations between $\Delta\omega_0$ (or Δm) and $\Delta\phi$
		0.5	Correct answer (Partial credit of 0.2 for the wrong sign.)
(e)	1.5	1.0	Modification of the equation with $f(h)$ and use of a proper approximation for the equation
		0.5	Correct answer
(f)	2.5	0.5	Use of a correct formula of Coulomb force
		0.3	Evaluation of c_3
		0.6	Use of the result in (e) for either $\Delta\omega_0$ or $\omega'^2_0 - \omega_0^2$
		0.6	Expression for d_0
		0.5	Correct answer
Total	10		

Experimental Competition

Apparatus and materials

1. List of available apparatus and materials

	Name	Quantity		Name	Quantity
A	Photogate timer	1	L	Philips screw driver	1
B	Photogate	1	M	Weight with a string	1
C	Connecting cable	1	N	Electronic balance	1
D	Mechanical "black box" (Black cylinder)	1	O	Stand with a ruler	1
E	Rotation stage	1	P	U-shaped support	1
F	Rubber pad	1	Q	C-clamp	1
G	Pulley	2		Ruler (0.50 m, 0.15 m)	1 each
H	Pin	2		Vernier calipers	1
I	U-shaped plate	1		Scissors	1
J	Screw	2		Thread	1
K	Allen (hexagonal, L-shaped) wrench	1		Spares (string, thread, pin, screw, Allen wrench)	

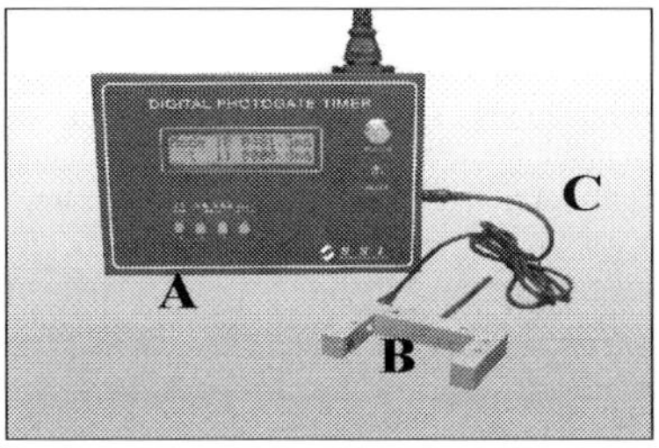

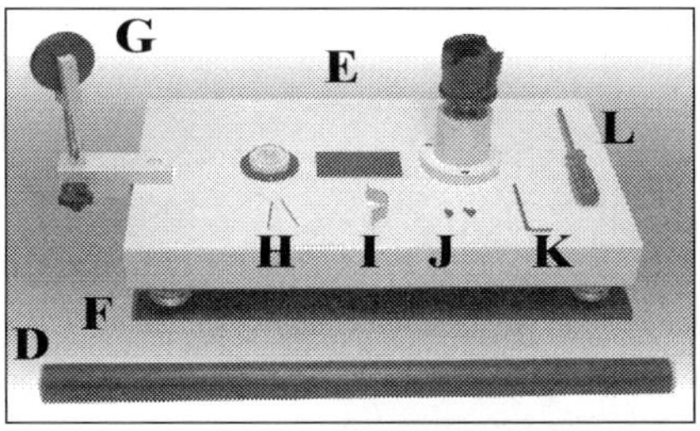

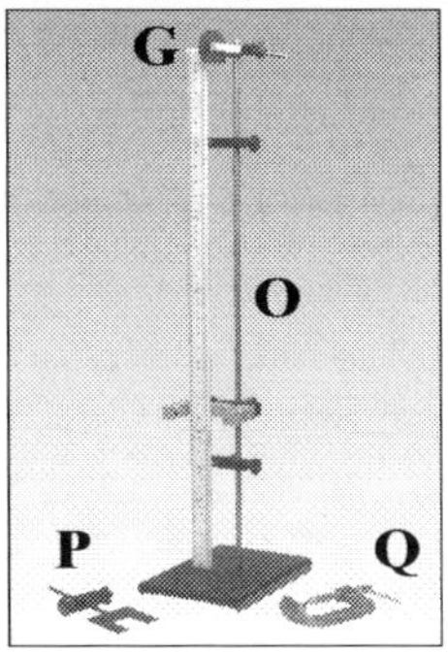

M

N

2. Instruction for the Photogate Timer

The Photogate consists of an infrared LED and a photodetector. By connecting the Photogate to the Photogate Timer, you can measure the time duration related to the blocking of the infrared light reaching the sensor.

- Be sure that the Photogate is connected to the Photogate Timer. Turn on the power by pushing the button labelled "POWER".
- To measure the time duration of a single blocking event, push the button labelled "GATE". Use this "GATE" mode for speed measurements.
- To measure the time interval between two or three successive blocking events, push the corresponding "PERIOD". Use this "PERIOD" mode for oscillation measurements.
- If "DELAY" button is pushed in, the Photogate Timer displays the result of each measurement for 5 seconds and then resets itself.
- If "DELAY" button is pushed out, the Photogate Timer displays the result of the previous measurement until the next measurement is completed.
- After any change of button position, press the "RESET" button once to activate the mode change.

Caution: Do not look directly into the Photogate. The invisible infrared light may be harmful to your eyes.

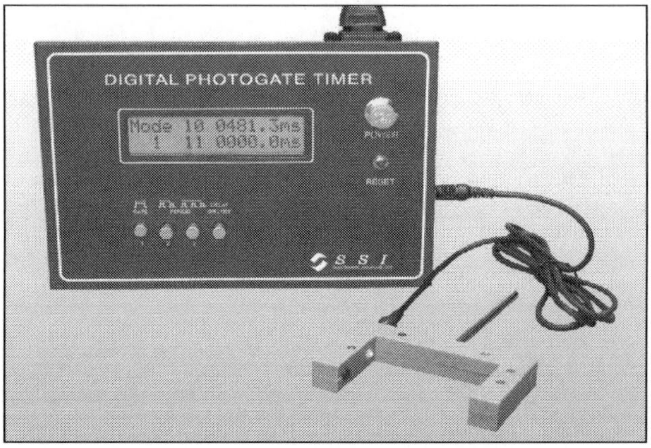

Photogate, Photogate Timer, and connection cable

3. Instruction for the Electronic Balance

- Adjust the bottom legs to set the balance stable. (Although there is a level indicator, setting the balance in a completely horizontal position is not necessary.)
- Without putting anything on the balance, turn it on by pressing the "On/Off" button.
- Place an object on the round weighing pan. Its mass will be displayed in grams.
- If there is nothing on the weighing pan, the balance will be turned off automatically in about 25 seconds.

Balance

4. Instruction for the Rotation Stage

- Adjust the bottom legs to set the rotation stage stable on a rubber pad in a near horizontal position.
- With a U-shaped plate and two screws, mount the Mechanical "Black Box" (black cylinder) on the top of the rotating stub. Use Allen (hexagonal, L-shaped) wrench to tighten the screws.
- The string attached to the weight is to be fixed to the screw on the side of the rotating stub. Use the Philips screw driver.

Caution: Do not look too closely at the Mechanical "Black Box" while it is rotating. Your eyes may get hurt.

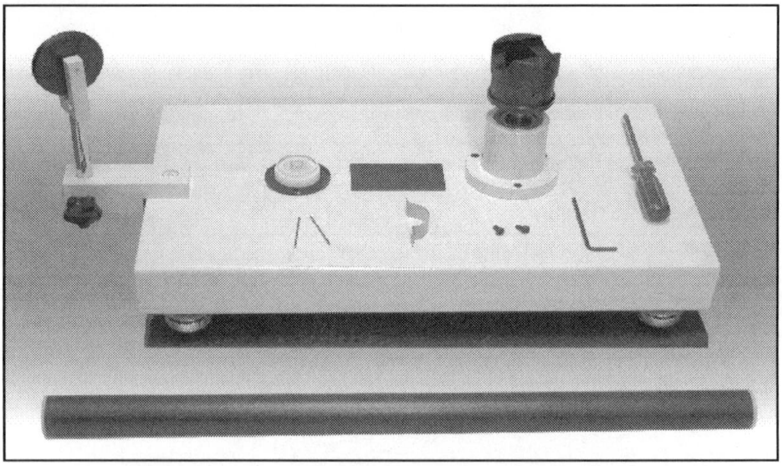

Mechanical "Black Box" and rotation stage

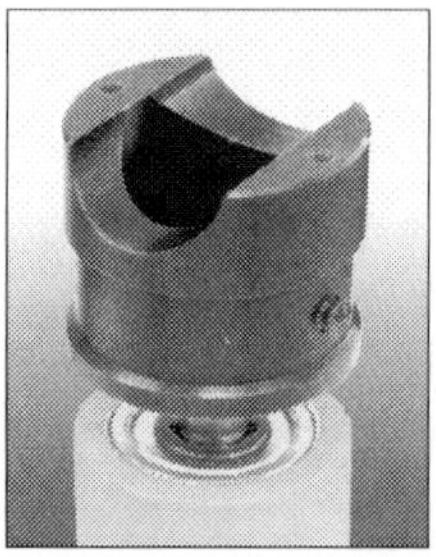

Rotating stub

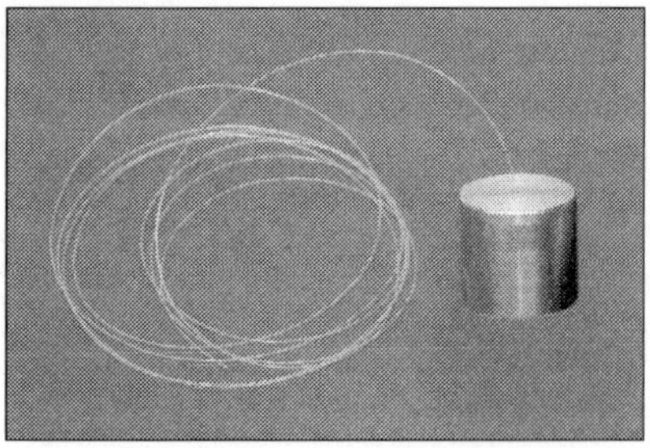

Weight with a string

Mechanical "Black Box"

[Question] ***<u>Find the mass of the ball and the spring constants of two springs in the Mechanical "Black Box".</u>***

General Information on the Mechanical "Black Box"

The Mechanical "Black Box" (MBB) consists of a solid ball attached to two springs in a black cylindrical tube as shown in Fig. 1. The two springs are fashioned from the same tightly wound spring with different number of turns. The masses and the lengths of the springs when they are not extended can be ignored. The tube is homogeneous and sealed with two identical end caps. The part of the end caps plugged into the tube is 5 mm long. The radius of the ball is 11 mm and the inner diameter of the tube is 23 mm. The gravitational acceleration is given as $g = 9.8\ \text{m/s}^2$. There is a finite friction between the ball and the inner walls of the tube.

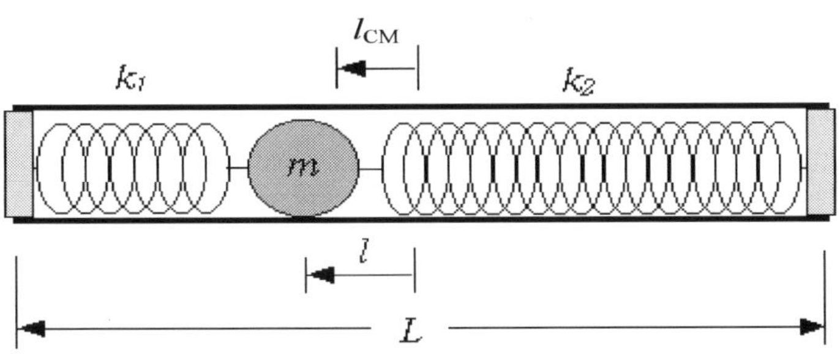

Fig. 1 Mechanical "Black Box" (not to scale)

The purpose of this experiment is to find out the mass m of the ball and the spring constants k_1 and k_2 of the springs without opening the MBB. The difficult aspect of this problem is that any single experiment cannot provide the mass m or the position l of the ball because the two quantities are interconnected. Here, l is the distance between the centers of the tube and the ball when the MBB lies horizontally in equilibrium when the friction is zero.

<u>The symbols listed below should be used to represent the physical quantities of interest.</u> If you need to use other physical quantities, use symbols different from those already assigned below to avoid confusion.

Assigned Physical Symbols

Mass of the ball: m
Radius of the ball: r (= 11 mm)
Mass of the MBB excluding the ball: M
Length of the black tube: L
Length of each end cap extending into the tube: δ(= 5.0 mm)
Distance from the center-of-mass of the MBB to the center of the tube: l_{CM}
Distance between the center of the ball and the center of the tube: x (or l at equilibrium when the MBB is horizontal)
Gravitational acceleration: g (= 9.8 m/s^2)
Mass of the weight attached to a string: m_o
Speed of the weight: v
Downward displacement of the weight: h
Radius of the rotating stub where the string is to be wound: R
Moments of inertia: I, I_o, I_1, I_2, and so on
Angular velocity and angular frequencies: ω, ω_1, ω_2, and so on
Periods of oscillation: T_1, T_2
Effective total spring constant: k
Spring constants of the two springs: k_1, k_2
Number of turns of the springs: N_1, N_2

Caution: Do not try to open the MBB. If you open it, you will be disqualified and your mark in the Experimental Competition will be zero.

Caution: Do not shake violently nor drop the MBB. The ball may be detached from the springs. If your MBB seems faulty, report to the proctors immediately. It will be replaced only once without affecting your mark. Any further replacement will cut down your mark by 0.5 points each time.

PART-A **Product of the mass and the position of the ball ($m \times l$)** (4.0 points)

l is the position of the center of the ball relative to that of the tube when the MBB lies horizontally in equilibrium as in Fig. 1. Find the value of the product of the mass m and the position l of the ball experimentally. You will need this to determine the value of m in ***PART-B***.

1. Suggest and justify, by using equations, a method allowing to obtain $m \times l$. (2.0 points)

2. Experimentally determine the value of $m \times l$. (2.0 points)

PART-B **The mass *m* of the ball** (10.0 points)

Figure 2 shows the MBB fixed horizontally on the rotating stub and a weight attached to one end of a string whose other end is wound on the rotating stub. When the weight falls, the string unwinds, and the MBB rotates. By combining the equation pertinent to this experiment with the one obtained in ***PART-A***, you can find an equation for m.

Between the ball and the inner walls of the cylindrical tube acts a frictional force. The physical mechanisms of the friction and the slipping of the ball under the rotational motion are complicated. To simplify the analysis, you may ignore the energy dissipation due to kinetic friction.

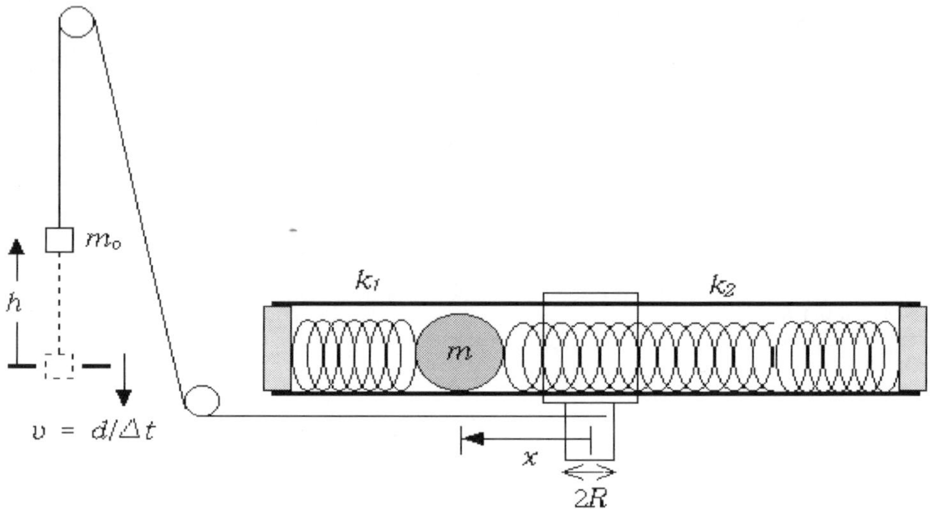

Fig. 2 Rotation of the Mechanical "Black Box" (not to scale)
The angular velocity ω of the MBB can be obtained from the speed v of the weight passing through the Photogate. x is the position of the ball relative to the rotation axis, and d is the length of the weight.

1. Measure the speed of the weight v for various values of downward displacement h of the weight. It is recommended to scan the whole range from $h = 1.0 \times 10^{-2}$ m to 4.0×10^{-1} m by measuring v just once at each h with an interval of 1.0×10^{-2} ~2.0×10^{-2} m. Plot the data on graph paper in a form that is suitable to find the value of m. After you get a general idea of the relation between v and h, you may repeat the measurement or add some data points, if necessary. When the MBB rotates slowly, the ball does not slip from its static equilibrium position because of the friction between the ball and the tube. When the MBB rotates sufficiently fast, the ball hits and actually stays at the end cap of the tube because the springs are weak. Identify the slow rotation region and the fast rotation region on the graph. (4.0 points)

2. Show your measurements are consistent with the fact that h is proportional to v^2 ($h = Cv^2$) in the slow rotation region. Show from your measurements that $h = Av^2 + B$ in the fast rotation region. (1.0 points)

3. The moment of inertia of a ball of radius r and mass m about the axis passing through its center is $2mr^2/5$. If the ball is displaced a distance a perpendicular to the axis, the moment of inertia increases by ma^2. Use the symbol I to represent the total moment of inertia of all the rotating bodies excluding the ball. Relate the coefficient C to the parameters of the MBB such as m, l, etc. (1.0 points)

4. Relate the coefficients A and B to the parameters of the MBB such as m, l, etc. (1.0 points)

5. Determine the value of m from your measurements and the results obtained in ***PART-A***. (3.0 points)

PART-C **The spring constants k_1 *and* k_2** (6.0 points)

In this part, you need to perform small oscillation experiments using the MBB as a rigid pendulum. There are two small holes at each end of the MBB. Two thin pins inserted into the holes can be used as the pivot of small oscillation. The U-shaped support is to be clamped to the stand and used to support the pivot. Note that the angular frequency ω of small oscillation is given as ω= [torque/(moment of inertia × angle)]$^{1/2}$. Here, the torque and the moment of inertia are with respect to the pivot. Similarly to ***PART-B***, consider two experimental conditions, shown in Fig. 3, to avoid the unknown moment of inertia I_o of the MBB excluding the ball.

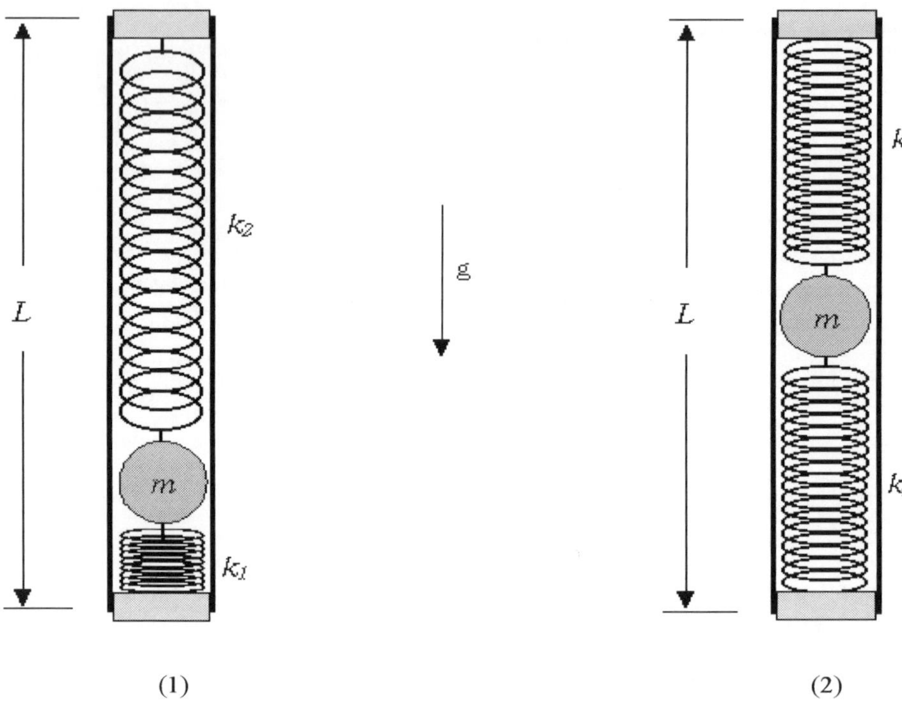

Fig. 3 Oscillation of the Mechanical "Black Box" (not to scale)
The periods of small oscillation, T_1 and T_2, for two configurations shown above can be measured using the Photogate. Two pins and a U-shaped support are supplied for this experiment.

1. Measure the periods T_1 and T_2 of small oscillation shown in Figs 3(1) and (2) and write down their values, respectively. (1.0 points)

2. Explain (by using equations) why the angular frequencies ω_1 and ω_2 of small oscillation of the configurations are different. Use the symbol I_o to represent the moment of inertia of the MBB excluding the ball for the axis perpendicular to the MBB at the end. Use the symbol Δl as the displacement of the ball from the horizontal equilibrium position. (1.0 points)

3. Evaluate Δl by eliminating I_o from the previous results. (1.0 points)

4. By combining the results of ***PART-C*** 1~3 and ***PART-B***, find and write down the value of the effective total spring constant k of the two-spring system. (2.0 points)

5. Obtain the respective values of k_1 and k_2. Write down their values. (1.0 points)

Solutions

PART-A Product of the mass and the position of the ball ($m \times l$) (4.0 points)

1. Suggest and justify, by using equations, a method allowing to obtain $m \times l$. (2.0 points)

$$m \times l = (M + m) \times l_{cm}$$

(Explanation) The lever rule is applied to the Mechanical "Black Box", shown in Fig. A-1, once the position of the center of mass of the whole system is found.

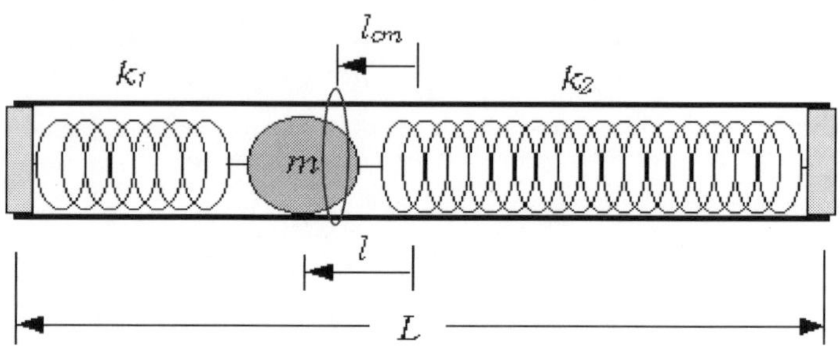

Fig. A-1 Experimental setup

2. Experimentally determine the value of $m \times l$. (2.0 points)

$$m \times l = 2.96 \times 10^{-3} \text{kg·m}$$

(Explanation) The measured quantities are

$$M + m = (1.411 \pm 0.0005) \times 10^{-1} \text{kg}$$

and

$$l_{cm} = (2.1 \pm 0.06) \times 10^{-2} \text{m} \quad \text{or} \quad 21 \pm 0.6 \text{ mm}.$$

Therefore

$$\begin{aligned} m \times l &= (M + m) \times l_{cm} \\ &= (1.411 \pm 0.0005) \times 10^{-1} \text{kg} \times (2.1 \pm 0.06) \times 10^{-2} \text{m} \\ &= (2.96 \pm 0.08) \times 10^{-3} \text{kg·m} \end{aligned}$$

PART-B The mass *m* of the ball (10.0 points)

1. Measure v for various values of h. Plot the data on a graph paper in a form that is suitable to find the value of m. Identify the slow rotation region and the fast rotation region on the graph. (4.0 points)

2. Show from your measurements that $h = Cv^2$ in the slow rotation region, and $h = Av^2 + B$ in the fast rotation region. (1.0 points)

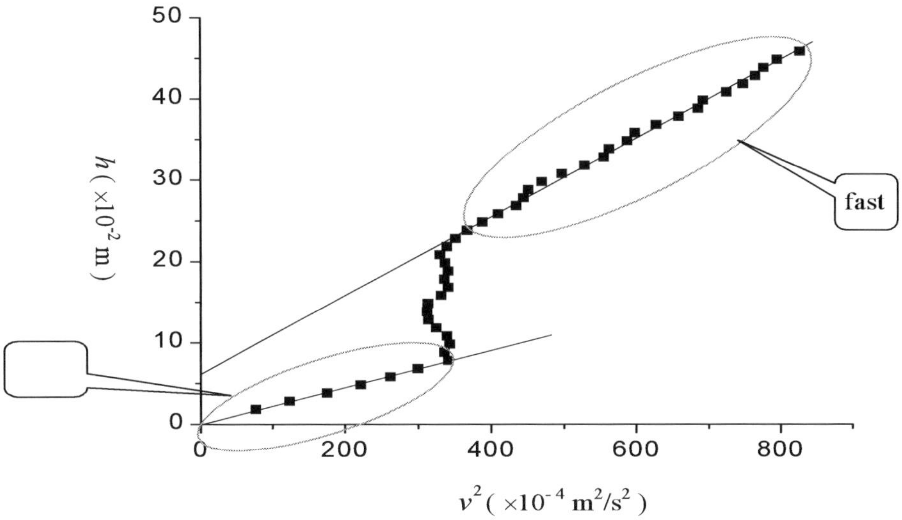

Fig. B-1 Experimental data

(Explanation) The measured data are

	h_1 ($\times 10^{-2}$ m) a)	Δt (ms)	h ($\times 10^{-2}$ m) b)	v ($\times 10^{-2}$ m/s) c)	v^2 ($\times 10^{-4}$ m²/s²)
1	25.5±0.1	269.4±0.05	1.8±0.1	8.75±0.02	76.6±0.2
2	26.5±0.1	235.7±0.05	2.8±0.1	11.12±0.02	123.7±0.3
3	27.5±0.1	197.9±0.05	3.8±0.1	13.24±0.03	175.3±0.6
4	28.5±0.1	176.0±0.05	4.8±0.1	14.89±0.03	221.7±0.6
5	29.5±0.1	161.8±0.05	5.8±0.1	16.19±0.03	262.1±0.7
6	30.5±0.1	151.4±0.05	6.8±0.1	17.31±0.03	299.6±0.7
7	31.5±0.1	141.8±0.05	7.8±0.1	18.48±0.04	342±1
8	32.5±0.1	142.9±0.05	8.8±0.1	18.33±0.04	336±1

9	33.5±0.1	141.4±0.05	9.8±0.1	18.53±0.04	343±1
10	34.5±0.1	142.2±0.05	10.8±0.1	18.42±0.04	339±1
11	35.5±0.1	145.4±0.05	11.8±0.1	18.02±0.04	325±1
12	36.5±0.1	147.8±0.05	12.8±0.1	17.73±0.04	314±1
13	37.5±0.1	148.3±0.05	13.8±0.1	17.67±0.04	312±1
14	38.5±0.1	148.0±0.05	14.8±0.1	17.70±0.04	313±1
15	39.5±0.1	143.9±0.05	15.8±0.1	18.21±0.04	332±1
16	40.5±0.1	141.9±0.05	16.8±0.1	18.46±0.04	341±1
17	41.5±0.1	142.9±0.05	17.8±0.1	18.33±0.04	336±1
18	42.5±0.1	141.9±0.05	18.8±0.1	18.46±0.04	341±1
19	43.5±0.1	142.8±0.05	19.8±0.1	18.35±0.04	337±1
20	44.5±0.1	144.3±0.05	20.8±0.1	18.16±0.04	330±1
21	45.5±0.1	142.2±0.05	21.8±0.1	18.42±0.04	339±1
22	46.5±0.1	139.8±0.05	22.8±0.1	18.74±0.04	351±1
23	47.5±0.1	136.7±0.05	23.8±0.1	19.17±0.04	368±1
24	48.5±0.1	133.0±0.05	24.8±0.1	19.70±0.04	388±1
25	49.5±0.1	129.5±0.05	25.8±0.1	20.23±0.04	409±1
26	50.5±0.1	125.7±0.05	26.8±0.1	20.84±0.04	434±1
27	51.5±0.1	124.3±0.05	27.8±0.1	21.08±0.04	444±1
28	52.5±0.1	123.4±0.05	28.8±0.1	21.23±0.04	451±1
29	53.5±0.1	120.9±0.05	29.8±0.1	21.67±0.04	470±1
30	54.5±0.1	117.5±0.05	30.8±0.1	22.30±0.04	497±1
31	55.5±0.1	114.0±0.05	31.8±0.1	22.98±0.04	528±1
32	56.5±0.1	111.2±0.05	32.8±0.1	23.56±0.05	555±2
33	57.5±0.1	110.5±0.05	33.8±0.1	23.71±0.05	562±2
34	58.5±0.1	108.1±0.05	34.8±0.1	24.24±0.05	588±2
35	59.5±0.1	107.1±0.05	35.8±0.1	24.46±0.05	598±2
36	60.5±0.1	104.6±0.05	36.8±0.1	25.05±0.05	628±2
37	61.5±0.1	102.1±0.05	37.8±0.1	25.66±0.05	658±2
38	62.5±0.1	100.1±0.05	38.8±0.1	26.17±0.05	685±2
39	63.5±0.1	99.6±0.05	39.8±0.1	26.31±0.05	692±2
40	64.5±0.1	97.3±0.05	40.8±0.1	26.93±0.05	725±2
41	65.5±0.1	95.8±0.05	41.8±0.1	27.35±0.05	748±2
42	66.5±0.1	94.7±0.05	42.8±0.1	27.67±0.05	766±2
43	67.5±0.1	94.0±0.05	43.8±0.1	27.87±0.06	777±2
44	68.5±0.1	92.9±0.05	44.8±0.1	28.20±0.06	795±2
45	69.5±0.1	91.1±0.05	45.8±0.1	28.76±0.06	827±2

where [a] h_1 is the reading of the top position of the weight before it starts to fall,

[b] h is the distance of fall of the weight which is obtained by $h = h_1 - h_2 + d/2$,

h_2 (= $(25\pm0.05)\times10^{-2}$ m) is the top position of the weight at the start of blocking of the photogate,

d (= $(2.62\pm0.005)\times10^{-2}$ m) is the length of the weight, and

[c] v is obtained from $v = d/\Delta t$.

3. Relate the coefficient C to the parameters of the MBB. (1.0 points)

$$h = C v^2, \text{ where } C = \{m_o + I/R^2 + m(l^2 + 2/5\, r^2)/R^2\}/2m_o g$$

(Explanation) The ball is at static equilibrium ($x = l$). When the speed of the weight is v, the increase in kinetic energy of the whole system is given by

$$\begin{aligned}\Delta K &= 1/2\, m_o v^2 + 1/2\, I\omega^2 + 1/2\, m(l^2 + 2/5\, r^2)\omega^2 \\ &= 1/2\, \{m_o + I/R^2 + m(l^2 + 2/5\, r^2)/R^2\}v^2,\end{aligned}$$

where ω (= v/R) is the angular velocity of the Mechanical "Black Box" and I is the *effective* moment of inertia of the whole system except the ball. Since the decrease in gravitational potential energy of the weight is

$$\Delta U = -\, m_o g h\,,$$

the energy conservation ($\Delta K + \Delta U = 0$) gives

$$\begin{aligned}h &= 1/2\, \{m_o + I/R^2 + m(l^2 + 2/5\, r^2)/R^2\}v^2/m_o g \\ &= C v^2, \quad \text{where } C = \{m_o + I/R^2 + m(l^2 + 2/5\, r^2)/R^2\}/2m_o g\end{aligned}$$

4. Relate the coefficients A and B to the parameters of the MBB. (1.0 points)

$$\begin{aligned}h = A v^2 + B, \text{ where } A &= [m_o + I/R^2 + m\{(L/2 - \delta - r)^2 + 2/5\, r^2\}/R^2]/2m_o g \\ \text{and } B &= [-k_1(L/2 - l - \delta - r)^2 \\ &\quad + k_2\{(L - 2\delta - 2r)^2 - (L/2 + l - \delta - r)^2\}]\,/2m_o g\end{aligned}$$

(Explanation) The ball stays at the end cap of the tube ($x = L/2 - \delta - r$). When the speed of the weight is v, the increase in kinetic energy of the whole system is given by

$$K = 1/2\ [m_o + I/R^2 + m\{(L/2 - \delta - r)^2 + 2/5\ r^2\}/R^2]v^2.$$

Since the increase in elastic potential energy of the springs is

$$\Delta U_e = 1/2\ [-k_1(L/2 - l - \delta - r)^2 + k_2\{(L - 2\delta - 2r)^2 - (L/2 + l - \delta - r)^2\}],$$

the energy conservation ($K + \Delta U + \Delta U_e = 0$) gives

$$h = 1/2\ [m_o + I/R^2 + m\{(L/2 - \delta - r)^2 + 2/5\ r^2\}/R^2]v^2/m_o g + \Delta U_e/m_o g = A v^2 + B,$$

where

$$A = [m_o + I/R^2 + m\{(L/2 - \delta - r)^2 + 2/5\ r^2\}/R^2]/2m_o g$$

and

$$B = [-k_1(L/2 - l - \delta - r)^2 + k_2\{(L - 2\delta - 2r)^2 - (L/2 + l - \delta - r)^2\}]/2m_o g.$$

5. Determine the value of m from your measurements and the results obtained in ***PART-A***. (3.0 points)

$$m = 6.2\times10^{-2}\ \text{kg}$$

(Explanation) From the results obtained in ***PART-B*** 3 and 4 we get

$$A - C = \frac{m}{2gm_o R^2}\left\{(L/2 - \delta - r)^2 - l^2\right\}.$$

The measured values are $L = (40.0\pm0.05)\times10^{-2}$ m

$m_o = (100.4\pm0.05)\times10^{-3}$ kg

$2R = (3.91\pm0.005)\times10^{-2}$ m

Therefore,

$$(L/2 - \delta - r)^2 = \{(20.0\pm0.03) - 0.5 - 1.1\}^2 \times10^{-4}\ \text{m}^2 = (338.6\pm0.8)\times10^{-4}\ \text{m}^2$$

and

$$2gm_oR^2 = 2\times980\times(100.4\pm0.05)\times(1.955\pm0.003)^2\times10^{-6}\text{kg·m}^3/\text{s}^2 = (752\pm2)\times10^{-6}\text{kg·m}^3/\text{s}^2.$$

The slopes of the two straight lines in the graph (Fig. B-1) of ***PART-B*** 1 are

$$A = 5.0 \pm 0.1 \text{s}^2/\text{m} \quad \text{and} \quad C = 2.4 \pm 0.1 \text{s}^2/\text{m},$$

respectively, and

$$A - C = 2.6 \pm 0.1 \text{s}^2/\text{m}.$$

Since we already obtained $m \times l = (M + m) \times l_{cm} = 2.96 \times 10^{-3}$kg·m from ***PART-A***, the equation

$$(338.6 \pm 0.8)m^2 - (752 \pm 2) \times 10^3 \times (0.026 \pm 0.001)m - (296 \pm 8)^2 = 0$$

or

$$(338.6 \pm 0.8)m^2 - (19600 \pm 800)m - (88000 \pm 3000) = 0$$

is resulted, where m is expressed in the unit of g.
The roots of this equation are

$$m = \frac{(9800 \pm 400) \pm \sqrt{(9800 \pm 400)^2 + (338.6 \pm 0.8) \times (88000 \pm 3000)}}{(338.6 \pm 0.8)}.$$

The physically meaningful positive root is

$$m = \frac{(9800 \pm 400) + \sqrt{(126000000 \pm 6000000)}}{(338.6 \pm 0.8)} = (62 \pm 2)\,\text{g} = (6.2 \pm 0.2) \times 10^{-2}\,\text{kg}.$$

PART-C **The spring constants k_1 *and* k_2 (6.0 points)**

1. Measure the periods T_1 and T_2 of small oscillation shown in Figs. 3 (1) and (2) and write down their values, respectively. (1.0 points)

$$T_1 = 1.1090\text{s} \quad \text{and} \quad T_2 = 1.0193\text{s}$$

(Explanation)

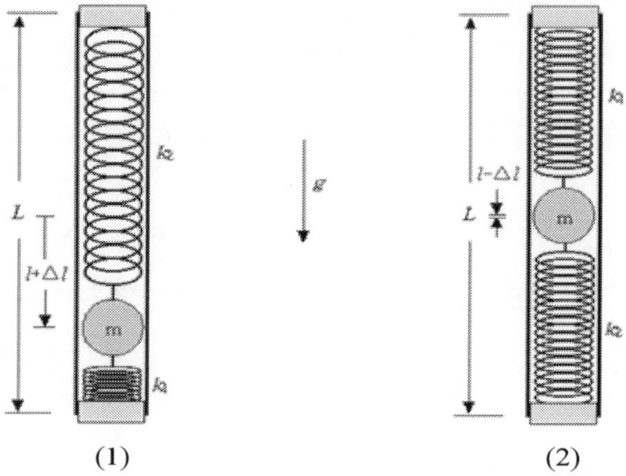

Fig. C-1 Small oscillation experimental set up

The measured periods are

	T_1 (s)		T_2 (s)
1	1.1085±0.00005	1	1.0194±0.00005
2	1.1092±0.00005	2	1.0194±0.00005
3	1.1089±0.00005	3	1.0193±0.00005
4	1.1085±0.00005	4	1.0191±0.00005
5	1.1094±0.00005	5	1.0192±0.00005
6	1.1090±0.00005	6	1.0194±0.00005
7	1.1088±0.00005	7	1.0194±0.00005
8	1.1090±0.00005	8	1.0191±0.00005
9	1.1092±0.00005	9	1.0192±0.00005
10	1.1094±0.00005	10	1.0193±0.00005

By averaging the10 measurements for each configuration, respectively, we get

$T_1 = 1.1090 \pm 0.0003$s and $T_2 = 1.0193 \pm 0.0001$s.

2. Explain, by using equations, why the angular frequencies ω_1 and ω_2 of small oscillation of the configurations are different. (1.0 points)

$$\omega_1 = \sqrt{\frac{Mg\,{}^{L}\!/_{2} + mg\left({}^{L}\!/_{2} + l + \Delta l\right)}{I_o + m\left\{\left({}^{L}\!/_{2} + l + \Delta l\right)^2 + \frac{2}{5}r^2\right\}}}$$

$$\omega_2 = \sqrt{\frac{Mg\,{}^{L}\!/_{2} + mg\left({}^{L}\!/_{2} - l + \Delta l\right)}{I_o + m\left\{\left({}^{L}\!/_{2} - l + \Delta l\right)^2 + \frac{2}{5}r^2\right\}}}$$

(Explanation) The moment of inertia of the Mechanical "Black Box" with respect to the pivot at the top of the tube is

$$I_1 = I_o + m\left\{\left({}^{L}\!/_{2} + l + \Delta l\right)^2 + \frac{2}{5}r^2\right\} \quad \text{or} \quad I_2 = I_o + m\left\{\left({}^{L}\!/_{2} - l + \Delta l\right)^2 + \frac{2}{5}r^2\right\}$$

depending on the orientation of the MBB as shown in Figs. C-1(1) and (2), respectively.

When the MBB is slightly tilted by an angle θ from vertical, the torque applied by the gravity is

$$\tau_1 = Mg\left({}^{L}\!/_{2}\right)\sin\theta + mg\left({}^{L}\!/_{2} + l + \Delta l\right)\sin\theta \approx \left\{Mg\left({}^{L}\!/_{2}\right) + mg\left({}^{L}\!/_{2} + l + \Delta l\right)\right\}\theta$$

or

$$\tau_2 = Mg\left({}^{L}\!/_{2}\right)\sin\theta + mg\left({}^{L}\!/_{2} - l + \Delta l\right)\sin\theta \approx \left\{Mg\left({}^{L}\!/_{2}\right) + mg\left({}^{L}\!/_{2} - l + \Delta l\right)\right\}\theta$$

depending on the orientation.

Therefore, the angular frequencies of oscillation become

$$\omega_1 = \sqrt{\frac{{}^{\tau_1}\!/_{\theta}}{I_1}} = \sqrt{\frac{Mg\,{}^{L}\!/_{2} + mg\left({}^{L}\!/_{2} + l + \Delta l\right)}{I_o + m\left\{\left({}^{L}\!/_{2} + l + \Delta l\right)^2 + \frac{2}{5}r^2\right\}}}$$

and

$$\omega_2 = \sqrt{\frac{{}^{\tau_2}\!/_{\theta}}{I_2}} = \sqrt{\frac{Mg\,{}^{L}\!/_{2} + mg\left({}^{L}\!/_{2} - l + \Delta l\right)}{I_o + m\left\{\left({}^{L}\!/_{2} - l + \Delta l\right)^2 + \frac{2}{5}r^2\right\}}}.$$

3. Evaluate Δl by eliminating I_o from the previous results. (1.0 points)

$$\Delta l = (7.2 \pm 0.9)\,\text{cm} = (7.2 \pm 0.9) \times 10^{-2}\,\text{m}$$

(Explanation) By rewriting the two expressions for the angular frequencies ω_1 and ω_2 as

$$Mg\,L/2 + mg\left(L/2 + l + \Delta l\right) = I_o\omega_1^2 + m\omega_1^2\left\{\left(L/2 + l + \Delta l\right)^2 + \frac{2}{5}r^2\right\}$$

and

$$Mg\,L/2 + mg\left(L/2 - l + \Delta l\right) = I_o\omega_2^2 + m\omega_2^2\left\{\left(L/2 - l + \Delta l\right)^2 + \frac{2}{5}r^2\right\}$$

one can eliminate the unknown moment of inertia I_o of the MBB without the ball. By eliminating the I_o one gets the equation for Δl

$$\left(\omega_2^2 - \omega_1^2\right)\left\{\frac{(M+m)gL}{2} + mg\Delta l\right\} + \left(\omega_1^2 + \omega_2^2\right)mgl = \omega_1^2\omega_2^2 m\left(L + 2\Delta l\right)(2l).$$

From the measured or given values we get,

$$\left(\omega_2^2 - \omega_1^2\right) = \left\{\left(\frac{2\pi}{T_2}\right)^2 - \left(\frac{2\pi}{T_1}\right)^2\right\} = \left(\frac{6.2832}{1.0193 \pm 0.0001}\right)^2 - \left(\frac{6.2832}{1.1090 \pm 0.0003}\right)^2$$

$$= 5.90 \pm 0.01\,\text{s}^{-2}$$

$$\frac{(M+m)gL}{2} = \frac{(141.1 \pm 0.05) \times 980 \times (40.0 \pm 0.05)}{2} = (27.66 \pm 0.04) \times 10^{-2}\,\text{kg·m}^2/\text{s}^2$$

$$\left(\omega_1^2 + \omega_2^2\right)mgl = \left\{\left(\frac{2\pi}{T_1}\right)^2 + \left(\frac{2\pi}{T_2}\right)^2\right\}(M+m)l_{cm}g$$

$$= \left\{\left(\frac{6.2832}{1.1090 \pm 0.0003}\right)^2 + \left(\frac{6.2832}{1.0193 \pm 0.0001}\right)^2\right\} \times (296 \pm 8) \times 980$$

$$= (203 \pm 5) \times 10^{-2}\ \text{kg·m}^2/\text{s}^4$$

$$\omega_1^2 \omega_2^2 ml = \left(\frac{2\pi}{T_1}\right)^2 \left(\frac{2\pi}{T_2}\right)^2 (M+m) l_{cm}$$

$$= \left(\frac{6.2832}{1.1090 \pm 0.0003}\right)^2 \left(\frac{6.2832}{1.0193 \pm 0.0001}\right)^2 \times (296 \pm 8)$$

$$= (3.6 \pm 0.1)\ \text{kg·m/s}^4.$$

Therefore, the equation we obtained in ***PART-C*** 3 becomes

$$(5.90 \pm 0.01)\{(27.66 \pm 0.04) \times 10^5 + (62 \pm 2) \times 980 \times \Delta l\} + (203 \pm 5) \times 10^5$$

$$= (7.2 \pm 0.2) \times 10^5 \times \{(40.0 \pm 0.05) + 2\Delta l\},$$

where Δl is expressed in the unit of cm. By solving the equation we get

$$\Delta l = (7.2 \pm 0.9)\ \text{cm} = (7.2 \pm 0.9) \times 10^{-2}\ \text{m}$$

4. Write down the value of the effective total spring constant k of the two-spring system. (2.0 points)

$$k = 9\ \text{N/m}$$

(Explanation) The effective total spring constant is

$$k \equiv \frac{mg}{\Delta l} = \frac{(62 \pm 2) \times 980}{7.2 \pm 0.9} = 9000 \pm 1000\ \text{dyne/cm} \quad \text{or} \quad 9 \pm 1\ \text{N/m}.$$

5. Obtain the respective values of k_1 and k_2. Write down their values. (1.0 point)

$k_1 = 5.7$ N/m
$k_2 = 3$ N/m

(Explanation) When the MBB is in equilibrium on a horizontal plane the force balance condition for the ball is that

$$\frac{L/2 - l - \delta - r}{L/2 + l - \delta - r} = \frac{N_1}{N_2} = \frac{k_2}{k_1}.$$

Since $k = k_1 + k_2$, we get

$$k_1 = \frac{k}{\dfrac{L/2 - l - \delta - r}{L/2 + l - \delta - r} + 1} = \frac{L/2 + l - \delta - r}{L - 2\delta - 2r} k$$

and

$$k_2 = k - k_1 = \frac{L/2 - l - \delta - r}{L - 2\delta - 2r} k.$$

From the measured or given values

$$\frac{L/2 + l - \delta - r}{L - 2\delta - 2r} = \frac{(20.0 \pm 0.03) + \left(\dfrac{296 \pm 8}{62 \pm 2}\right) - 0.5 - 1.1}{(40.0 \pm 0.05) - 1.0 - 2.2} = 0.63 \pm 0.005.$$

Therefore,

$$k_1 = (0.63 \pm 0.005) \times (9000 \pm 1000) = 5700 \pm 600 \text{ dyne/cm} \quad \text{or} \quad 5.7 \pm 0.6 \text{N/m},$$

and

$$k_2 = (9000 \pm 1000) - (5700 \pm 600) = 3000 \pm 1000 \text{ dyne/cm} \quad \text{or} \quad 3 \pm 1 \text{N/m}.$$

36TH

INTERNATIONAL PHYSICS OLYMPIAD

SALAMANCA, SPAIN

2005

Th 1 AN ILL FATED SATELLITE

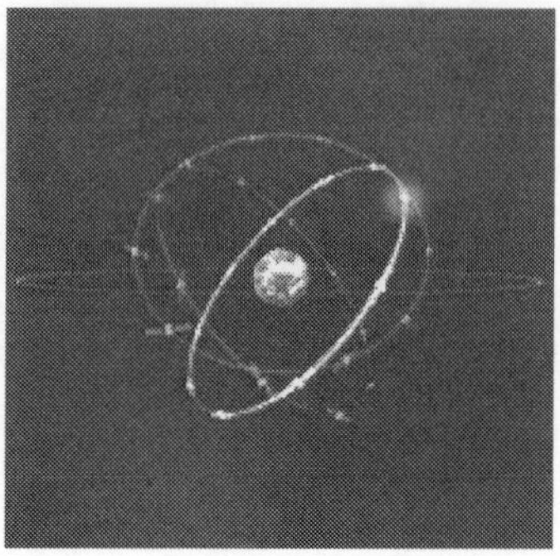
Image: ESA

The most frequent orbital manoeuvres performed by spacecraft consist of velocity variations along the direction of flight, namely accelerations to reach higher orbits or brakings done to initiate re-entering in the atmosphere. In this problem we will study the orbital variations when the engine thrust is applied in a radial direction.

To obtain numerical values use: Earth radius $R_T = 6.37 \cdot 10^6$ m, Earth surface gravity $g = 9.81\,\text{m/s}^2$, and take the length of the sidereal day to be $T_0 = 24.0$ h.

We consider a geosynchronous[1] communications satellite of mass m placed in an equatorial circular orbit of radius r_0. These satellites have an "apogee engine" which provides the tangential thrusts needed to reach the final orbit.

Marks are indicated at the beginning of each subquestion, in parenthesis.

Question 1

1.1 (0.3) Compute the numerical value of r_0.

1.2 (0.3+0.1) Give the analytical expression of the velocity v_0 of the satellite as a function of g, R_T, and r_0, and calculate its numerical value.

1.3 (0.4+0.4) Obtain the expressions of its angular momentum L_0 and mechanical energy E_0, as functions of v_0, m, g and R_T.

Once this geosynchronous circular orbit has been reached (see Figure F-1), the satellite has been stabilised in the desired location, and is being readied to do its work, an error by the ground controllers causes the apogee engine to be fired again. The thrust happens to be directed towards the Earth and, despite the quick reaction of the ground crew to shut the engine off, an unwanted velocity variation Δv is imparted on the satellite. We characterize this boost by the parameter $\beta = \Delta v / v_0$. The duration of the engine burn is always negligible with respect to any other orbital times, so that it can be considered as instantaneous.

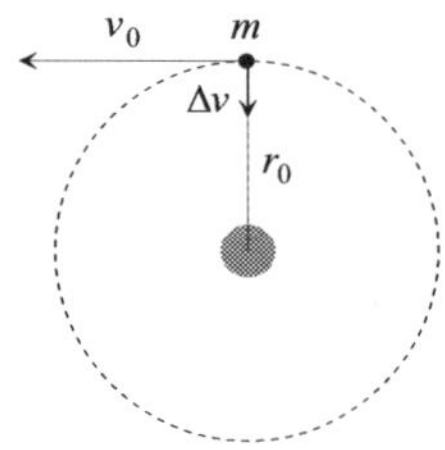

F-1

Question 2

Suppose $\beta < 1$.

2.1 (0.4+0.5) Determine the parameters of the new orbit[2], *semi-latus-rectum* l and *eccentricity* ε, in terms of r_0 and β.

2.2 (1.0) Calculate the angle α between the major axis of the new orbit and the position vector at the accidental misfire.

2.3 (1.0+0.2) Give the analytical expressions of the perigee r_{min} and apogee r_{max} distances to the Earth centre, as functions of r_0 and β, and calculate their numerical values for $\beta = 1/4$.

2.4 (0.5+0.2) Determine the period of the new orbit, T, as a function of T_0 and β, and calculate its numerical value for $\beta = 1/4$.

[1] Its revolution period is T_0.

[2] See the "hint".

Question 3

3.1 (0.5) Calculate the minimum boost parameter, β_{esc}, needed for the satellite to escape Earth gravity.

3.2 (1.0) Determine in this case the closest approach of the satellite to the Earth centre in the new trajectory, r'_{min}, as a function of r_0.

Question 4

Suppose $\beta > \beta_{esc}$.

4.1 (1.0) Determine the residual velocity at the infinity, v_∞, as a function of v_0 and β.

4.2 (1.0) Obtain the "impact parameter" b of the asymptotic escape direction in terms of r_0 and β. (See Figure F-2).

4.3 (1.0+0.2) Determine the angle ϕ of the asymptotic escape direction in terms of β. Calculate its numerical value for $\beta = \frac{3}{2}\beta_{esc}$.

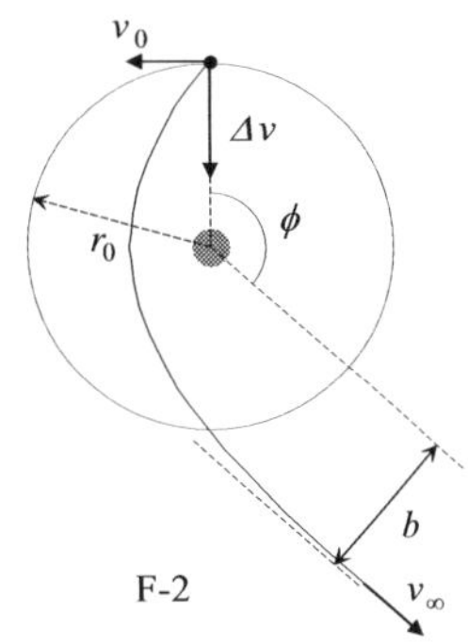

F-2

HINT

Under the action of central forces obeying the inverse-square law, bodies follow trajectories described by ellipses, parabolas or hyperbolas. In the approximation $m << M$ the gravitating mass M is at one of the focuses. Taking the origin at this focus, the general polar equation of these curves can be written as (see Figure F-3)

$$r(\theta) = \frac{l}{1 - \varepsilon \cos\theta}$$

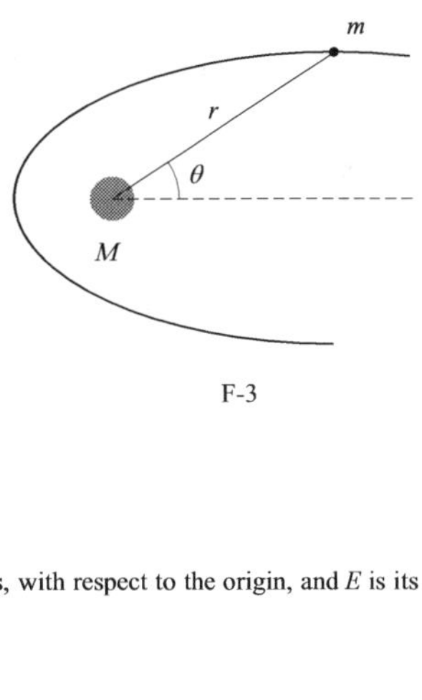

F-3

where l is a positive constant named the *semi-latus-rectum* and ε is the *eccentricity* of the curve. In terms of constants of motion:

$$l = \frac{L^2}{GMm^2} \qquad \text{and} \qquad \varepsilon = \left(1 + \frac{2EL^2}{G^2M^2m^3}\right)^{1/2}$$

where G is the Newton constant, L is the modulus of the angular momentum of the orbiting mass, with respect to the origin, and E is its mechanical energy, with zero potential energy at infinity.

We may have the following cases:

i) If $0 \le \varepsilon < 1$, the curve is an ellipse (circumference for $\varepsilon = 0$).

ii) If $\varepsilon = 1$, the curve is a parabola.

iii) If $\varepsilon > 1$, the curve is a hyperbola.

Th 2 ABSOLUTE MEASUREMENTS OF ELECTRICAL QUANTITIES

The technological and scientific transformations underwent during the XIX century produced a compelling need of universally accepted standards for the electrical quantities. It was thought the new absolute units should only rely on the standards of length, mass and time established after the French Revolution. An intensive experimental work to settle the values of these units was developed from 1861 until 1912. We propose here three case studies.

Marks are indicated at the beginning of each subquestion, in parenthesis.

Determination of the ohm (Kelvin)

A closed circular coil of N turns, radius a and total resistance R is rotated with uniform angular velocity ω about a vertical diameter in a horizontal magnetic field $\vec{B}_0 = B_0 \vec{i}$.

1. (0.5+1.0) Compute the electromotive force ε induced in the coil, and also the mean power[1] $\langle P \rangle$ required for maintaining the coil in motion. Neglect the coil self inductance.

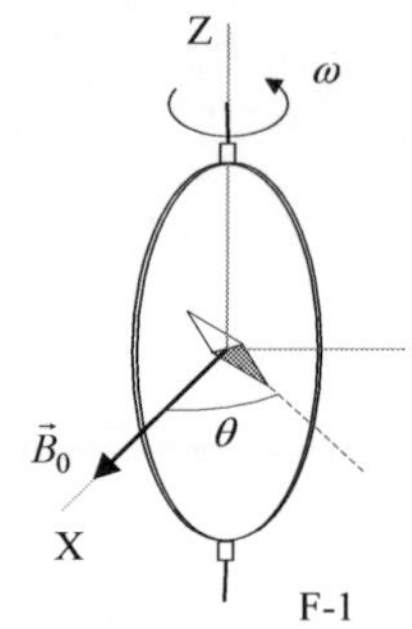

F-1

A small magnetic needle is placed at the center of the coil, as shown in Figure F-1. It is free to turn slowly around the Z axis in a horizontal plane, but it cannot follow the rapid rotation of the coil.

2. (2.0) Once the stationary regime is reached, the needle will set at a direction making a small angle θ with $\vec{B}_0$. Compute the resistance R of the coil in terms of this angle and the other parameters of the system.

Lord Kelvin used this method in the 1860s to set the absolute standard for the ohm. To avoid the rotating coil, Lorenz devised an alternative method used by Lord Rayleigh and Ms. Sidgwick, that we analyze in the next paragraphs.

Determination of the ohm (Rayleigh, Sidgwick).

The experimental setup is shown in Figure F-2. It consists of two identical metal disks D and D' of radius b mounted on the conducting shaft SS'. A motor rotates the set at an angular velocity ω, which can be adjusted for measuring R. Two identical coils C and C' (of radius a and with N turns each) surround the disks. They are connected in such a form that the current I flows through them in opposite directions. The whole apparatus serves to measure the resistance R.

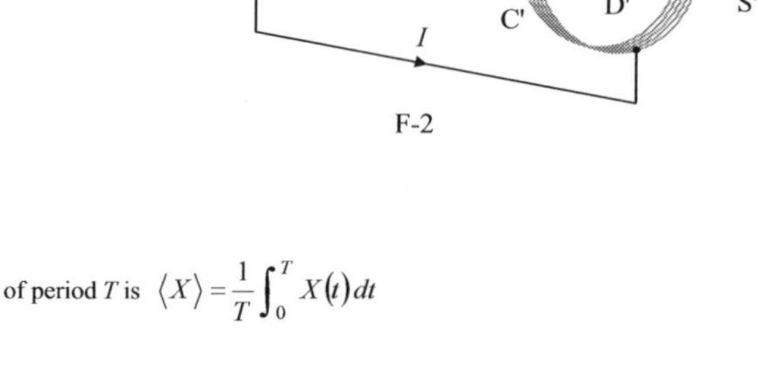

F-2

[1] The mean value $\langle X \rangle$ of a quantity $X(t)$ in a periodic system of period T is $\langle X \rangle = \frac{1}{T}\int_0^T X(t)\,dt$

You may need one or more of these integrals:

$$\int_0^{2\pi} \sin x\,dx = \int_0^{2\pi} \cos x\,dx = \int_0^{2\pi} \sin x \cos x\,dx = 0, \quad \int_0^{2\pi} \sin^2 x\,dx = \int_0^{2\pi} \cos^2 x\,dx = \pi, \quad \text{and later} \quad \int x^n\,dx = \frac{1}{n+1}x^{n+1}$$

3. (2.0) Assume that the current I flowing through the coils C and C' creates a uniform magnetic field B around D and D', equal to the one at the centre of the coil. Compute[1] the electromotive force $\mathcal{E}$ induced between the rims 1 and 4, assuming that the distance between the coils is much larger than the radius of the coils and that $a >> b$.

The disks are connected to the circuit by brush contacts at their rims 1 and 4. The galvanometer G detects the flow of current through the circuit 1-2-3-4.

4. (0.5) The resistance R is measured when G reads zero. Give R in terms of the physical parameters of the system.

Determination of the ampere

Passing a current through two conductors and measuring the force between them provides an absolute determination of the current itself. The "Current Balance" designed by Lord Kelvin in 1882 exploits this method. It consists of six identical single turn coils $C_1 \ldots C_6$ of radius a, connected in series. As shown in Figure F-3, the fixed coils C_1, C_3, C_4, and C_6 are on two horizontal planes separated by a small distance $2h$. The coils C_2 and C_5 are carried on balance arms of length d, and they are, in equilibrium, equidistant from both planes.

The current I flows through the various coils in such a direction that the magnetic force on C_2 is upwards while that on C_5 is downwards. A mass m at a distance x from the fulcrum O is required to restore the balance to the equilibrium position described above when the current flows through the circuit.

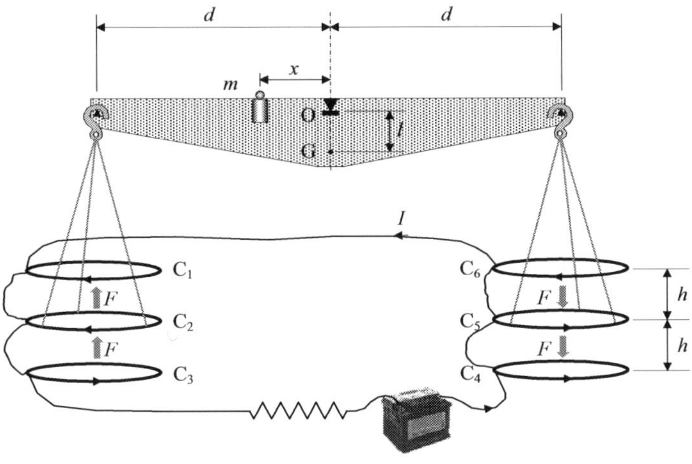

F-3

5. (1.0) Compute the force F on C_2 due to the magnetic interaction with C_1. For simplicity assume that the force per unit length is the one corresponding to two long, straight wires carrying parallel currents.

6. (1.0) The current I is measured when the balance is in equilibrium. Give the value of I in terms of the physical parameters of the system. The dimensions of the apparatus are such that we can neglect the mutual effects of the coils on the left and on the right.

Let M be the mass of the balance (except for m and the hanging parts), G its centre of mass and l the distance $\overline{OG}$.

7. (2.0) The balance equilibrium is stable against deviations producing small changes δz in the height of C_2 and $-\delta z$ in C_5. Compute[2] the maximum value $\delta z_{\max}$ so that the balance still returns towards the equilibrium position when it is released.

---

[2] Consider that the coils centres remain approximately aligned.

Use the approximations $\frac{1}{1 \pm \beta} \approx 1 \mp \beta + \beta^2$ or $\frac{1}{1 \pm \beta^2} \approx 1 \mp \beta^2$ for $\beta << 1$, and $\sin\theta \approx \tan\theta$ for small θ.

Th 3 NEUTRONS IN A GRAVITATIONAL FIELD

In the familiar classical world, an elastic bouncing ball on the Earth's surface is an ideal example for perpetual motion. The ball is trapped: it can not go below the surface or above its turning point. It will remain bounded in this state, turning down and bouncing up once and again, forever. Only air drag or inelastic bounces could stop the process and will be ignored in the following.

A group of physicists from the Institute Laue - Langevin in Grenoble reported[1] in 2002 experimental evidence on the behaviour of neutrons in the gravitational field of the Earth. In the experiment, neutrons moving to the right were allowed to fall towards a horizontal crystal surface acting as a neutron mirror, where they bounced back elastically up to the initial height once and again.

The setup of the experiment is sketched in Figure F-1. It consists of the opening W, the neutron mirror M (at height $z = 0$), the neutron absorber A (at height $z = H$ and with length L) and the neutron detector D. The beam of neutrons flies with constant horizontal velocity component v_x from W to D through the cavity between A and M. All the neutrons that reach the surface of A are absorbed and disappear from the experiment. Those that reach the surface of M are reflected elastically. The detector D counts the transmission rate $N(H)$, that is, the total number of neutrons that reach D per unit time.

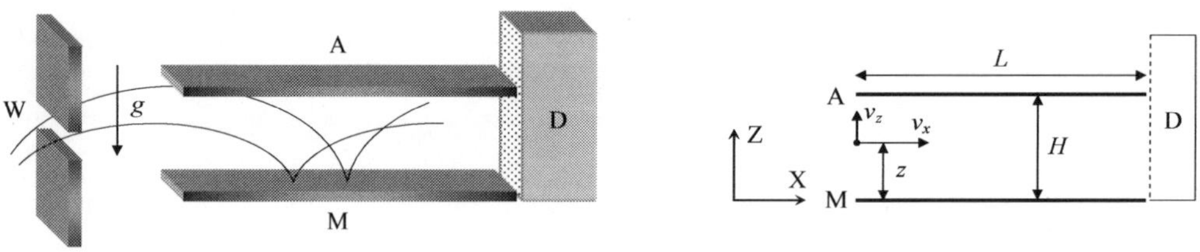

F-1

Marks are indicated at the beginning of each subquestion, in parenthesis.

The neutrons enter the cavity with a wide range of positive and negative vertical velocities, v_z. Once in the cavity, they fly between the mirror below and the absorber above.

1. (1.5) Compute classically the range of vertical velocities $v_z(z)$ of the neutrons that, entering at a height z, can arrive at the detector D. Assume that L is much larger than any other length in the problem.

2. (1.5) Calculate classically the minimum length L_c of the cavity to ensure that all neutrons outside the previous velocity range, regardless of the values of z, are absorbed by A. Use $v_x = 10\ \text{m s}^{-1}$ and $H = 50\ \mu\text{m}$.

The neutron transmission rate $N(H)$ is measured at D. We expect that it increases monotonically with H.

3. (2.5) Compute the classical rate $N_c(H)$ assuming that neutrons arrive at the cavity with vertical velocity v_z and at height z, being all the values of v_z and z equally probable. Give the answer in terms of ρ, the constant number of neutrons per unit time, per unit vertical velocity, per unit height, that enter the cavity with vertical velocity v_z and at height z.

---

1 V. V. Nesvizhevsky et al. "Quantum states of neutrons in the Earth's gravitational field." Nature, **415** (2002) 297. Phys Rev D 67, 102002 (2003).

The experimental results obtained by the Grenoble group disagree with the above classical predictions, showing instead that the value of $N(H)$ experiences sharp increases when H crosses some critical heights H_1, H_2 ... (Figure F-2 shows a sketch). In other words, the experiment showed that the vertical motion of neutrons bouncing on the mirror is quantized. In the language that Bohr and Sommerfeld used to obtain the energy levels of the hydrogen atom, this can be written as: "The action S of these neutrons along the vertical direction is an integer multiple of the Planck action constant h". Here S is given by

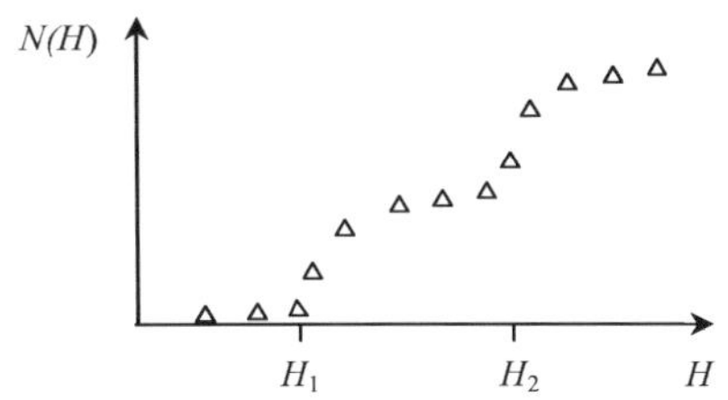

F-2

$$S = \int p_z(z)dz = nh, \quad n = 1, 2, 3 \ldots \qquad \text{(Bohr-Sommerfeld quantization rule)}$$

where p_z is the vertical component of the classical momentum, and the integral covers a whole bouncing cycle. Only neutrons with these values of S are allowed in the cavity.

4. (2.5) Compute the turning heights H_n and energy levels E_n (associated to the vertical motion) using the Bohr-Sommerfeld quantization condition. Give the numerical result for H_1 in μm and for E_1 in eV.

The uniform initial distribution ρ of neutrons at the entrance changes, during the flight through a long cavity, into the step-like distribution detected at D (see Figure F-2). From now on, we consider for simplicity the case of a long cavity with $H < H_2$. Classically, all neutrons with energies in the range considered in question 1 were allowed through it, while quantum mechanically only neutrons in the energy level E_1 are permitted. According to the time-energy Heisenberg uncertainty principle, this reshuffling requires a minimum time of flight. The uncertainty of the vertical motion energy will be significant if the cavity length is small. This phenomenon will give rise to the widening of the energy levels.

5. (2.0) Estimate the minimum time of flight t_q and the minimum length L_q of the cavity needed to observe the first sharp increase in the number of neutrons at D. Use $v_x = 10 \text{ m s}^{-1}$.

Data:

Planck action constant	$h = 6.63 \cdot 10^{-34}$ J s
Speed of light in vacuum	$c = 3.00 \cdot 10^{8}$ m s^{-1}
Elementary charge	$e = 1.60 \cdot 10^{-19}$ C
Neutron mass	$M = 1.67 \cdot 10^{-27}$ kg
Acceleration of gravity on Earth	$g = 9.81$ m s^{-2}
If necessary, use the expression:	$\int (1-x)^{1/2} dx = -\frac{2(1-x)^{3/2}}{3}$

Th1 AN ILL FATED SATELLITE
SOLUTION

1.1 and 1.2

$$\left.\begin{array}{l} G\dfrac{M_T m}{r_0^2} = m\dfrac{v_0^2}{r_0} \\ v_0 = \dfrac{2\pi r_0}{T_0} \\ g = \dfrac{GM_T}{R_T^2} \end{array}\right\} \Rightarrow \left\{\begin{array}{l} r_0 = \left(\dfrac{g R_T^2 T_0^2}{4\pi^2}\right)^{1/3} \Rightarrow \boxed{r_0 = 4.22\cdot 10^7 \text{ m}} \\ \boxed{v_0 = R_T\sqrt{\dfrac{g}{r_0}}} \Rightarrow \boxed{v_0 = 3.07\cdot 10^3 \text{ m/s}} \end{array}\right.$$

1.3

$$L_0 = r_0 m v_0 = \frac{g R_T^2}{v_0^2} m v_0 \Rightarrow \boxed{L_0 = \frac{m g R_T^2}{v_0}}$$

$$E_0 = \frac{1}{2}mv_0^2 - G\frac{M_T m}{r_0} = \frac{1}{2}mv_0^2 - \frac{g R_T^2 m}{r_0} = \frac{1}{2}mv_0^2 - mv_0^2 \Rightarrow \boxed{E_0 = -\frac{1}{2}mv_0^2}$$

2.1

The value of the *semi-latus-rectum* l is obtained taking into account that the orbital angular momentum is the same in both orbits. That is

$$l = \frac{L_0^2}{G M_T m^2} = \frac{m^2 g^2 R_T^4}{v_0^2}\frac{1}{g R_T^2 m^2} = \frac{g R_T^2}{v_0^2} = r_0 \Rightarrow \boxed{l = r_0}$$

The eccentricity value is

$$\varepsilon^2 = 1 + \frac{2 E L_0^2}{G^2 M_T^2 m^3}$$

where E is the new satellite mechanical energy

$$E = \frac{1}{2}m\left(v_0^2 + \Delta v^2\right) - G\frac{M_T m}{r_0} = \frac{1}{2}m\Delta v^2 + E_0 = \frac{1}{2}m\Delta v^2 - \frac{1}{2}mv_0^2$$

that is

$$E = \frac{1}{2}mv_0^2\left(\frac{\Delta v^2}{v_0^2} - 1\right) = \frac{1}{2}mv_0^2\left(\beta^2 - 1\right)$$

Combining both, one gets $\boxed{\varepsilon = \beta}$

This is an elliptical trajectory because $\varepsilon = \beta < 1$.

2.2

The initial and final orbits cross at P, where the satellite engine fired instantaneously (see Figure 4). At this point

$$r(\theta = \alpha) = r_0 = \frac{r_0}{1-\beta\cos\alpha} \quad \Rightarrow \quad \boxed{\alpha = \frac{\pi}{2}}$$

2.3

From the trajectory expression one immediately obtains that the maximum and minimum values of r correspond to $\theta = 0$ and $\theta = \pi$ respectively (see Figure 4). Hence, they are given by

$$r_{max} = \frac{l}{1-\varepsilon} \qquad r_{min} = \frac{l}{1+\varepsilon}$$

that is

$$\boxed{r_{\max} = \frac{r_0}{1-\beta}} \quad \text{and} \quad \boxed{r_{\min} = \frac{r_0}{1+\beta}}$$

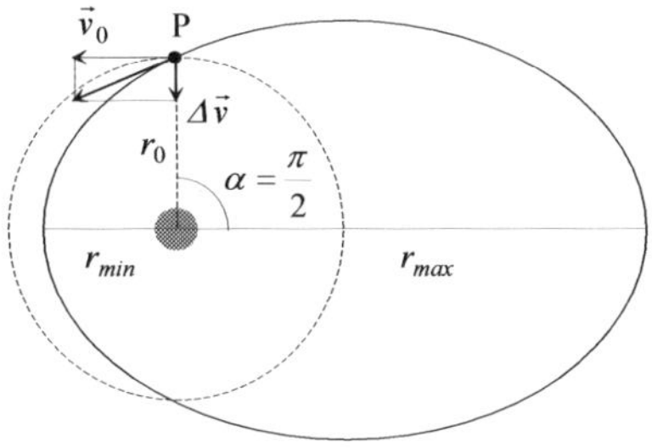

Figure 4

For $\beta = 1/4$, one gets

$$\boxed{r_{max} = 5.63 \cdot 10^7 \text{ m}; \quad r_{min} = 3.38 \cdot 10^7 \text{ m}}$$

The distances r_{max} and r_{min} can also be obtained from mechanical energy and angular momentum conservation, taking into account that $\vec{r}$ and $\vec{v}$ are orthogonal at apogee and at perigee

$$E = \frac{1}{2}mv_0^2\left(\beta^2 - 1\right) = \frac{1}{2}mv^2 - \frac{gR_T^2 m}{r}$$

$$L_0 = \frac{mgR_T^2}{v_0} = m\,v\,r$$

What remains of them, after eliminating v, is a second-degree equation whose solutions are r_{max} and r_{min}.

2.4

By the Third Kepler Law, the period T in the new orbit satisfies that

$$\frac{T^2}{a^3} = \frac{T_0^2}{r_0^3}$$

where a, the semi-major axis of the ellipse, is given by

$$a = \frac{r_{max} + r_{min}}{2} = \frac{r_0}{1-\beta^2}$$

Therefore

$$\boxed{T = T_0\left(1-\beta^2\right)^{-3/2}}$$

For β = 1/4 $\qquad \boxed{T = T_0\left(\frac{15}{16}\right)^{-3/2} = 26.4\,\text{h}}$

3.1

Only if the satellite follows an open trajectory it can escape from the Earth gravity attraction. Then, the orbit eccentricity has to be equal or larger than one. The minimum boost corresponds to a parabolic trajectory, with $\varepsilon = 1$

$$\varepsilon = \beta \qquad \Rightarrow \qquad \boxed{\beta_{esc} = 1}$$

This can also be obtained by using that the total satellite energy has to be zero to reach infinity ($E_p = 0$) without residual velocity ($E_k = 0$)

$$E = \frac{1}{2} m v_0^2 \left(\beta_{esc}^2 - 1\right) = 0 \quad \Rightarrow \qquad \beta_{esc} = 1$$

This also arises from $T = \infty$ or from $r_{max} = \infty$.

3.2

Due to $\varepsilon = \beta_{esc} = 1$, the polar parabola equation is

$$r = \frac{l}{1 - \cos\theta}$$

where the semi-latus-rectum continues to be $l = r_0$. The minimum Earth - satellite distance corresponds to $\theta = \pi$, where

$$\boxed{r'_{min} = \frac{r_0}{2}}$$

This also arises from energy conservation (for $E = 0$) and from the equality between the angular momenta (L_0) at the initial point P and at maximum approximation, where $\vec{r}$ and $\vec{v}$ are orthogonal.

4.1

If the satellite escapes to infinity with residual velocity v_∞, by energy conservation

$$E = \frac{1}{2} m v_0^2 \left(\beta^2 - 1\right) = \frac{1}{2} m v_\infty^2 \quad \Rightarrow$$

$$\boxed{v_\infty = v_0 \left(\beta^2 - 1\right)^{1/2}}$$

4.2

As $\varepsilon = \beta > \beta_{esc} = 1$ the satellite trajectory will be a hyperbola.

The satellite angular momentum is the same at P than at the point where its residual velocity is v_∞ (Figure 5), thus

$$m v_0 r_0 = m v_\infty b$$

So

$$b = r_0 \frac{v_0}{v_\infty} \quad \Rightarrow \quad \boxed{b = r_0 \left(\beta^2 - 1\right)^{-1/2}}$$

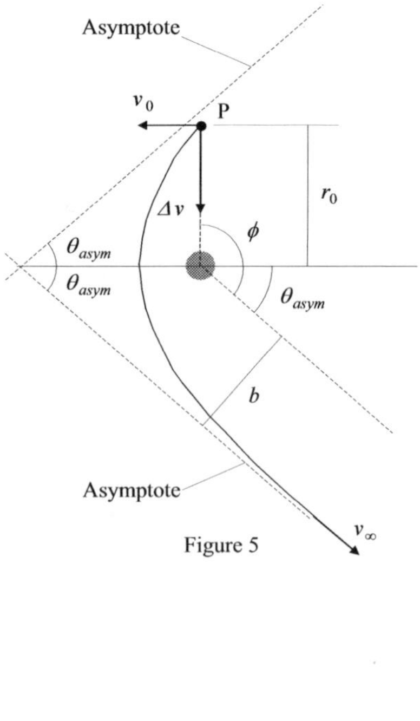

Figure 5

4.3

The angle between each asymptote and the hyperbola axis is that appearing in its polar equation in the limit $r \to \infty$. This is the angle for which the equation denominator vanishes

$$1-\beta\cos\theta_{asym}=0 \quad \Rightarrow \quad \theta_{asym}=\cos^{-1}\left(\frac{1}{\beta}\right)$$

According to Figure 5

$$\phi=\frac{\pi}{2}+\theta_{asym} \quad \Rightarrow \quad \boxed{\phi=\frac{\pi}{2}+\cos^{-1}\left(\frac{1}{\beta}\right)}$$

For $\beta=\frac{3}{2}\beta_{esc}=\frac{3}{2}$, one gets $\boxed{\phi=138^{o}=2.41\,\text{rad}}$

Th 1 ANSWER SHEET

Question	Basic formulas and ideas used	Analytical results	Numerical results	Marking guideline
1.1	$G\frac{M_T m}{r_0^2} = m\frac{v_0^2}{r_0}$ $v_0 = \frac{2\pi r_0}{T_0}$		$r_0 = 4.22 \cdot 10^7$ m	0.3
1.2	$g = \frac{GM_T}{R_T^2}$	$v_0 = R_T\sqrt{\frac{g}{r_0}}$	$v_0 = 3.07 \cdot 10^3$ m/s	0.3 + 0.1
1.3	$\vec{L} = m\vec{r} \times \vec{v}$ $E = \frac{1}{2}mv^2 - G\frac{Mm}{r}$	$L_0 = \frac{mgR_T^2}{v_0}$ $E_0 = -\frac{1}{2}mv_0^2$		0.4 0.4
2.1	Hint on the conical curves	$l = r_0$ $\varepsilon = \beta$		0.4 0.5
2.2			$\alpha = \frac{\pi}{2}$	1.0
2.3	Results of 2.1, or conservation of E and L	$r_{max} = \frac{r_0}{1-\beta}$ $r_{min} = \frac{r_0}{1+\beta}$	$r_{max} = 5.63 \cdot 10^7$ m $r_{min} = 3.38 \cdot 10^7$ m	1.0 + 0.2
2.4	Third Kepler's Law	$T = T_0\left(1-\beta^2\right)^{-3/2}$	$T = 26.4$ h	0.5 + 0.2
3.1	$\varepsilon = 1,\ E = 0,\ T = \infty$ or $r_{max} = \infty$		$\beta_{esc} = 1$	0.5
3.2	$\varepsilon = 1$ and results of 2.1	$r'_{min} = \frac{r_0}{2}$		1.0
4.1	Conservation of E	$v_\infty = v_0\left(\beta^2 - 1\right)^{1/2}$		1.0
4.2	Conservation of L	$b = r_0\left(\beta^2 - 1\right)^{-1/2}$		1.0
4.3	Hint on the conical curves	$\phi = \frac{\pi}{2} + \cos^{-1}\left(\frac{1}{\beta}\right)$	$\phi = 138^\circ = 2.41$ rad	1.0 + 0.2

Th 2 ABSOLUTE MEASUREMENTS OF ELECTRICAL QUANTITIES

SOLUTION

1. After some time t, the normal to the coil plane makes an angle ωt with the magnetic field $\vec{B}_0 = B_0\vec{i}$. Then, the magnetic flux through the coil is

$$\phi = N\ \vec{B}_0 \cdot \vec{S}$$

where the vector surface $\vec{S}$ is given by $\vec{S} = \pi a^2 \left(\cos\omega t\,\vec{i} + \sin\omega t\,\vec{j}\right)$

Therefore $\phi = N\pi a^2 B_0 \cos\omega t$

The induced electromotive force is

$$\varepsilon = -\frac{d\phi}{dt} \quad \Rightarrow \quad \boxed{\varepsilon = N\pi a^2 B_0 \omega \sin\omega t}$$

The instantaneous power is $P = \varepsilon^2/\mathrm{R}$, therefore

$$\boxed{\langle P\rangle = \frac{\left(N\,\pi\,a^2 B_0\omega\right)^2}{2R}}$$

where we used $<\sin^2\omega\,t> = \frac{1}{T}\int_0^T \sin^2\omega t\,dt = \frac{1}{2}$

2. The total field at the center the coil at the instant t is

$$\vec{B}_t = \vec{B}_0 + \vec{B}_i$$

where $\vec{B}_i$ is the magnetic field due to the induced current $\vec{B}_i = B_i\left(\cos\omega t\,\vec{i} + \sin\omega t\,\vec{j}\right)$

with $B_i = \frac{\mu_0 N I}{2a}$ and $I = \varepsilon / R$

Therefore $B_i = \frac{\mu_0 N^2 \pi a B_0 \omega}{2R}\sin\omega t$

The mean values of its components are

$$\langle B_{ix}\rangle = \frac{\mu_0 N^2 \pi a B_0 \omega}{2R}\langle \sin\omega t\cos\omega t\rangle = 0$$

$$\langle B_{iy}\rangle = \frac{\mu_0 N^2 \pi a B_0 \omega}{2R}\langle \sin^2\omega t\rangle = \frac{\mu_0 N^2 \pi a B_0 \omega}{4R}$$

And the mean value of the total magnetic field is

$$\langle \vec{B}_t\rangle = B_0\vec{i} + \frac{\mu_0 N^2 \pi a B_0 \omega}{4R}\vec{j}$$

The needle orients along the mean field, therefore

$$\tan\theta = \frac{\mu_0 N^2 \pi a \omega}{4R}$$

Finally, the resistance of the coil measured by this procedure, in terms of θ, is

$$\boxed{R = \frac{\mu_0 N^2 \pi a \omega}{4 \tan\theta}}$$

3. The force on a unit positive charge in a disk is radial and its modulus is

$$\left|\vec{v} \times \vec{B}\right| = v B = \omega r B$$

where B is the magnetic field at the center of the coil

$$B = N \frac{\mu_0 I}{2a}$$

Then, the electromotive force (e.m.f.) induced on each disk by the magnetic field B is

$$\varepsilon_D = \varepsilon_{D'} = B\omega \int_0^b r\,dr = \frac{1}{2} B\omega b^2$$

Finally, the induced e.m.f. between 1 and 4 is $\varepsilon = \varepsilon_D + \varepsilon_{D'}$

$$\boxed{\varepsilon = N \frac{\mu_0 b^2 \omega I}{2a}}$$

4. When the reading of G vanishes, $I_G = 0$ and Kirchoff laws give an immediate answer. Then we have

$$\varepsilon = I R \qquad \Rightarrow \qquad \boxed{R = N \frac{\mu_0 b^2 \omega}{2a}}$$

5. The force per unit length f between two indefinite parallel straight wires separated by a distance h is.

$$f = \frac{\mu_0}{2\pi} \frac{I_1 I_2}{h}$$

for $I_1 = I_2 = I$ and length $2\pi a$, the force F induced on C_2 by the neighbor coils C_1 is

$$\boxed{F = \frac{\mu_0 a}{h} I^2}$$

6. In equilibrium

$$mgx = 4Fd$$

Then

$$mgx = \frac{4\mu_0 ad}{h} I^2 \qquad (1)$$

so that

$$\boxed{I = \left(\frac{mghx}{4\mu_0 ad}\right)^{1/2}}$$

7. The balance comes back towards the equilibrium position for a little angular deviation $\delta\varphi$ if the gravity torques with respect to the fulcrum O are greater than the magnetic torques.

$$Mgl\sin\delta\varphi + mgx\cos\delta\varphi > 2\mu_0 a I^2\left(\frac{1}{h-\delta z}+\frac{1}{h+\delta z}\right)d\cos\delta\varphi$$

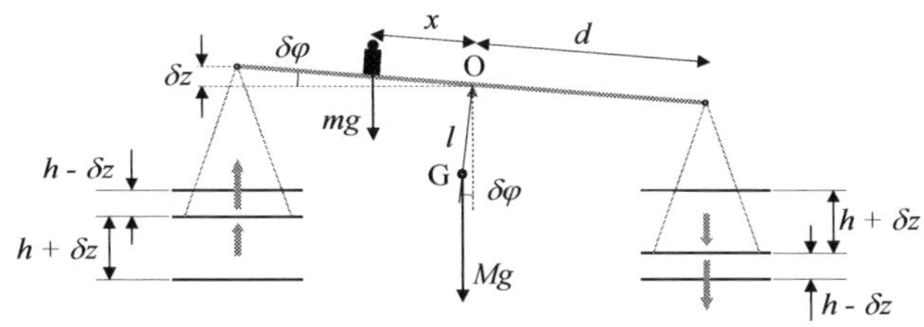

Therefore, using the suggested approximation

$$Mgl\sin\delta\varphi + mgx\cos\delta\varphi > \frac{4\mu_0 a d I^2}{h}\left(1+\frac{\delta z^2}{h^2}\right)\cos\delta\varphi$$

Taking into account the equilibrium condition (1), one obtains

$$M\,g\,l\,sin\,\delta\varphi > m\,g\,x\frac{\delta z^2}{h^2}cos\,\delta\varphi$$

Finally, for $\tan\delta\varphi \approx \sin\delta\varphi = \dfrac{\delta z}{d}$

$$\delta z < \frac{Mlh^2}{mxd} \quad\Rightarrow\quad \boxed{\delta z_{\max} = \frac{Mlh^2}{mxd}}$$

Th 2 ANSWER SHEET

Question	Basic formulas and ideas used	Analytical results	Marking guideline
1	$\Phi = N\,\vec{B}_0 \cdot \vec{S}$ $\varepsilon = -\dfrac{d\Phi}{dt}$ $P = \dfrac{\varepsilon^2}{R}$	$\varepsilon = N\pi a^2 B_0 \omega \sin\omega t$ $\langle P\rangle = \dfrac{(N\pi a^2 B_0\omega)^2}{2R}$	0.5 1.0
2	$\vec{B} = \vec{B}_0 + \vec{B}_i$ $B_i = \dfrac{\mu_0 N}{2a} I$ $\tan\theta = \dfrac{\langle B_y\rangle}{\langle B_x\rangle}$	$R = \dfrac{\mu_0 N^2 \pi a\omega}{4\tan\theta}$	2.0
3	$\vec{E} = \vec{v}\times\vec{B}$ $v = \omega r$ $B = N\dfrac{\mu_0 I}{2a}$ $\varepsilon = \int_0^b \vec{E}\,d\vec{r}$	$\varepsilon = N\dfrac{\mu_0 b^2 \omega I}{2a}$	2.0
4	$\varepsilon = RI$	$R = N\dfrac{\mu_0 b^2 \omega}{2a}$	0,5
5	$f = \dfrac{\mu_0}{2\pi}\dfrac{I I'}{h}$	$F = \dfrac{\mu_0\, a}{h} I^2$	1.0
6	$mgx = 4Fd$	$I = \left(\dfrac{mghx}{4\mu_0 ad}\right)^{1/2}$	1.0
7	$\Gamma_{grav} > \Gamma_{mag}$	$\delta z_{\max} = \dfrac{Mlh^2}{mxd}$	2.0

Th3 QUANTUM EFFECTS OF GRAVITY
SOLUTION

1. The only neutrons that will survive absorption at A are those that cannot cross H. Their turning points will be below H. So that, for a neutron entering to the cavity at height z with vertical velocity v_z, conservation of energy implies

$$\frac{1}{2}M v_z^2 + Mgz \le MgH \qquad \Rightarrow \qquad \boxed{-\sqrt{2g(H-z)} \le v_z(z) \le \sqrt{2g(H-z)}}$$

2. The cavity should be long enough to ensure the absorption of all neutrons with velocities outside the allowed range. Therefore, neutrons have to reach its maximum height at least once within the cavity. The longest required length corresponds to neutrons that enter at $z = H$ with $v_z = 0$ (see the figure). Calling t_f to their time of fall

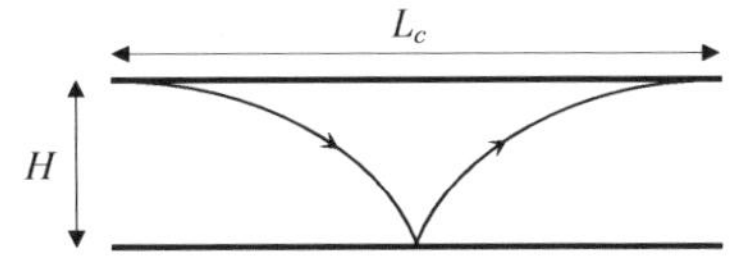

$$\left.\begin{aligned} L_c &= v_x 2t_f \\ H &= \frac{1}{2}g t_f^2 \end{aligned}\right\} \Rightarrow \qquad \boxed{L_c = 2v_x\sqrt{\frac{2H}{g}}} \qquad \boxed{L_c = 6.4\text{ cm}}$$

3. The rate of transmitted neutrons entering at a given height z, per unit height, is proportional to the range of allowed velocities at that height, ρ being the proportionality constant

$$\frac{dN_c(z)}{dz} = \rho\left[v_{z,\max}(z) - v_{z,\min}(z)\right] = 2\rho\sqrt{2g(H-z)}$$

The total number of transmitted neutrons is obtained by adding the neutrons entering at all possible heights. Calling $y = z/H$

$$N_c(H) = \int_0^H dN_c(z) = \int_0^H 2\rho\sqrt{2g(H-z)}dz = 2\rho\sqrt{2g}\,H^{3/2}\int_0^1 (1-y)^{1/2}dy = 2\rho\sqrt{2g}\,H^{3/2}\left[-\frac{2}{3}(1-y)^{3/2}\right]_0^1$$

$$\Rightarrow \qquad \boxed{N_c(H) = \frac{4}{3}\rho\sqrt{2g}\,H^{3/2}}$$

4. For a neutron falling from a height H, the action over a bouncing cycle is twice the action during the fall or the ascent

$$S = 2\int_0^H p_z dz = 2M\sqrt{2g}\,H^{3/2}\int_0^1 (1-y)^{1/2}dy = \frac{4}{3}M\sqrt{2g}\,H^{3/2}$$

Using the BS quantization condition

$$S = \frac{4}{3}M\sqrt{2g}\,H^{3/2} = nh \qquad \Rightarrow \qquad \boxed{H_n = \left(\frac{9h^2}{32M^2 g}\right)^{1/3} n^{2/3}}$$

The corresponding energy levels (associated to the vertical motion) are

$$E_n = MgH_n \qquad \Rightarrow \qquad \boxed{E_n = \left(\frac{9Mg^2h^2}{32}\right)^{1/3} n^{2/3}}$$

Numerical values for the first level:

$$H_1 = \left(\frac{9h^2}{32 M^2 g} \right)^{1/3} = 1.65 \times 10^{-5} \text{ m} \qquad \boxed{H_1 = 16.5 \ \mu\text{m}}$$

$$E_1 = M g H_1 = 2.71 \times 10^{-31} \text{ J} = 1.69 \times 10^{-12} \text{ eV} \qquad \boxed{E_1 = 1.69 \text{ peV}}$$

Note that H_1 is of the same order than the given cavity height, $H = 50$ μm. This opens up the possibility for observing the spatial quantization when varying H.

5. The uncertainty principle says that the minimum time Δt and the minimum energy ΔE satisfy the relation $\Delta E \, \Delta t \geq \hbar$. During this time, the neutrons move to the right a distance

$$\Delta x = v_x \Delta t \geq v_x \frac{\hbar}{\Delta E}$$

Now, the minimum neutron energy allowed in the cavity is E_1, so that $\Delta E \approx E_1$. Therefore, an estimation of the minimum time and the minimum length required is

$$\boxed{t_q \approx \frac{\hbar}{E_1} = 0.4 \cdot 10^{-3} \text{ s} = 0.4 \text{ ms}} \qquad \boxed{L_q \approx v_x \frac{\hbar}{E_1} = 4 \cdot 10^{-3} \text{ m} = 4 \text{ mm}}$$

Th 3 ANSWER SHEET

Question	Basic formulas used	Analytical results	Numerical results	Marking guideline
1	$\frac{1}{2}Mv_z^2 + Mgz \le MgH$	$-\sqrt{2g(H-z)} \le v_z(z) \le \sqrt{2g(H-z)}$		1.5
2	$L_c = v_x 2t_f$ $H = \frac{1}{2}gt_f^2$	$L_c = 2v_x\sqrt{\frac{2H}{g}}$	$L_c = 6.4$ cm	1.3 + 0.2
3	$\frac{dN_c}{dz} = \rho\left[v_{z,\max} - v_{z,\min}\right]$ $N_c(H) = \int_0^H dN_c(z)$	$N_c(H) = \frac{4}{3}\rho\sqrt{2g}\,H^{3/2}$		2.5
4	$S = 2\int_0^H p_z dz = nh$	$H_n = \left(\frac{9h^2}{32M^2g}\right)^{1/3} n^{2/3}$ $E_n = \left(\frac{9Mg^2h^2}{32}\right)^{1/3} n^{2/3}$	$H_1 = 16.5$ μm $E_1 = 1.69$ peV	1.6 + 0.2 0.5 + 0.2
5	$\Delta E\,\Delta t \ge \hbar$ $\Delta E \approx E_1$ $\Delta x = v_x \Delta t$	$t_q \approx \frac{\hbar}{E_1}$ $L_q \approx v_x \frac{\hbar}{E_1}$	$t_q \approx 0.4$ ms $L_q \approx 4$ mm	1.3 + 0.2 0.3 + 0.2

PLANCK'S CONSTANT IN THE LIGHT OF AN INCANDESCENT LAMP

In 1900 Planck introduced the hypothesis that light is emitted by matter in the form of quanta of energy $h\nu$. In 1905 Einstein extended this idea proposing that once emitted, the energy quantum remains intact as a quantum of light (that later received the name photon). Ordinary light is composed of an enormous number of photons on each wave front. They remain masked in the wave, just as individual atoms are in bulk matter, but h – the Planck's constant – reveals their presence. The purpose of this experiment is to measure Planck's constant.

A body not only emits, it can also absorb radiation arriving from outside. Black body is the name given to a body that can absorb all radiation incident upon it, for any wavelength. It is a full radiator. Referring to electromagnetic radiation, black bodies absorb everything, reflect nothing, and emit everything. Real bodies are not completely black; the ratio between the energy emitted by a body and the one that would be emitted by a black body at the same temperature, is called emissivity, ε, usually depending on the wavelength.

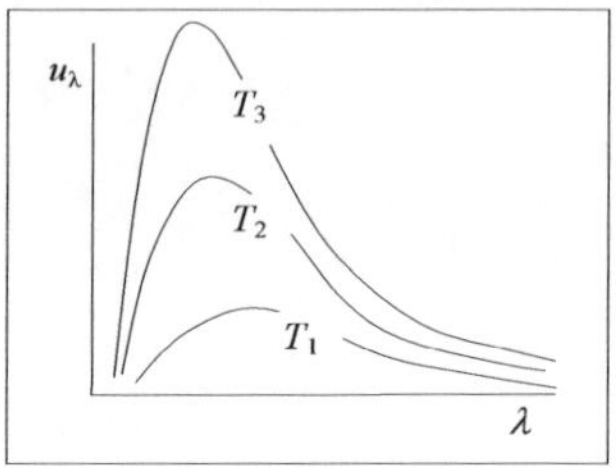

F-1

Planck found that the power density radiated by a body at absolute temperature T in the form of electromagnetic radiation of wavelength λ can be written as

$$u_\lambda = \varepsilon \frac{c_1}{\lambda^5 \left(e^{c_2/\lambda T} - 1\right)} \qquad (1)$$

where c_1 and c_2 are constants. In this question we ask you to determine c_2 experimentally, which is proportional to h.

For emission at small λ, far at left of the maxima in Figure F-1, it is permissible to drop the -1 from the denominator of Eq. (1), that reduces to

$$u_\lambda = \varepsilon \frac{c_1}{\lambda^5 e^{c_2/\lambda T}} \qquad (2)$$

The basic elements of this experimental question are sketched in Fig. F-2.

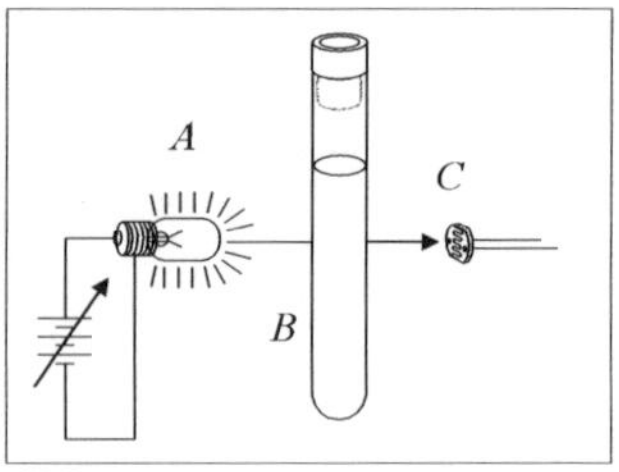

F-2

- The emitter body is the tungsten filament of an incandescent lamp A that emits a wide range of λ's, and whose luminosity can be varied.
- The test tube B contains a liquid filter that only transmits a thin band of the visible spectrum around a value λ_0 (see Fig. F-3). More information on the filter properties will be found in page 5.
- Finally, the transmitted radiation falls upon a photo resistor C (also known as LDR, the acronym of Light Dependent Resistor). Some properties of the LDR will be described in page 6.

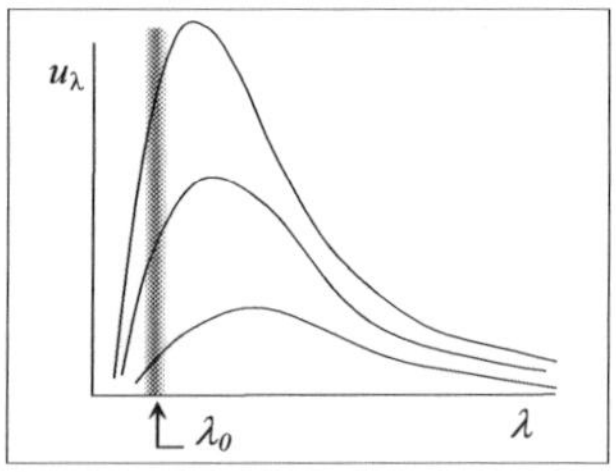

F-3

The LDR resistance R depends on its illumination, E, which is proportional to the filament power energy density

$$\left.\begin{array}{l} E \propto u_{\lambda_0} \\ R \propto E^{-\gamma} \end{array}\right\} \Rightarrow R \propto u_{\lambda_0}^{-\gamma}$$

where the dimensionless parameter γ is a property of the LDR that will be determined in the experiment. For this setup we finally obtain a relation between the LDR resistance R and the filament temperature T

$$R = c_3 e^{c_2 \gamma / \lambda_0 T} \qquad (3)$$

that we will use in page 6. In this relation c_3 is an unknown proportionality constant. By measuring R as a function on T one can obtain c_2, the objective of this experimental question.

DESCRIPTION OF THE APPARATUS

The components of the apparatus are shown in Fig. F-4, which also includes some indications for its setup. Check now that all the components are available, but refrain for making any manipulation on them until reading the instructions in the next page.

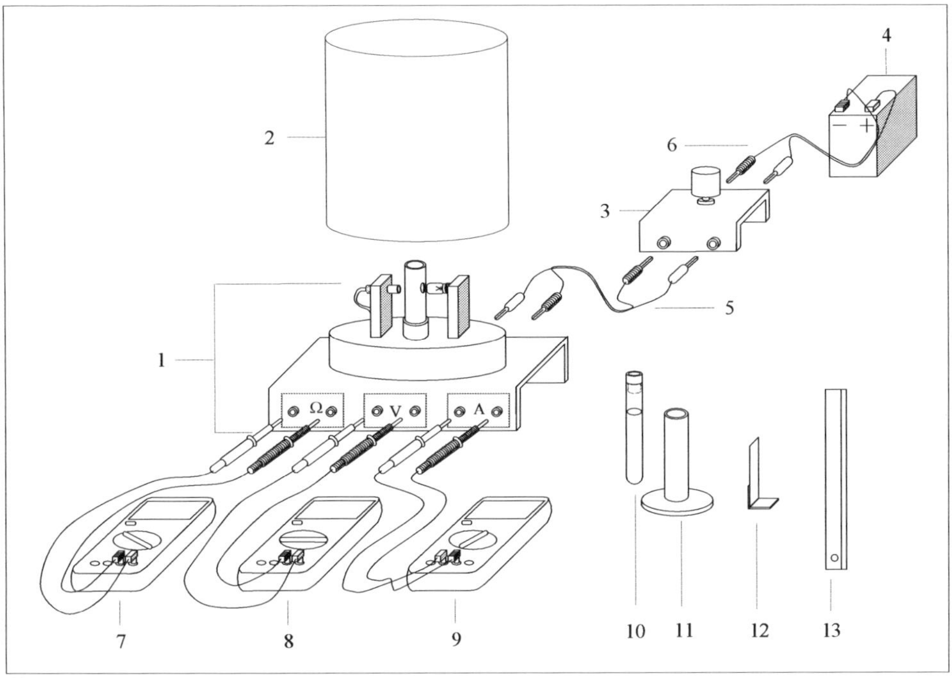

F-4

EQUIPMENT:

1. Platform. It has a disk on the top that holds a support for the LDR, a support for the tube and a support for an electric lamp of 12 V, 0.1 A.
2. Protecting cover.
3. 10 turns and 1 kΩ potentiometer.
4. 12 V battery.
5. Red and black wires with plugs at both ends to connect platform to potentiometer.
6. Red and black wires with plugs at one end and sockets for the battery at the other end.
7. Multimeter to work as ohmmeter.
8. Multimeter to work as voltmeter.
9. Multimeter to work as ammeter.
10. Test tube with liquid filter.
11. Stand for the test tube.
12. Grey filter.
13. Ruler.

An abridged set of instructions for the use of multimeters, along with information on the least squares method, is provided in a separate page.

SETTING UP THE EQUIPMENT

Follow these instructions:

- Carefully make the electric connections as indicated in Fig. F-4, but do not plug the wires 6 to the potentiometer.
- By looking at Fig. F-5, follow the steps indicated below:

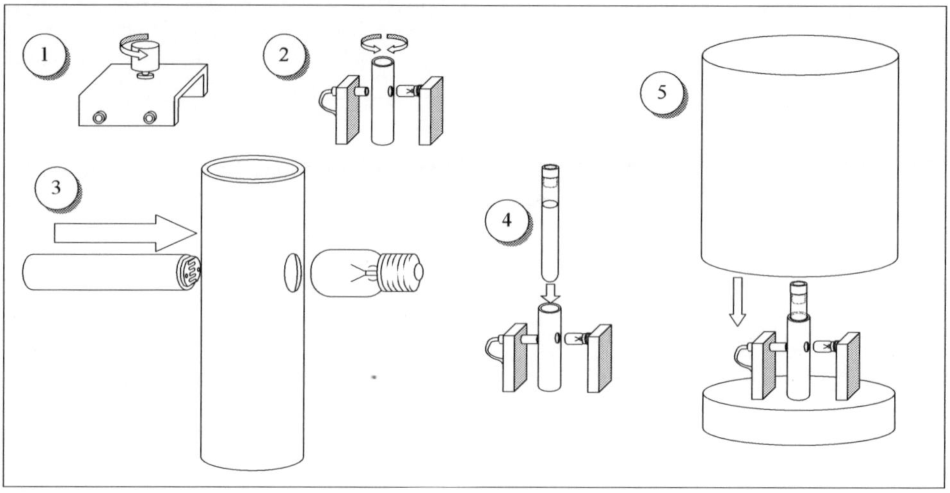

F-5

1. Turn the potentiometer knob anticlockwise until reaching the end.
2. Turn slowly the support for the test tube so that one of the lateral holes is in front of the lamp and the other in front of the LDR.
3. Bring the LDR nearer to the test tube support until making a light touch with its lateral hole. It is advisable to orient the LDR surface as indicated in Fig. F-5.
4. Insert the test tube into its support.
5. Put the cover onto the platform to protect from the outside light. Be sure to keep the LDR in total darkness for at least 10 minutes before starting the measurements of its resistance. This is a cautionary step, as the resistance value at darkness is not reached instantaneously.

Task 1

Draw in Answer Sheet 1 the complete electric circuits in the boxes and between the boxes, when the circuit is fully connected. Please, take into account the indications contained in Fig. F-4 to make the drawings.

Measurement of the filament temperature

The electric resistance R_B of a conducting filament can be given as

$$R_B = \rho \frac{l}{S} \tag{4}$$

where ρ is the resistivity of the conductor, l is the length and S the cross section of the filament.

This resistance depends on the temperature due to different causes such as:

- Metal resistivity increases with temperature. For tungsten and for temperatures in the range 300 K to 3655 K, it can be given by the empirical expression, valid in SI units,

$$T = 3.05 \cdot 10^8 \rho^{0.83} \tag{5}$$

- Thermal dilatation modifies the filament's length and section. However, its effects on the filament resistance will be negligible small in this experiment.

  From (4) and (5) and neglecting dilatations one gets

$$T = a R_B^{0.83} \tag{6}$$

- Therefore, to get T it is necessary to determine a. This can be achieved by measuring the filament resistance $R_{B,0}$ at ambient temperature T_0.

Task 2

a) Measure with the multimeter the ambient temperature T_0.

b) It is not a good idea to use the ohmmeter to measure the filament resistance $R_{B,0}$ at T_0 because it introduces a small unknown current that increases the filament temperature. Instead, to find $R_{B,0}$ connect the battery to the potentiometer and make a sufficient number of current readings for voltages from the lowest values attainable up to 1 V. (It will prove useful to make at least 15 readings below 100 mV.) At the end, leave the potentiometer in the initial position and disconnect one of the cables from battery to potentiometer.

   Find R_B for each pair of values of V and I, translate these values into the Table for Task 2,b) in the Answer Sheets. Indicate there the lowest voltage that you can experimentally attain. Draw a graph and represent R_B in the vertical axis against I.

c) After inspecting the graphics obtained at b), select an appropriate range of values to make a linear fit to the data suitable for extrapolating to the ordinate at the origin, $R_{B,0}$. Write the selected values in the Table for Task 2, c) in the Answer Sheets. Finally, obtain $R_{B,0}$ and $\Delta R_{B,0}$.

d) Compute the numerical values of a and Δa for $R_{B,0}$ in Ω and T_0 in K using (6).

OPTICAL PROPERTIES OF THE FILTER

The liquid filter in the test tube is an aqueous solution of copper sulphate (II) and Orange (II) aniline dye. The purpose of the salt is to absorb the infrared radiation emitted by the filament.

The filter transmittance (transmitted intensity/incident intensity) is shown in Figure F-6 versus the wavelength.

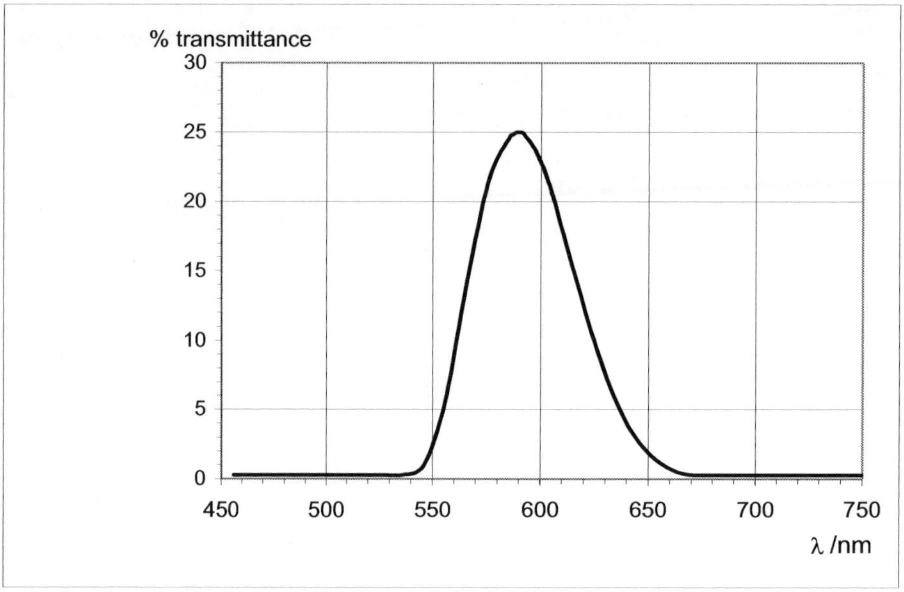

F-6

Task 3

Determine λ_0 and $\Delta\lambda$ from Fig. F-6.

Note: $2\,\Delta\lambda$ is the total width at half height and λ_0 the wavelength at the maximum.

PROPERTIES OF THE LDR

The material which composes the LDR is non conducting in darkness conditions. By illuminating it some charge carriers are activated allowing some flow of electric current through it. In terms of the resistance of the LDR one can write the following relation

$$R = bE^{-\gamma} \qquad (7)$$

where b is a constant that depends on the composition and geometry of the LDR and γ is a dimensionless parameter that measures the variation of the resistance with the illumination E produced by the incident radiation. Theoretically, an ideal LDR would have $\gamma = 1$, however many factors intervene, so that in the real case $\gamma < 1$.

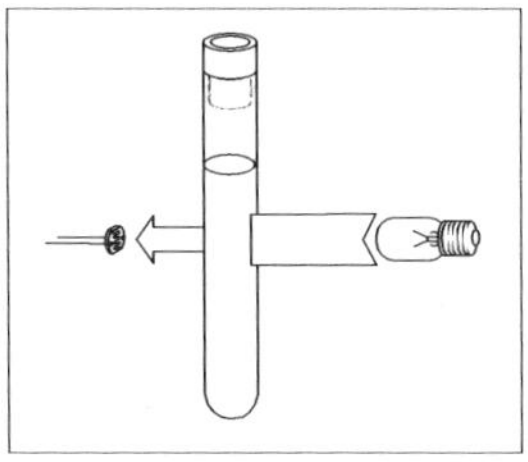

F-7

It is necessary to determine γ. This is achieved by measuring a pair R and E (Fig. F-7) and then introducing between the lamp and the tube the grey filter F (Fig. F-8) whose transmittance is known to be 51.2 %, and we consider free of error. This produces an illumination $E' = 0.51\ E$. After measuring the resistance R' corresponding to this illumination, we have

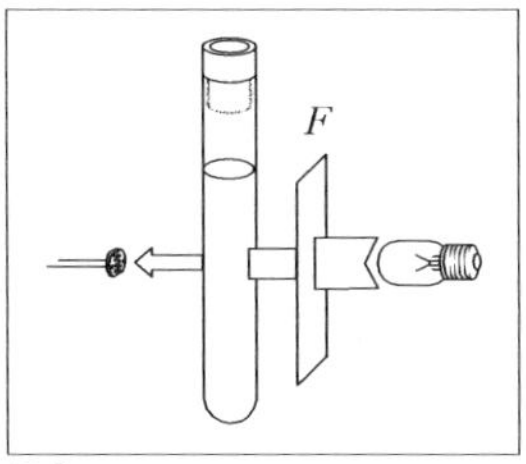

F-8

$$R = bE^{-\gamma} \quad ; \qquad R' = b\,(0.512\,E)^{-\gamma}$$

From this

$$\ln\frac{R}{R'} = \gamma \ln 0.512 \qquad (8)$$

Do not carry out this procedure until arriving at part b) of task 4 below.

Task 4

a) Check that the LDR remained in complete darkness for at least 10 minutes before starting this part. Connect the battery to the potentiometer and, rotating the knob very slowly, increase the lamp voltage. Read the pairs of values of V and I for V in the range between 9.50 V and 11.50 V, and obtain the corresponding LDR resistances R. (It will be useful to make at least 12 readings). Translate all these values to a table in the Answer Sheet. To deal with the delay in the LDR response, we recommend the following procedure: Once arrived at $V > 9.5$ V, wait 10 min approximately before making the first reading. Then wait 5 min for the second one, and so on. Before doing any further calculation go to next step.

b) Once obtained the lowest value of the resistance R, open the protecting cover, put the grey filter as indicated in F-9, cover again - as soon as possible - the platform and record the new LDR resistance R'. Using these data in (8) compute γ and $\Delta\gamma$.

c) Modify Eq. (3) to display a linear dependence of ln R on $R_B^{-0.83}$. Write down that equation there and label it as (9).

d) Using now the data from a), work out a table that will serve to plot Eq. (9).

e) Make the graphics plot and, knowing that $c_2 = hc/k$, compute h and Δh by any method (you are allowed to use statistical functions of the calculators provided by the organization).

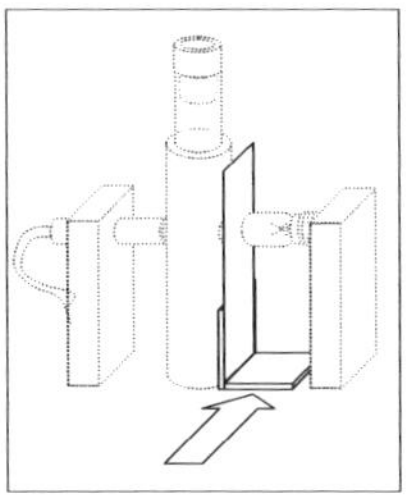

F-9

(Speed of light, $c = 2.998 \cdot 10^8$ m s^{-1} ; Boltzmann constant, $k = 1.381 \cdot 10^{-23}$ J K^{-1})

PLANCK'S CONSTANT IN THE LIGHT OF AN INCANDESCENT LAMP

SOLUTION

TASK 1

Draw the electric connections in the boxes and between boxes below.

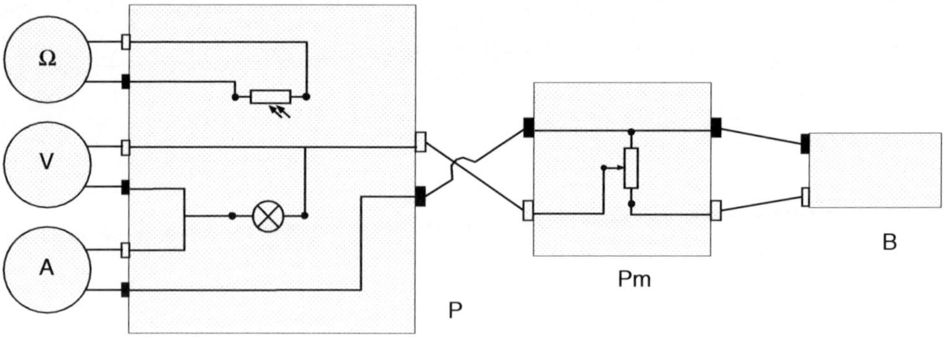

Photoresistor	
Incandescent Bulb	
Potentiometer	
Red socket	▯
Black socket	▮

Ω	Ohmmeter
V	Voltmeter
A	Ammeter
P	Platform
Pm	Potentiometer
B	Battery

TASK 2

a)

$t_0 =$ 24 ºC	$T_0 =$ 297 K	$\Delta T_0 = 1$ K

b)

V /mV	I / mA	R_B /Ω
21.9	1.87	11.7
30.5	2.58	11.8
34.9	2.95	11.8
37.0	3.12	11.9
40.1	3.37	11.9
43.0	3.60	11.9
47.6	3.97	12.0
51.1	4.24	12.1
55.3	4.56	12.1
58.3	4.79	12.2
61.3	5.02	12.2
65.5	5.33	12.3
67.5	5.47	12.3
73.0	5.88	12.4
80.9	6.42	12.6
85.6	6.73	12.7
89.0	6.96	12.8
95.1	7.36	12.9
111.9	8.38	13.4
130.2	9.37	13.9
181.8	11.67	15.6
220	13.04	16.9
307	15.29	20.1
447	17.68	25.1
590	19.8	29.8
730	21.5	33.9
860	23.2	37.1
960	24.4	39.3

$V_{min} =$ 9.2 mV *
* This is a characteristic of your apparatus. You can´t go below it.

We represent R_B in the vertical axis against I

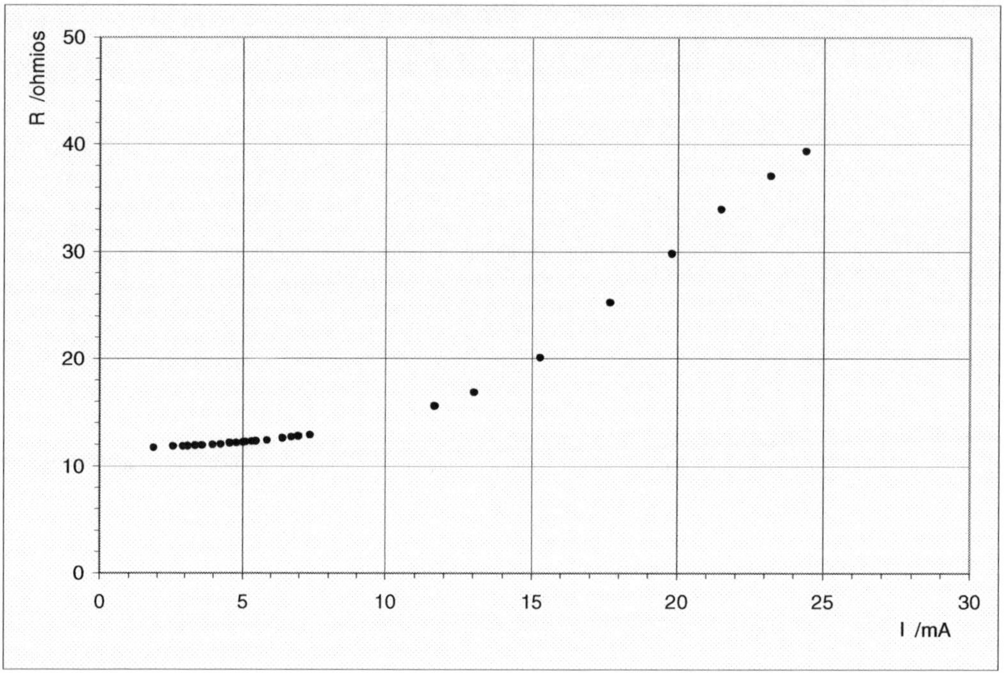

In order to work out R_{B0} , we choose the first ten readings.

TASK 2

c)

V /mV	I / mA	R_B /Ω
21.9 ± 0.1	1.87± 0.01	11.7± 0.1
30.5± 0.1	2.58± 0.01	11.8± 0.1
34.9± 0.1	2.95± 0.01	11.8± 0.1
37.0± 0.1	3.12± 0.01	11.9± 0.1
40.1± 0.1	3.37± 0.01	11.9± 0.1
43.0± 0.1	3.60± 0.01	11.9± 0.1
47.6± 0.1	3.97± 0.01	12.0± 0.1
51.1± 0.1	4.24± 0.01	12.1± 0.1
55.3± 0.1	4.56± 0.01	12.1± 0.1
58.3± 0.1	4.79± 0.01	12.2± 0.1

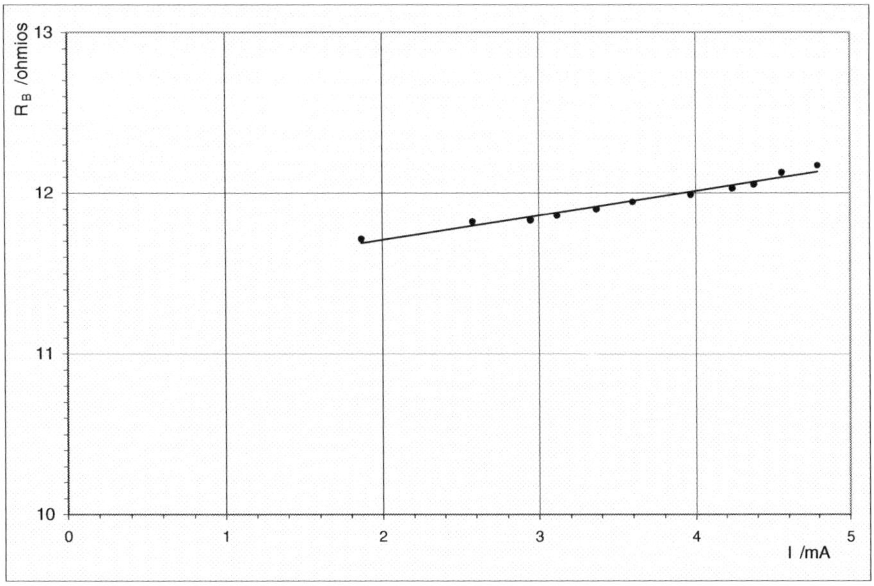

Error for R_B (We work out the error for first value, as example).

$$\Delta R_B = R_B\sqrt{\left(\frac{\Delta V}{V}\right)^2+\left(\frac{\Delta I}{I}\right)^2} = 11.71\sqrt{\left(\frac{0.1}{21.9}\right)^2+\left(\frac{0.01}{1.87}\right)^2} = 0.1$$

We have worked out R_{B0} by the least squares.

$R_{B0} = 11.4$

slope $= m = 0.167$

$\sum I^2 = 130.38$

$\sum I = 35.05$

$n = 10$

$$\text{For axis } X: \sigma_I = \sqrt{\frac{\sum \Delta I^2}{n}} = 0.01$$

$$\text{For axis } Y: \sigma_{R_B} = \sqrt{\frac{\sum \Delta R_B^2}{n}} = 0.047$$

$$\sigma = \sqrt{\sigma_{R_B}^2 + m^2\sigma_I^2} = \sqrt{0.1^2 + 0.167^2\cdot 0.01^2} = 0.1$$

$$\Delta R_{B0} = \sqrt{\frac{\sigma^2\sum I^2}{n\sum I^2 - \left(\sum I\right)^2}} = \sqrt{\frac{0.1^2\times 130.38}{10\cdot 130.38 - 35.05^2}} = 0.13$$

R_{B0} = 11,4Ω	ΔR_{B0} = 0.1Ω

d) $T = aR^{0.83}$; $a = \dfrac{T_0}{R_0^{0.83}}$; $a = \dfrac{297}{11.4^{0.83}} = 39.40$

Working out the error for two methods:

<u>Method A</u>

$\ln a = \ln T_0 - 0.83 \ln R_{B0}$; $\Delta a = a\left(\dfrac{\Delta T_0}{T_0} + 0.83\dfrac{\Delta R_{B0}}{R_{B0}}\right)$; $\Delta a = 39.40\left(\dfrac{1}{297} + 0.83\dfrac{0.1}{11.40}\right) = 0.419 = 0.4$

<u>Method B</u>

Higher value of a: $a_{\max} = \dfrac{T_0 + \Delta T_0}{(R_0 - \Delta R_0)^{0.83}} = \dfrac{297+1}{(11.4 - 0.1)^{0.83}} = 39.8255$

Smaller value of a: $a_{\min} = \dfrac{T_0 - \Delta T_0}{(R_0 + \Delta R_0)^{0.83}} = \dfrac{297-1}{(11.4 + 0.1)^{0.83}} = 38.9863$

$$\Delta a = \frac{a_{\max} - a_{\min}}{2} = \frac{39.8255 - 38.9863}{2} = 0.419 = 0.4$$

$a = 39.4$	$\Delta a = 0.4$

TASK 3

Because of $2\Delta\lambda = 620 - 565$; $\Delta\lambda = 28$ nm

$\lambda_0 = 590$ nm	$\Delta\lambda = 28$ nm

TASK 4

a)

V /V	I / mA	R /kΩ
9.48	85.5	8.77
9.73	86.8	8.11
9.83	87.3	7.90
100.1	88.2	7.49
10.25	89.4	7.00
10.41	90.2	6.67
10.61	91.2	6.35
10.72	91.8	6.16
10.82	92.2	6.01
10.97	93.0	5.77
11.03	93.3	5.69
11.27	94.5	5.35
11.42	95.1	5.17
11.50	95.5	5.07

b)

Because of $\ln\frac{R}{R'}=\gamma\ln 0.512$; $\gamma=\ln\frac{R}{R'}\Big/\ln 0.512=\ln\frac{5.07}{8.11}\Big/\ln 0.512=0.702$

For working out $\Delta\gamma$ we know that:

$R\pm\Delta R=5.07\pm 0.01$ kΩ

$R'\pm\Delta R'=8.11\pm 0.01$ kΩ

Transmittance, $t=51.2$ %

Working out the error for two methods:

<u>Method A</u>

$$\gamma=\frac{\ln R/R'}{\ln t};\quad \Delta\gamma=\frac{1}{\ln t}\left(\frac{\Delta R}{R}+\frac{\Delta R'}{R'}\right)=\frac{1}{\ln 0.512}\left(\frac{0.01}{5.07}+\frac{0.01}{8.11}\right)=0.00479;\ \Delta\gamma=0.005$$

<u>Method B</u>

Higher value of γ: $\gamma_{\max}=\ln\frac{R-\Delta R}{R'+\Delta R'}\Big/\ln\gamma=\ln\frac{5.07-0.01}{8.11+0.01}\Big/\ln 0.512=0.70654$

Smaller value of γ: $\gamma_{\max}=\ln\frac{R+\Delta R}{R'-\Delta R'}\Big/\ln\gamma=\ln\frac{5.07+0.01}{8.11-0.01}\Big/\ln 0.512=0.69696$

$$\Delta\gamma=\frac{\gamma_{\max}-\gamma_{\min}}{2}=\frac{0.70654-0.69696}{2}=0.00479;\quad \Delta\gamma=0.005$$

$R=5.07$ kΩ	$\gamma=0.702$
$R'=8.11$ kΩ	$\Delta\gamma=0.005$

c)

We know that $R=c_3e^{\frac{c_2\gamma}{\lambda_0 T}}$ (3)

then $\ln R=\ln c_3+\frac{c_2\gamma}{\lambda_0 T}$

Because of $T=aR_B^{0.83}$ (6)

consequently $\ln R=\ln c_3+\frac{c_2\gamma}{\lambda_0 a}R_B^{-0.83}$

$$\ln R=\ln c_3+\frac{c_2\gamma}{\lambda_0 a}R_B^{-0.83}\quad \text{Eq.(9)}$$

d)

V/V	I / mA	R_B / Ω	T / K	$R_B^{-0.83}$ (S.I.)	R / kΩ	ln R
9.48± 0.01	85.5± 0.1	110.9± 0.2	1962± 18	(2.008± 0.004)10^{-2}	8.77± 0.01	2.171± 0.001
9.73± 0.01	86.8± 0.1	112.1± 0.2	1980± 18	(1.990± 0.004)10^{-2}	8.11± 0.01	2.093± 0.001
9.83± 0.01	87.3± 0.1	112.6± 0.2	1987± 18	(1.983± 0.004)10^{-2}	7.90± 0.01	2.067± 0.001
10.01± 0.01	88.2± 0.1	113.5± 0.2	2000± 18	(1.970± 0.004)10^{-2}	7.49± 0.01	2.014± 0.001
10.25± 0.01	89.4± 0.1	114.7± 0.2	2018± 18	(1.952± 0.003)10^{-2}	7.00± 0.01	1.946± 0.001
10.41± 0.01	90.2± 0.1	115.4± 0.2	2028± 18	(1.943± 0.003)10^{-2}	6.67± 0.01	1.894± 0.002
10.61± 0.01	91.2± 0.1	116.3± 0.2	2041± 18	(1.930± 0.003)10^{-2}	6.35± 0.01	1.849± 0.002
10.72± 0.01	91.8± 0.1	116.8± 0.2	2049± 19	(1.923± 0.003)10^{-2}	6.16± 0.01	1.818± 0.002
10.82± 0.01	92.2± 0.1	117.4± 0.2	2057± 19	(1.915± 0.003)10^{-2}	6.01± 0.01	1.793± 0.002
10.97± 0.01	93.0± 0.1	118.0± 0.2	2066± 19	(1.907± 0.003)10^{-2}	5.77± 0.01	1.753± 0.002
11.03± 0.01	93.3± 0.1	118.2± 0.2	2069± 19	(1.904± 0.003)10^{-2}	5.69± 0.01	1.739± 0.002
11.27± 0.01	94.5± 0.1	119.3± 0.2	2085± 19	(1.890± 0.003)10^{-2}	5.35± 0.01	1.677± 0.002
11.42± 0.01	95.1± 0.1	120.1± 0.2	2096± 19	(1.880± 0.003)10^{-2}	5.15± 0.01	1.639± 0.002
11.50± 0.01	95.5± 0.1	120.4± 0.2	2101± 19	(1.875± 0.003)10^{-2}	5.07± 0.01	1.623± 0.002
			unnecessary			

We work out the errors for all the first row, as example.

Error for R_B: $\Delta R_B = R_B\sqrt{\left(\frac{\Delta V}{V}\right)^2 + \left(\frac{\Delta I}{I}\right)^2} = 110.9\sqrt{\left(\frac{0.01}{9.48}\right)^2 + \left(\frac{0.1}{85.5}\right)^2} = 0.2\ \Omega$

Error for T: $\Delta T = T\left(\frac{\Delta a}{a} + 0.83\frac{\Delta R_B}{R_B}\right)$; $\Delta T = 1962\left(\frac{0.3}{39.4} + 0.83\frac{0.2}{110.9}\right) = 18\text{K}$

Error for $R_B^{-0.83}$:

$$x = R_B^{-0.83}\ ;\ \ln x = -0.83 \ln R_B\ ;\ \Delta x = x \cdot 0.83\frac{\Delta R_B}{R_B}\ ;\ \Delta\left(R_B^{-0.83}\right) = R_B^{-0.83}\frac{\Delta R_B}{R_B}$$

$$\Delta\left(R_B^{-0.83}\right) = 0.020077\frac{0.2}{110.9} \approx 0.004 \times 10^{-2}$$

Error for ln$R$: $\Delta \ln R = \frac{\Delta R}{R}$; $\Delta \ln R = \frac{0.01}{8.77} = 0.001$

e)

We plot ln$R$ versus $R_B^{-0.83}$.

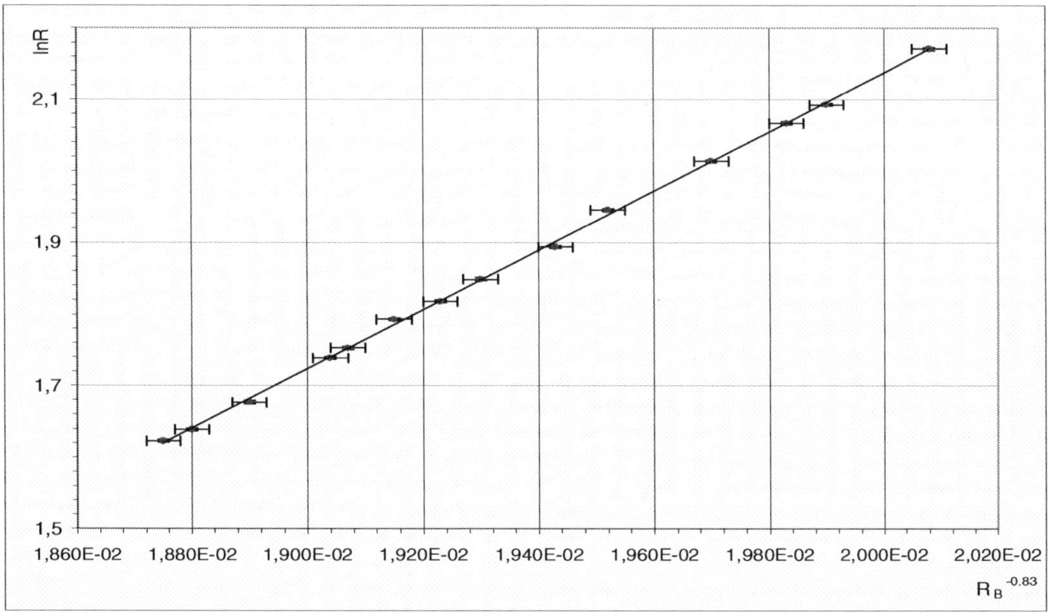

By theleastsquares

Slope = $m = 414{,}6717$

$$\sum \left(R_B^{-0.83}\right)^2 = 5.23559\times 10^{-3}$$

$$\sum \left(R_B^{-0.83}\right) = 0.27068$$

$n = 14$

For axis X: $\sigma_{R_B^{-0.83}} = \sqrt{\dfrac{\sum \Delta\left(R_B^{-0.83}\right)^2}{n}} = 0.003\times 10^{-2}$

For axis Y: $\sigma_{\ln R} = \sqrt{\dfrac{\sum \Delta(\ln R)^2}{n}} = 0.002$

$$\sigma = \sqrt{\sigma_{\ln R}^2 + m^2\sigma_{R_B^{-0.83}}^2} = \sqrt{0.002^2 + 414.672^2\cdot\left(0.003\times 10^{-2}\right)^2} = 0.0126$$

$$\Delta m = \sqrt{\frac{n\sigma^2}{n\sum\left(R_B^{-0.83}\right)^2 - \left(\sum R_B^{-0.83}\right)^2}} = \sqrt{\frac{14\cdot 0{,}0126^2}{14\cdot 5.23559\times 10^{-3} - (0.27068)^2}} = 8.295$$

Because of

$$m = \frac{c_2\gamma}{\lambda_0 a}$$

and

$$c_2 = \frac{hc}{k}$$

then

$$h = \frac{mk\lambda_0 a}{c\gamma}$$

$$h = \frac{414.67 \cdot 1.381\times 10^{-23} \cdot 590\times 10^{-9} \cdot 39.4}{2.998\times 10^{8} \cdot 0.702} = 6.33\times 10^{-34}$$

$$\Delta h = h\sqrt{\left(\frac{\Delta m}{m}\right)^2 + \left(\frac{\Delta k}{k}\right)^2 + \left(\frac{\Delta \lambda_0}{\lambda_0}\right)^2 + \left(\frac{\Delta a}{a}\right)^2 + \left(\frac{\Delta \gamma}{\gamma}\right)^2}$$

$$\Delta h = 6.34\times 10^{-34}\sqrt{\left(\frac{8.3}{415}\right)^2 + 0 + \left(\frac{28}{590}\right)^2 + \left(\frac{0.3}{39.4}\right)^2 + 0 + \left(\frac{0.01}{0.70}\right)^2} = 0.34\times 10^{-34}$$

$h = 6.3\times 10^{34}$ J · s	$\Delta h = 0.3\times 10^{34}$ J · s

37TH

INTERNATIONAL PHYSICS OLYMPIAD

SINGAPORE, SINGAPORE

2006

Theory Question 1: Gravity in a Neutron Interferometer

Enter all your answers into the **Answer Script**.

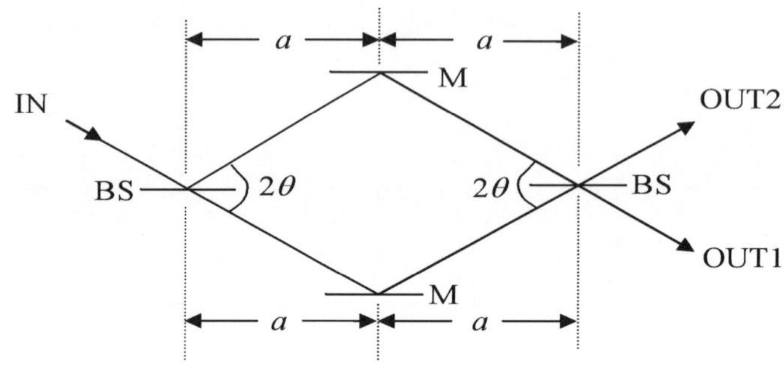

BS - Beam Splitters M - Mirror

Figure 1a

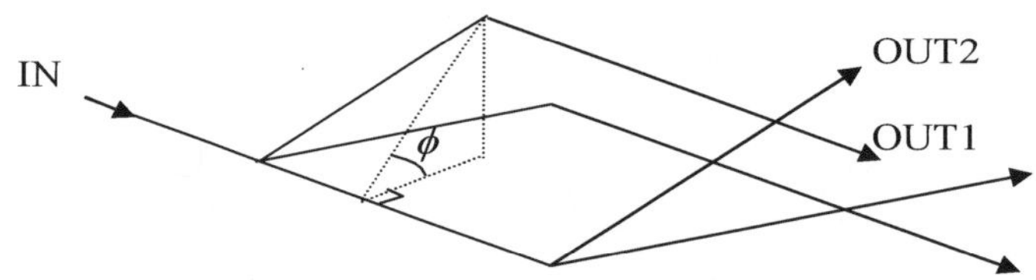

Figure 1b

Physical situation We consider the situation of the famous neutron-interferometer experiment by Collela, Overhauser and Werner, but idealize the set-up inasmuch as we shall assume perfect beam splitters and mirrors within the interferometer. The experiment studies the effect of the gravitational pull on the de Broglie waves of neutrons.

The symbolic representation of this interferometer in analogy to an optical interferometer is shown in Figure 1a. The neutrons enter the interferometer through the IN port and follow the two paths shown. The neutrons are detected at either one of the two output ports, OUT1 or OUT2. The two paths enclose a diamond-shaped area, which is typically a few cm^2 in size.

The neutron de Broglie waves (of typical wavelength of 10^{-10} m) interfere such that all neutrons emerge from the output port OUT1 if the interferometer plane is horizontal. But when the interferometer is tilted around the axis of the incoming neutron beam by angle ϕ (Figure 1b), one observes a ϕ dependent redistribution of the neutrons between the two output ports OUT1 and OUT2.

Geometry For $\phi = 0°$ the interferometer plane is horizontal; for $\phi = 90°$ the plane is vertical with the output ports above the tilt axis.

1.1 (1.0) How large is the diamond-shaped area A enclosed by the two paths of the interferometer?

1.2 (1.0) What is the height H of output port OUT1 above the horizontal plane of the tilt axis?

Express A and H in terms of a, θ, and ϕ.

Optical path length The optical path length N_{opt} (a number) is the ratio of the geometrical path length (a distance) and the wavelength λ. If λ changes along the path, N_{opt} is obtained by integrating λ^{-1} along the path.

1.3 (3.0) What is the difference ΔN_{opt} in the optical path lengths of the two paths when the interferometer has been tilted by angle ϕ? Express your answer in terms of a, θ, and ϕ as well as the neutron mass M, the de Broglie wavelength λ_0 of the incoming neutrons, the gravitational acceleration g, and Planck's constant h.

1.4 (1.0) Introduce the volume parameter

$$V = \frac{h^2}{gM^2}$$

and express ΔN_{opt} solely in terms of A, V, λ_0, and ϕ. State the value of V for $M = 1.675 \times 10^{-27}$ kg, $g = 9.800$ m s^{-2}, and $h = 6.626 \times 10^{-34}$ J s.

1.5 (2.0) How many cycles — from high intensity to low intensity and back to high intensity — are completed by output port OUT1 when ϕ is increased from $\phi = -90°$ to $\phi = 90°$?

Experimental data The interferometer of an actual experiment was characterized by $a = 3.600$ cm and $\theta = 22.10°$, and 19.00 full cycles were observed.

1.6 (1.0) How large was λ_0 in this experiment?

1.7 (1.0) If one observed 30.00 full cycles in another experiment of the same kind that uses neutrons with $\lambda_0 = 0.2000$ nm, how large would be the area A?

Hint: If $|\alpha x| << 1$, it is permissible to replace $(1+x)^{\alpha}$ by $1+\alpha x$.

Theory Question 2: Watching a Rod in Motion

Enter all your answers into the **Answer Script**.

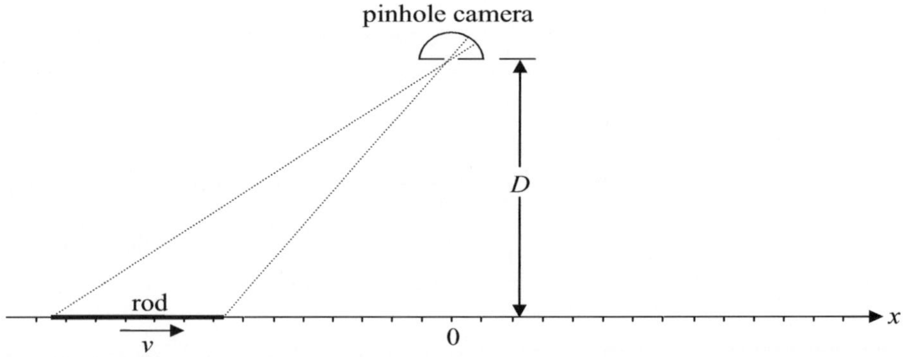

Physical situation A pinhole camera, with the pinhole at $x = 0$ and at distance D from the x axis, takes pictures of a rod, by opening the pinhole for a very short time. There are equidistant marks along the x axis by which the *apparent length* of the rod, as it is seen on the picture, can be determined from the pictures taken by the pinhole camera. On a picture of the rod *at rest*, its length is L. However, the rod is *not* at rest, but is moving with constant velocity υ along the x axis.

Basic relations A picture taken by the pinhole camera shows a tiny segment of the rod at position $\tilde{x}$.

2.1 (0.6) What is the *actual position* x of this segment at the time when the picture is taken? State your answer in terms of $\tilde{x}$, D, L, υ, and the speed of light c=3.00×10^{8} ms^{-1}. Employ the quantities

$$\beta = \frac{\upsilon}{c} \text{ and } \gamma = \frac{1}{\sqrt{1-\beta^2}}$$

if they help to simplify your result.

2.2 (0.9) Find also the corresponding inverse relation, that is: express $\tilde{x}$ in terms of x, D, L, υ, and c.

Note: The *actual position* is the position in the frame in which the camera is at rest

Apparent length of the rod The pinhole camera takes a picture at the instant when the actual position of the center of the rod is at some point x_0.

2.3 (1.5) In terms of the given variables, determine the apparent length of the rod on this picture.

2.4 (1.5) Check one of the boxes in the **Answer Script** to indicate how the apparent length changes with time.

Symmetric picture One pinhole-camera picture shows both ends of the rod at the same distance from the pinhole.

2.5 (0.8) Determine the apparent length of the rod on this picture.

2.6 (1.0) What is the actual position of the middle of the rod at the time when this picture is taken?

2.7 (1.2) Where does the picture show the image of the middle of the rod?

Very early and very late pictures The pinhole camera took one picture very early, when the rod was very far away and approaching, and takes another picture very late, when the rod is very far away and receding. On one of the pictures the apparent length is 1.00 m, on the other picture it is 3.00 m.

2.8 (0.5) Check the box in the **Answer Script** to indicate which length is seen on which picture.

2.9 (1.0) Determine the velocity υ.

2.10 (0.6) Determine the length L of the rod at rest.

2.11 (0.4) Infer the apparent length on the symmetric picture.

Theory Question 3

This question consists of five independent parts. Each of them asks for an estimate of an order of magnitude only, not for a precise answer. Enter all your answers into the **Answer Script**.

Digital Camera Consider a digital camera with a square CCD chip with linear dimension L = 35 mm having N_p = 5 Mpix (1 Mpix = 10^6 pixels). The lens of this camera has a focal length of f = 38 mm. The well known sequence of numbers (2, 2.8, 4, 5.6, 8, 11, 16, 22) that appear on the lens refer to the so called F-number, which is denoted by $F\#$ and defined as the ratio of the focal length and the diameter D of the aperture of the lens, $F\# = f / D$.

3.1 (1.0) Find the best possible spatial resolution $\Delta x_{\min}$, at the chip, of the camera as limited by the lens. Express your result in terms of the wavelength λ and the F-number $F\#$ and give the numerical value for λ = 500 nm.

3.2 (0.5) Find the necessary number N of Mpix that the CCD chip should possess in order to match this optimal resolution.

3.3 (0.5) Sometimes, photographers try to use a camera at the smallest practical aperture. Suppose that we now have a camera of N_0 = 16 Mpix, with the chip size and focal length as given above. Which value is to be chosen for $F\#$ such that the image quality is not limited by the optics?

3.4 (0.5) Knowing that the human eye has an approximate angular resolution of ϕ = 2 arcmin and that a typical photo printer will print a minimum of 300 dpi (dots per inch), at what minimal distance z should you hold the printed page from your eyes so that you do not see the individual dots?

Data 1 inch = 25.4 mm
1 arcmin = 2.91×10^{-4} rad

Hard-boiled egg An egg, taken directly from the fridge at temperature $T_0 = 4°C$, is dropped into a pot with water that is kept boiling at temperature T_1.

3.5 (0.5) How large is the amount of energy U that is needed to get the egg coagulated?

3.6 (0.5) How large is the heat flow J that is flowing into the egg?

3.7 (0.5) How large is the heat power P transferred to the egg?

3.8 (0.5) For how long do you need to cook the egg so that it is hard-boiled?

Hint You may use the simplified form of Fourier's Law $J = \kappa \Delta T / \Delta r$, where ΔT is the temperature difference associated with Δr, the typical length scale of the problem. The heat flow J is in units of $\mathrm{W\,m^{-2}}$.

Data Mass density of the egg: $\mu = 10^3\,\mathrm{kg\,m^{-3}}$
Specific heat capacity of the egg: $C = 4.2\,\mathrm{J\,K^{-1}\,g^{-1}}$
Radius of the egg: $R = 2.5$ cm
Coagulation temperature of albumen (egg protein): $T_c = 65°C$
Heat transport coefficient: $\kappa = 0.64\,\mathrm{W\,K^{-1}\,m^{-1}}$ (assumed to be the same for liquid and solid albumen)

Lightning An oversimplified model of lightning is presented. Lightning is caused by the build-up of electrostatic charge in clouds. As a consequence, the bottom of the cloud usually gets positively charged and the top gets negatively charged, and the ground below the cloud gets negatively charged. When the corresponding electric field exceeds the breakdown strength value of air, a disruptive discharge occurs: this is lightning.

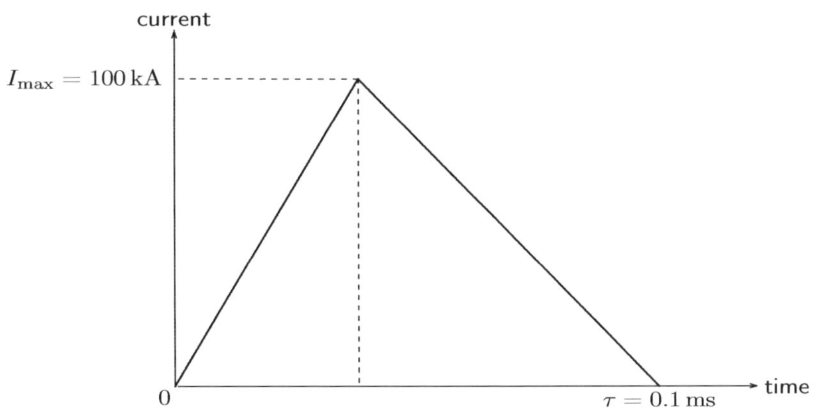

Idealized current pulse flowing between the cloud and the ground during lightning.

Answer the following questions with the aid of this simplified curve for the current as a function of time and these data:

Distance between the bottom of the cloud and the ground: $h = 1$ km;
Breakdown electric field of humid air: $E_0 = 300\,\text{kV m}^{-1}$;
Total number of lightning striking Earth per year: 32×10^6;
Total human population: 6.5×10^9 people.

3.9 (0.5) What is the total charge Q released by lightning?

3.10 (0.5) What is the average current I flowing between the bottom of the cloud and the ground during lightning?

3.11 (1.0) Imagine that the energy of all storms of one year is collected and equally shared among all people. For how long could you continuously light up a 100 W light bulb for your share?

Capillary Vessels Let us regard blood as an incompressible viscous fluid with mass density μ similar to that of water and dynamic viscosity $\eta = 4.5$ g m^{-1} s^{-1}. We model blood vessels as circular straight pipes with radius r and length L and describe the blood flow by Poiseuille's law,

$$\Delta p = RD,$$

the Fluid Dynamics analog of Ohm's law in Electricity. Here Δp is the pressure difference between the entrance and the exit of the blood vessel, $D = S\upsilon$ is the volume flow through the cross-sectional area S of the blood vessel and υ is the blood velocity. The hydraulic resistance R is given by

$$R = \frac{8\eta L}{\pi r^4}.$$

For the systemic blood circulation (the one flowing from the left ventricle to the right auricle of the heart), the blood flow is $D \approx 100$ cm^{3}s^{-1} for a man at rest. Answer the following questions under the assumption that all capillary vessels are connected in parallel and that each of them has radius $r = 4$ μm and length $L = 1$ mm and operates under a pressure difference $\Delta p = 1$ kPa.

3.12 (1.0) How many capillary vessels are in the human body?

3.13 (0.5) How large is the velocity υ with which blood is flowing through a capillary vessel?

Skyscraper At the bottom of a 1000 m high skyscraper, the outside temperature is T_{bot}= 30°C. The objective is to estimate the outside temperature T_{top} at the top. Consider a thin slab of air (ideal nitrogen gas with adiabatic coefficient γ = 7/5) rising slowly to height z where the pressure is lower, and assume that this slab expands adiabatically so that its temperature drops to the temperature of the surrounding air.

3.14 (0.5) How is the fractional change in temperature dT/T related to dp/p, the fractional change in pressure?

3.15 (0.5) Express the pressure difference dp in terms of dz, the change in height.

3.16 (1.0) What is the resulting temperature at the top of the building?

Data Boltzmann constant: $k = 1.38 \times 10^{-23}$ J K^{-1}
Mass of a nitrogen molecule: $m = 4.65 \times 10^{-26}$ kg
Gravitational acceleration: $g = 9.80$ m s^{-2}

SOLUTIONS to Theory Question 1

Geometry Each side of the diamond has length $L = \dfrac{a}{\cos\theta}$ and the distance between parallel sides is $D = \dfrac{a}{\cos\theta}\sin(2\theta) = 2a\sin\theta$. The area is the product thereof, $A = LD$, giving

1.1
$$A = 2a^2\tan\theta\,.$$

The height H by which a tilt of ϕ lifts OUT1 above IN is $H = D\sin\phi$ or

1.2
$$H = 2a\sin\theta\,\sin\phi\,.$$

Optical path length Only the two parallel lines for IN and OUT1 matter, each having length L. With the de Broglie wavelength λ_0 on the IN side and λ_1 on the OUT1 side, we have

$$\Delta N_{\text{opt}} = \frac{L}{\lambda_0} - \frac{L}{\lambda_1} = \frac{a}{\lambda_0\cos\theta}\left(1 - \frac{\lambda_0}{\lambda_1}\right).$$

The momentum is h/λ_0 or h/λ_1, respectively, and the statement of energy conservation reads

$$\frac{1}{2M}\left(\frac{h}{\lambda_0}\right)^2 = \frac{1}{2M}\left(\frac{h}{\lambda_1}\right)^2 + MgH\,,$$

which implies

$$\frac{\lambda_0}{\lambda_1} = \sqrt{1 - 2\frac{gM^2}{h^2}\lambda_0^2 H}\,.$$

Upon recognizing that $(gM^2/h^2)\lambda_0^2 H$ is of the order of 10^{-7}, this simplifies to

$$\frac{\lambda_0}{\lambda_1} = 1 - \frac{gM^2}{h^2}\lambda_0^2 H\,,$$

and we get

$$\Delta N_{\text{opt}} = \frac{a}{\lambda_0\cos\theta}\frac{gM^2}{h^2}\lambda_0^2 H$$

or

1.3
$$\Delta N_{\text{opt}} = 2\frac{gM^2}{h^2}a^2\lambda_0 \tan\theta \, \sin\phi \,.$$

A more compact way of writing this is

1.4
$$\Delta N_{\text{opt}} = \frac{\lambda_0 A}{V} \sin\phi \,,$$

where

1.4
$$V = 0.1597 \times 10^{-13}\,\text{m}^3 = 0.1597\,\text{nm}\,\text{cm}^2$$

is the numerical value for the volume parameter V.

There is constructive interference (high intensity in OUT1) when the optical path lengths of the two paths differ by an integer, $\Delta N_{\text{opt}} = 0, \pm 1, \pm 2, \ldots$, and we have destructive interference (low intensity in OUT1) when they differ by an integer plus half, $\Delta N_{\text{opt}} = \pm\frac{1}{2}, \pm\frac{3}{2}, \pm\frac{5}{2}, \ldots$. Changing ϕ from $\phi = -90°$ to $\phi = 90°$ gives

$$\Delta N_{\text{opt}}\Big|_{\phi=-90°}^{\phi=90°} = \frac{2\lambda_0 A}{V} \,,$$

which tell us that

1.5
$$\sharp \text{ of cycles} = \frac{2\lambda_0 A}{V} \,.$$

Experimental data For $a = 3.6\,\text{cm}$ and $\theta = 22.1°$ we have $A = 10.53\,\text{cm}^2$, so that

1.6
$$\lambda_0 = \frac{19 \times 0.1597}{2 \times 10.53}\,\text{nm} = 0.1441\,\text{nm} \,.$$

And 30 full cycles for $\lambda_0 = 0.2\,\text{nm}$ correspond to an area

1.7
$$A = \frac{30 \times 0.1597}{2 \times 0.2}\,\text{cm}^2 = 11.98\,\text{cm}^2 \,.$$

SOLUTIONS to Theory Question 2

Basic relations Position $\tilde{x}$ shows up on the picture if light was emitted from there at an instant that is earlier than the instant of the picture taking by the light travel time T that is given by

$$T = \sqrt{D^2 + \tilde{x}^2}\Big/c\,.$$

During the lapse of T the respective segment of the rod has moved the distance vT, so that its actual position x at the time of the picture taking is

2.1

$$x = \tilde{x} + \beta\sqrt{D^2 + \tilde{x}^2}\,.$$

Upon solving for $\tilde{x}$ we find

2.2

$$\tilde{x} = \gamma^2 x - \beta\gamma\sqrt{D^2 + (\gamma x)^2}\,.$$

Apparent length of the rod Owing to the Lorentz contraction, the actual length of the moving rod is L/γ, so that the actual positions of the two ends of the rod are

$$x_\pm = x_0 \pm \frac{L}{2\gamma} \text{ for the } \left\{\begin{array}{c}\text{front end}\\ \text{rear end}\end{array}\right\} \text{ of the rod.}$$

The picture taken by the pinhole camera shows the images of the rod ends at

$$\tilde{x}_\pm = \gamma\left(\gamma x_0 \pm \frac{L}{2}\right) - \beta\gamma\sqrt{D^2 + \left(\gamma x_0 \pm \frac{L}{2}\right)^2}\,.$$

The apparent length $\tilde{L}(x_0) = \tilde{x}_+ - \tilde{x}_-$ is therefore

2.3

$$\tilde{L}(x_0) = \gamma L + \beta\gamma\sqrt{D^2 + \left(\gamma x_0 - \frac{L}{2}\right)^2} - \beta\gamma\sqrt{D^2 + \left(\gamma x_0 + \frac{L}{2}\right)^2}\,.$$

Since the rod moves with the constant speed v, we have $\dfrac{\mathrm{d}x_0}{\mathrm{d}t} = v$ and therefore the question is whether $\tilde{L}(x_0)$ increases or decreases when x_0 increases. We sketch the two square root terms:

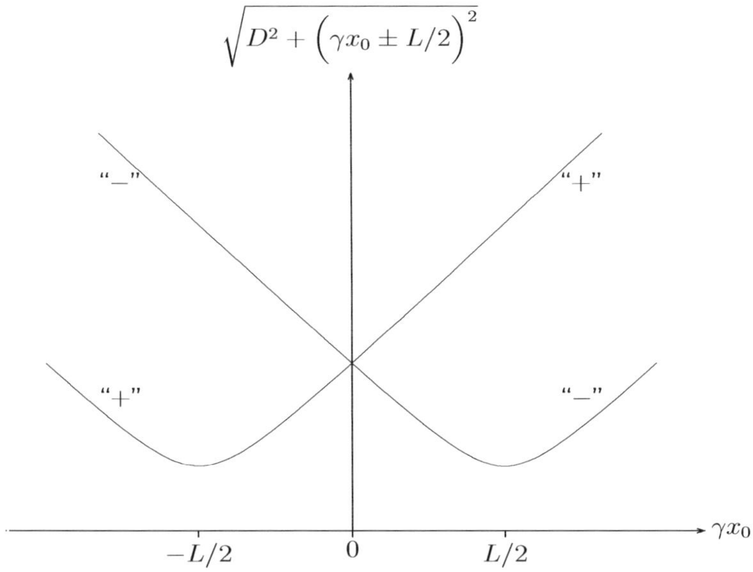

The difference of the square roots with "−" and "+" appears in the expression for $\tilde{L}(x_0)$, and this difference clearly decreases when x_0 increases.

2.4 The apparent length decreases all the time.

Symmetric picture For symmetry reasons, the apparent length on the symmetric picture is the actual length of the moving rod, because the light from the two ends was emitted simultaneously to reach the pinhole at the same time, that is:

2.5
$$\tilde{L} = \frac{L}{\gamma}.$$

The apparent endpoint positions are such that $\tilde{x}_- = -\tilde{x}_+$, or

$$0 = \tilde{x}_+ + \tilde{x}_- = 2\gamma^2 x_0 - \beta\gamma\sqrt{D^2 + \left(\gamma x_0 + \frac{L}{2}\right)^2} - \beta\gamma\sqrt{D^2 + \left(\gamma x_0 - \frac{L}{2}\right)^2}.$$

In conjunction with

$$\frac{L}{\gamma} = \tilde{x}_+ - \tilde{x}_- = \gamma L - \beta\gamma\sqrt{D^2 + \left(\gamma x_0 + \frac{L}{2}\right)^2} + \beta\gamma\sqrt{D^2 + \left(\gamma x_0 - \frac{L}{2}\right)^2}$$

this tells us that

$$\sqrt{D^2 + \left(\gamma x_0 \pm \frac{L}{2}\right)^2} = \frac{2\gamma^2 x_0 \pm (\gamma L - L/\gamma)}{2\beta\gamma} = \frac{\gamma x_0}{\beta} \pm \frac{\beta L}{2}.$$

As they should, both the version with the upper signs and the version with the lower signs give the same answer for x_0, namely

2.6

$$x_0 = \beta\sqrt{D^2 + \left(\frac{L}{2\gamma}\right)^2}.$$

The image of the middle of the rod on the symmetric picture is, therefore, located at

$$\begin{aligned}\tilde{x}_0 &= \gamma^2 x_0 - \beta\gamma\sqrt{D^2 + (\gamma x_0)^2} \\ &= \beta\gamma\left(\sqrt{(\gamma D)^2 + \left(\frac{L}{2}\right)^2} - \sqrt{(\gamma D)^2 + \left(\frac{\beta L}{2}\right)^2}\right),\end{aligned}$$

which is at a distance $\ell = \tilde{x}_+ - \tilde{x}_0 = \dfrac{L}{2\gamma} - \tilde{x}_0$ from the image of the front end, that is

2.7

$$\ell = \frac{L}{2\gamma} - \beta\gamma\sqrt{(\gamma D)^2 + \left(\frac{L}{2}\right)^2} + \beta\gamma\sqrt{(\gamma D)^2 + \left(\frac{\beta L}{2}\right)^2}$$

or

$$\ell = \frac{L}{2\gamma}\left[1 - \frac{\dfrac{\beta L}{2}}{\sqrt{(\gamma D)^2 + \left(\dfrac{L}{2}\right)^2} + \sqrt{(\gamma D)^2 + \left(\dfrac{\beta L}{2}\right)^2}}\right].$$

Very early and very late pictures At the very early time, we have a very large negative value for x_0, so that the apparent length on the very early picture is

$$\tilde{L}_{\text{early}} = \tilde{L}(x_0 \to -\infty) = (1+\beta)\gamma L = \sqrt{\frac{1+\beta}{1-\beta}}\,L.$$

Likewise, at the very late time, we have a very large positive value for x_0, so that the apparent length on the very late picture is

$$\tilde{L}_{\text{late}} = \tilde{L}(x_0 \to \infty) = (1-\beta)\gamma L = \sqrt{\frac{1-\beta}{1+\beta}}\, L\,.$$

It follows that $\tilde{L}_{\text{early}} > \tilde{L}_{\text{late}}$, that is:

2.8	The apparent length is 3 m on the early picture and 1 m on the late picture.

Further, we have

$$\beta = \frac{\tilde{L}_{\text{early}} - \tilde{L}_{\text{late}}}{\tilde{L}_{\text{early}} + \tilde{L}_{\text{late}}}\,,$$

so that $\beta = \frac{1}{2}$ and the velocity is

2.9	$v = \frac{c}{2}\,.$

It follows that $\gamma = \frac{\tilde{L}_{\text{early}} + \tilde{L}_{\text{late}}}{2\sqrt{\tilde{L}_{\text{early}}\tilde{L}_{\text{late}}}} = \frac{2}{\sqrt{3}} = 1.1547$. Combined with

2.10	$L = \sqrt{\tilde{L}_{\text{early}}\tilde{L}_{\text{late}}} = 1.73\,\text{m}\,,$

this gives the length on the symmetric picture as

2.11	$\tilde{L} = \frac{2\tilde{L}_{\text{early}}\tilde{L}_{\text{late}}}{\tilde{L}_{\text{early}} + \tilde{L}_{\text{late}}} = 1.50\,\text{m}\,.$

SOLUTIONS to Theory Question 3

Digital Camera Two factors limit the resolution of the camera as a photographic tool: the diffraction by the aperture and the pixel size. For diffraction, the inherent angular resolution θ_R is the ratio of the wavelength λ of the light and the aperture D of the camera,

$$\theta_R = 1.22\frac{\lambda}{D},$$

where the standard factor of 1.22 reflects the circular shape of the aperture. When taking a picture, the object is generally sufficiently far away from the photographer for the image to form in the focal plane of the camera where the CCD chip should thus be placed. The Rayleigh diffraction criterion then states that two image points can be resolved if they are separated by more than

3.1

$$\Delta x = f\theta_R = 1.22\lambda\, F\sharp,$$

which gives

$$\Delta x = 1.22\ \mu\text{m}$$

if we choose the largest possible aperture (or smallest value $F\sharp = 2$) and assume $\lambda = 500\,\text{nm}$ for the typical wavelength of daylight

The digital resolution is given by the distance ℓ between the center of two neighboring pixels. For our 5 Mpix camera this distance is roughly

$$\ell = \frac{L}{\sqrt{N_p}} = 15.65\,\mu\text{m}.$$

Ideally we should match the optical and the digital resolution so that neither aspect is overspecified. Taking the given optical resolution in the expression for the digital resolution, we obtain

3.2

$$N = \left(\frac{L}{\Delta x}\right)^2 \approx 823\ \text{Mpix}.$$

Now looking for the unknown optimal aperture, we note that we should have $\ell \geq \Delta x$, that is: $F\sharp \leq F_0$ with

$$F_0 = \frac{L}{1.22\lambda\sqrt{N_0}} = 2\sqrt{\frac{N}{N_0}} = 14.34.$$

Since this $F\sharp$ value is not available, we choose the nearest value that has a higher optical resolution,

3.3
$$F_0 = 11\,.$$

When looking at a picture at distance z from the eye, the (small) subtended angle between two neighboring dots is $\phi = \ell/z$ where, as above, ℓ is the distance between neighboring dots. Accordingly,

3.4
$$z = \frac{\ell}{\phi} = \frac{2.54 \times 10^{-2}/300\,\text{dpi}}{5.82 \times 10^{-4}\,\text{rad}} = 14.55\,\text{cm} \approx 15\,\text{cm}\,.$$

Hard-boiled egg All of the egg has to reach coagulation temperature. This means that the increase in temperature is

$$\Delta T = T_c - T_0 = 65\,°\text{C} - 4\,°\text{C} = 61\,°\text{C}\,.$$

Thus the minimum amount of energy that we need to get into the egg such that all of it has coagulated is given by $U = \mu V C \Delta T$ where $V = 4\pi R^3/3$ is the egg volume. We thus find

3.5
$$U = \mu \frac{4\pi R^3}{3} C(T_c - T_0) = 16768\ \text{J}\,.$$

The simplified equation for heat flow then allows us to calculate how much energy has flown into the egg through the surface per unit time. To get an approximate value for the time we assume that the center of the egg is at the initial temperature $T = 4\,°\text{C}$. The typical length scale is $\Delta r = R$, and the temperature difference associated with it is $\Delta T = T_1 - T_0$ where $T_1 = 100\,°\text{C}$ (boiling water). We thus get

3.6
$$J = \kappa(T_1 - T_0)/R = 2458\,\text{W}\,\text{m}^{-2}\,.$$

Heat is transferred from the boiling water to the egg through the surface of the egg. This gives

3.7
$$P = 4\pi R^2 J = 4\pi\kappa R(T_1 - T_0) \approx 19.3\,\mathrm{W}$$

for the amount of energy transferred to the egg per unit time. From this we get an estimate for the time τ required for the necessary amount of heat to flow into the egg all the way to the center:

3.8
$$\tau = \frac{U}{P} = \frac{\mu C R^2}{3\kappa}\frac{T_c - T_0}{T_1 - T_0} = \frac{16768}{19.3} = 869\ \mathrm{s} \approx 14.5\,\mathrm{min}\,.$$

Lightning The total charge Q is just the area under the curve of the figure. Because of the triangular shape, we immediately get

3.9
$$Q = \frac{I_0\tau}{2} = 5\,\mathrm{C}\,.$$

The average current is

3.10
$$I = Q/\tau = \frac{I_0}{2} = 50\,\mathrm{kA}\,,$$

simply half the maximal value.

Since the bottom of the cloud gets negatively charged and the ground positively charged, the situation is essentially that of a giant parallel-plate capacitor. The amount of energy stored just before lightning occurs is $QE_0h/2$ where E_0h is the voltage difference between the bottom of the cloud and the ground, and lightning releases this energy. Altogether we thus get for one lightning the energy $QE_0h/2 = 7.5 \times 10^8\,\mathrm{J}$. It follows that you could light up the 100 W bulb for the duration

3.11
$$t = \frac{32 \times 10^6}{6.5 \times 10^9} \times \frac{7.5 \times 10^8\,\mathrm{J}}{100\,\mathrm{W}} \approx 10\,\mathrm{h}\,.$$

Capillary Vessels Considering *all* capillaries, one has

$$R_{\mathrm{all}} = \frac{\Delta p}{D} = 10^7\,\mathrm{Pa\,m^{-3}\,s}\,.$$

All capillaries are assumed to be connected in parallel. The analogy between Poiseuille's and Ohm's laws then gives the hydraulic resistance R of one capillary as

$$\frac{1}{R_{\text{all}}} = \frac{N}{R} .$$

We thus get

$$N = \frac{R}{R_{\text{all}}}$$

for the number of capillary vessels in the human body. Now calculate R using Poiseuille's law,

$$R = \frac{8\eta L}{\pi r^4} \approx 4.5 \times 10^{16}\,\text{kg}\,\text{m}^{-4}\,\text{s}^{-1} ,$$

and arrive at

3.12
$$N \approx \frac{4.5 \times 10^{16}}{10^7} = 4.5 \times 10^9 .$$

The volume flow is $D = S_{\text{all}}v$ where $S_{\text{all}} = N\pi r^2$ is the *total* cross-sectional area associated with all capillary vessels. We then get

3.13
$$v = \frac{D}{N\pi r^2} = \frac{r^2 \Delta p}{8\eta L} = 0.44\,\text{mm}\,\text{s}^{-1} ,$$

where the second expression is found by alternatively considering one capillary vessel by itself.

Skyscraper When the slab is at height z above the ground, the air in the slab has pressure $p(z)$ and temperature $T(z)$ and the slab has volume $V(z) = Ah(z)$ where A is the cross-sectional area and $h(z)$ is the thickness of the slab. At any given height z, we combine the ideal gas law

$$pV = NkT \qquad (N \text{ is the number of molecules in the slab})$$

with the adiabatic law

$$pV^\gamma = \text{const} \quad \text{or} \quad (pV)^\gamma \propto p^{\gamma-1}$$

to conclude that $p^{\gamma-1} \propto T^\gamma$. Upon differentiation this gives $(\gamma-1)\dfrac{dp}{p} = \gamma\dfrac{dT}{T}$, so that

3.14
$$\frac{dT}{T} = (1 - 1/\gamma)\frac{dp}{p}\,.$$

Since the slab is not accelerated, the weight must be balanced by the force that results from the difference in pressure at the top and bottom of the slab. Taking downward forces as positive, we have the net force

$$0 = Nmg + A[p(z+h) - p(z)] = \frac{pV}{kT}mg + \frac{V}{h}\frac{dp}{dz}h\,,$$

so that $\dfrac{dp}{dz} = -\dfrac{mg}{k}\dfrac{p}{T}$ or

3.15
$$dp = -\frac{mg}{k}\frac{p}{T}dz\,.$$

Taken together, the two expressions say that

$$dT = -(1 - 1/\gamma)\frac{mg}{k}dz$$

and therefore we have

$$T_{\text{top}} = T_{\text{bot}} - (1 - 1/\gamma)\frac{mgH}{k}$$

for a building of height H, which gives

3.16
$$T_{\text{top}} = 20.6\,^\circ\text{C}$$

for $H = 1\,\text{km}$ and $T_{\text{bot}} = 30\,^\circ\text{C}$.

List of apparatus and materials

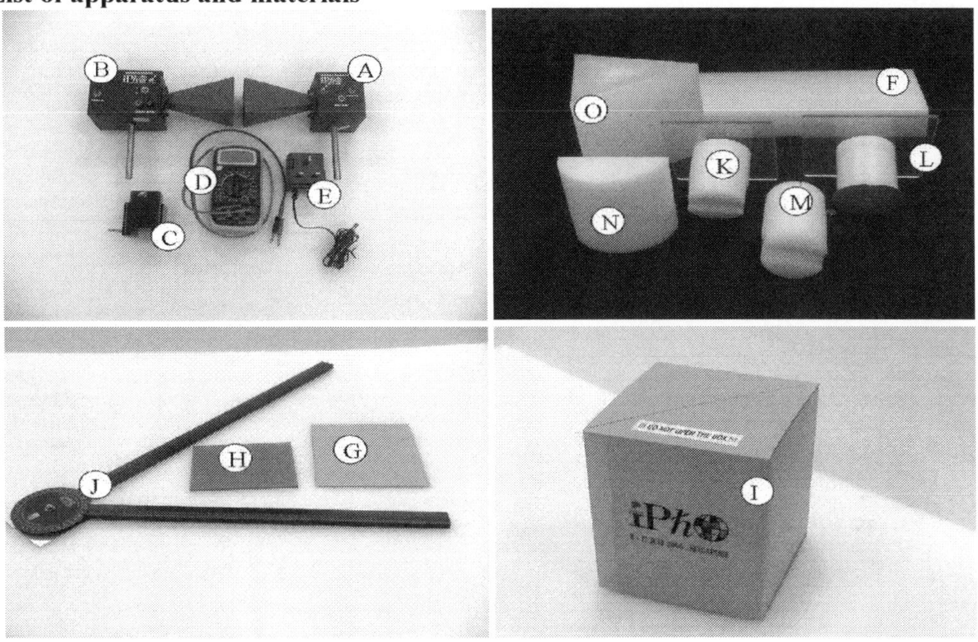

Label	Component	Quantity	Label	Component	Quantity
Ⓐ	Microwave transmitter	1	Ⓘ	Lattice structure in a black box	1
Ⓑ	Microwave receiver	1	Ⓙ	Goniometer	1
Ⓒ	Transmitter/receiver holder	2	Ⓚ	Prism holder	1
Ⓓ	Digital multimeter	1	Ⓛ	Rotating table	1
Ⓔ	DC power supply for transmitter	1	Ⓜ	Lens/reflector holder	1
Ⓕ	Slab as a "Thin film" sample	1	Ⓝ	Plano-cylindrical lens	1
Ⓖ	Reflector (silver metal sheet)	1	Ⓞ	Wax prism	2
Ⓗ	Beam splitter (blue Perspex)	1		Blu-Tack	1 pack
	Vernier caliper (provided separately)			30 cm ruler (provided separately)	

Caution:

- **The output power of the microwave transmitter is well within standard safety levels. Nevertheless, one should never look directly into the microwave horn at close range when the transmitter is on.**
- **Do not open the box containing the lattice Ⓘ.**
- **The wax prisms Ⓞ are fragile (used in Part 3).**

Note:

- ***It is important to note that the microwave receiver output (CURRENT) is proportional to the AMPLITUDE of the microwave.***
- ***Always use LO gain setting of the microwave receiver.***
- ***Do not change the range of the multimeter during the data collection.***
- ***Place the unused components away from the experiment to minimize interference.***
- ***Always use the component labels (Ⓐ, Ⓑ, Ⓒ,...) to indicate the components in all your drawings.***

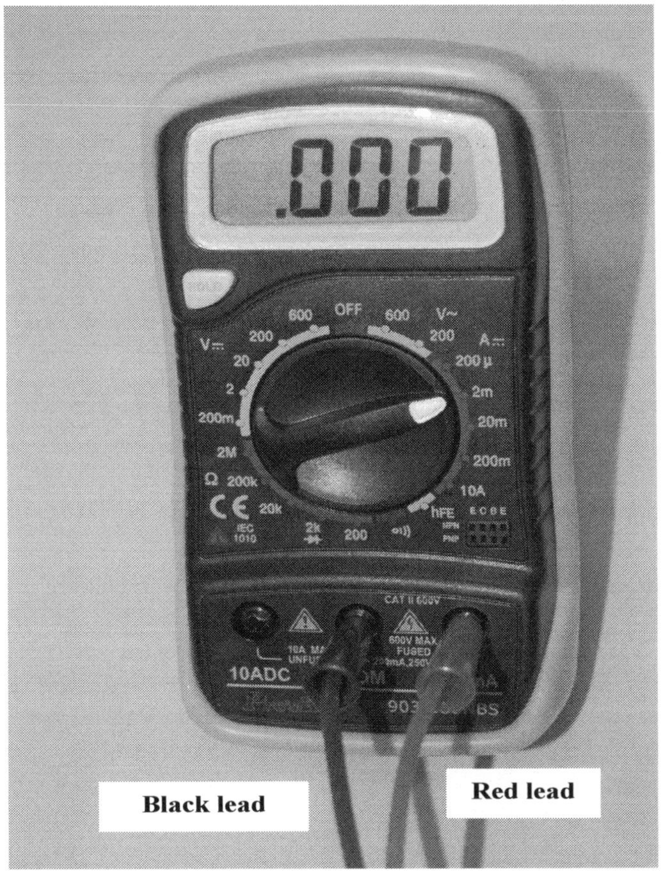

The digital multimeter should be used with the two leads connected as shown in the diagram. You should use the "2m" current setting in this experiment.

Part 1: Michelson interferometer

1.1. Introduction

In a Michelson interferometer, a beam splitter sends an incoming electromagnetic (EM) wave along two separate paths, and then brings the constituent waves back together after reflection so that they superpose, forming an interference pattern. Figure 1.1 illustrates the setup for a Michelson interferometer. An incident wave travels from the transmitter to the receiver along two different paths. These two waves superpose and interfere at the receiver. The strength of signal at the receiver depends on the phase difference between the two waves, which can be varied by changing the optical path difference.

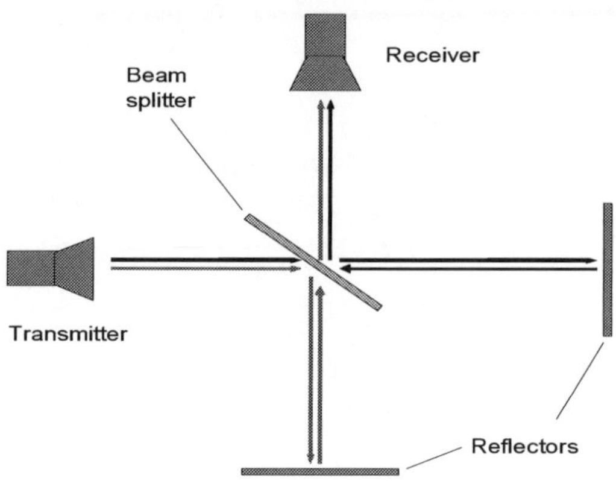

Figure 1.1: Schematic diagram of a Michelson interferometer.

1.2. List of components

1) Microwave transmitter Ⓐ with holder Ⓒ
2) Microwave receiver Ⓑ with holder Ⓒ
3) Goniometer Ⓙ
4) 2 reflectors: reflector Ⓖ with holder Ⓜ and thin film Ⓕ acting as a reflector.
5) Beam splitter Ⓗ with rotating table Ⓛ acting as a holder
6) Digital multimeter Ⓓ

1.3. Task: Determination of wavelength of the microwave [2 marks]

Using only the experimental components listed in Section 1.2, set up a Michelson interferometer experiment to determine the wavelength λ of the microwave in air. Record your data and determine λ in such a way that the uncertainty is ≤ 0.02 cm.

Note that the "thin film" is partially transmissive, so make sure you do not stand or move behind it as this might affect your results.

Part 2: "Thin film" interference

2.1. Introduction

A beam of EM wave incident on a dielectric thin film splits into two beams, as shown in Figure 2.1. Beam A is reflected from the top surface of the film whereas beam B is reflected from the bottom surface of the film. The superposition of beams A and B results in the so called thin film interference.

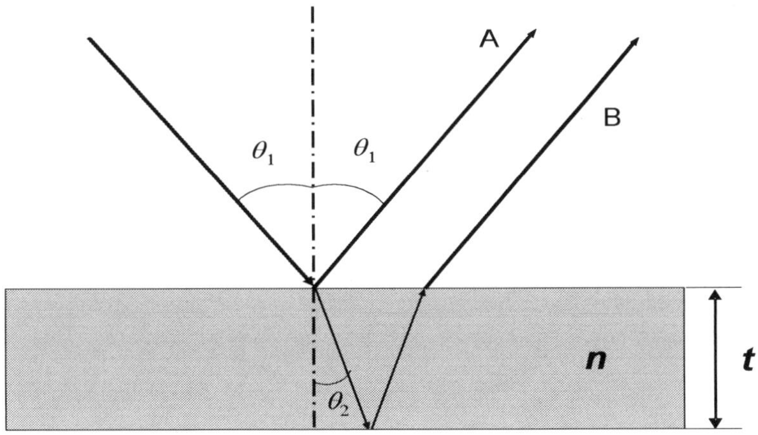

Figure 2.1: Schematic of thin film interference.

The difference in the optical path lengths of beam A and B leads to constructive or destructive interference. The resultant EM wave intensity I depends on the path difference of the two interfering beams which in turn depends on the angle of incidence, θ_1, of the incident beam, wavelength λ of the radiation, and the thickness t and refractive index n of the thin film. Thus, the refractive index n of the thin film can be determined from I-θ_1 plot, using values of t and λ.

2.2. List of components

1) Microwave transmitter Ⓐ with holder Ⓒ
2) Microwave receiver Ⓑ with holder Ⓒ
3) Plano-cylindrical lens Ⓝ with holder Ⓜ
4) Goniometer Ⓙ
5) Rotating table Ⓛ
6) Digital multimeter Ⓓ
7) Polymer slab acting as a "thin film" sample Ⓕ
8) Vernier caliper

2.3. Tasks: Determination of refractive index of polymer slab [6 marks]

1) Derive expressions for the conditions of constructive and destructive interferences in terms of θ_1, t, λ and n.

[1 mark]

2) Using only the experimental components listed in Section 2.2, set up an experiment to measure the receiver output S as a function of the angle of incidence θ_1 in the range from 40° to 75°. Sketch your experimental setup, clearly showing the angles of incidence and reflection and the position of the film on the rotating table. Mark all components using the labels given on page 2. Tabulate your data. Plot the receiver output S versus the angle of incidence θ_1. Determine accurately the angles corresponding to constructive and destructive interferences.

[3 marks]

3) Assuming that the refractive index of air is 1.00, determine the order of interference m and the refractive index of the polymer slab n. Write the values of m and n on the answer sheet.

[1.5 marks]

4) Carry out error analysis for your results and estimate the uncertainty of n. Write the value of the uncertainty Δn on the answer sheet.

[0.5 marks]

Note:

- ***The lens should be placed in front of the microwave transmitter with the planar surface facing the transmitter to obtain a quasi-parallel microwave beam. The distance between the planar surface of the lens and the aperture of transmitter horn should be 3 cm.***
- ***For best results, maximize the distance between the transmitter and receiver.***
- ***Deviations of the microwave emitted by transmitter from a plane wave may cause extra peaks in the observed pattern. In the prescribed range from 40° to 75°, only one maximum and one minimum exist due to interference.***

Part 3: Frustrated Total Internal Reflection

3.1. Introduction

The phenomenon of total internal reflection (TIR) may occur when the plane wave travels from an optically dense medium to less dense medium. However, instead of TIR at the interface as predicted by geometrical optics, the incoming wave in reality penetrates into the less dense medium and travels for some distance parallel to the interface before being scattered back to the denser medium (see Figure 3.1). This effect can be described by a shift D of the reflected beam, known as the Goos-Hänchen shift.

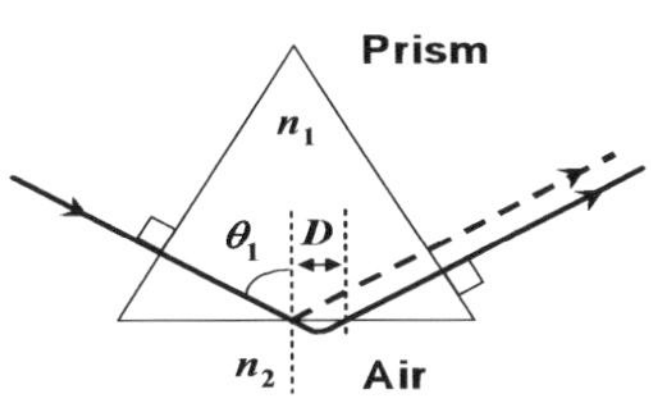

Figure 3.1: A sketch illustrating an EM wave undergoing total internal reflection in a prism. The shift D parallel to the surface in air represents the Goos-Hänchen shift

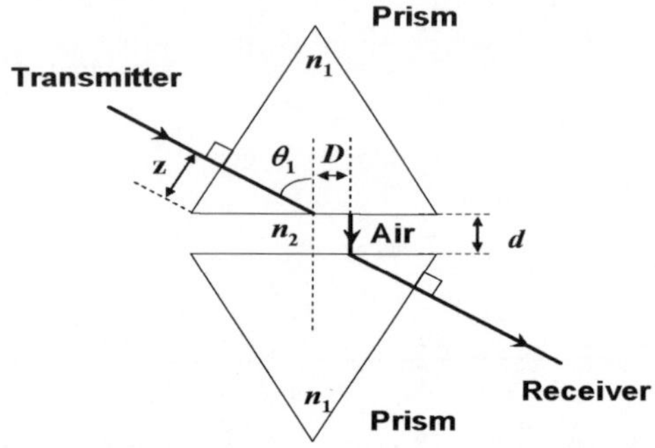

Figure 1.2: A sketch of the experimental setup showing the prisms and the air gap of distance d. The shift D parallel to the surface in air represents the Goos-Hänchen shift. z is the distance from the tip of the prism to the central axis of the transmitter.

If another medium of refractive index n_1 (i.e. made of the same material as the first medium) is placed at a small distance d to the first medium as shown in Figure 3.2, tunneling of the EM wave through the second medium occurs. This intriguing phenomenon is known as the *frustrated total internal reflection* (FTIR). The intensity of the transmitted wave, I_t, decreases exponentially with the distance d:

$$I_t = I_0 \exp(-2\gamma d) \tag{3.1}$$

where I_0 is the intensity of the incident wave and γ is:

$$\gamma = \frac{2\pi}{\lambda}\sqrt{\frac{n_1^2}{n_2^2}\sin^2\theta_1 - 1} \tag{3.2}$$

where λ is the wavelength of EM wave in medium 2 and n_2 is the refractive index of medium 2 (assume that the refractive index of medium 2, air, is 1.00).

3.2. List of components

1) Microwave transmitter Ⓐ with holder Ⓒ

2) Microwave receiver Ⓑ with holder Ⓒ
3) Plano-cylindrical lens Ⓝ with holder Ⓜ
4) 2 equilateral wax prisms Ⓞ with holder Ⓚ and rotating table Ⓛ acting as a holder
5) Digital multimeter Ⓓ
6) Goniometer Ⓙ
7) Ruler

3.3. Description of the Experiment

Using only the list of components described in Section 3.2, set up an experiment to investigate the variation of the intensity I_t as a function of the air gap separation d in FTIR. For consistent results, please take note of the following:

- Use one arm of the goniometer for this experiment.
- Choose the prism surfaces carefully so that they are parallel to each other.
- The distance from the centre of the curved surface of the lens should be 2 cm from the surface of the prism.
- Place the detector such that its horn is in contact with the face of the prism.
- For each value of d, adjust the position of the microwave receiver along the prism surface to obtain the maximum signal.
- Make sure that the digital multi-meter is on the 2mA range. Collect data starting from $d = 0.6$ cm. Discontinue the measurements when the reading of the multimeter falls below 0.20 mA.

3.4. Tasks: Determination of refractive index of prism material [6 marks]

Task 1

Sketch your final experimental setup and mark all components using the labels given at page 2. In your sketch, record the value of the distance z (see Figure 3.2), the distance from the tip of the prism to the central axis of the transmitter.

[1 Mark]

Task 2

Perform your experiment and tabulate your data. Perform this task twice.

[2.1 Marks]

Task 3

(a) By plotting appropriate graphs, determine the refractive index, n_1, of the prism with error analysis.

(b) Write the refractive index n_1, and its uncertainty Δn_1, of the prism in the answer sheet provided.

[2.9 Marks]

Part 4: Microwave diffraction of a metal-rod lattice: Bragg reflection

4.1. Introduction

Bragg's Law

The lattice structure of a real crystal can be examined using Bragg's Law,

$$2d \sin \theta = m\lambda \tag{4.1}$$

where d refers to the distance between a set of parallel crystal planes that "reflect" the X-ray; m is the order of diffraction and θ is the angle between the incident X-ray beam and the crystal planes. Bragg's law is also commonly known as Bragg's reflection or X-ray diffraction.

Metal-rod lattice

Because the wavelength of the X-ray is comparable to the lattice constant of the crystal, traditional Bragg's diffraction experiment is performed using X-ray. For microwave, however, diffraction occurs in lattice structures with much larger lattice constant, which can be measured easily with a ruler.

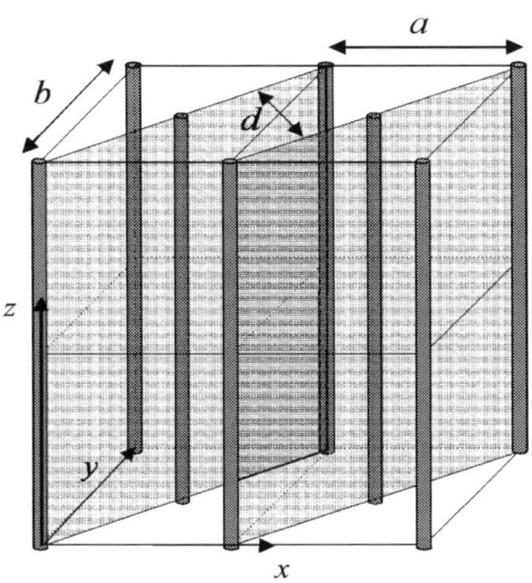

Figure 4.1: A metal-rod lattice of lattice constants a and b, and interplanar spacing d.

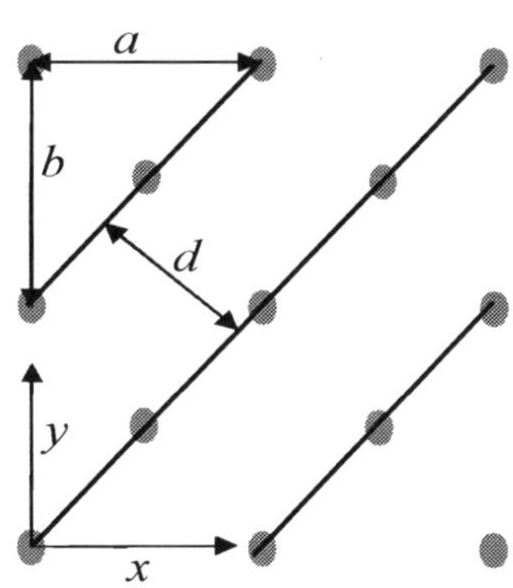

Figure 4.2: Top-view of the metal-rod lattice shown in Fig. 4.1 (not to scale). The lines denote diagonal planes of the lattice.

In this experiment, the Bragg law is used to measure the lattice constant of a lattice made of metal rods. An example of such metal-rod lattice is shown in Fig. 4.1, where the metal rods are shown as thick vertical lines. The lattice planes along the diagonal direction of the xy-plane are shown as shaded planes. Fig. 4.2 shows the top-view (looking down along the z-axis) of the metal-rod lattice, where the points represent the rods and the lines denote the diagonal lattice planes.

4.2. List of components

1) Microwave transmitter Ⓐ with holder Ⓒ
2) Microwave receiver Ⓑ with holder Ⓒ
3) Plano-cylindrical lens Ⓝ with holder Ⓜ
4) Sealed box containing a metal-rod lattice Ⓘ
5) Rotating table Ⓛ
6) Digital multimeter Ⓓ
7) Goniometer Ⓙ

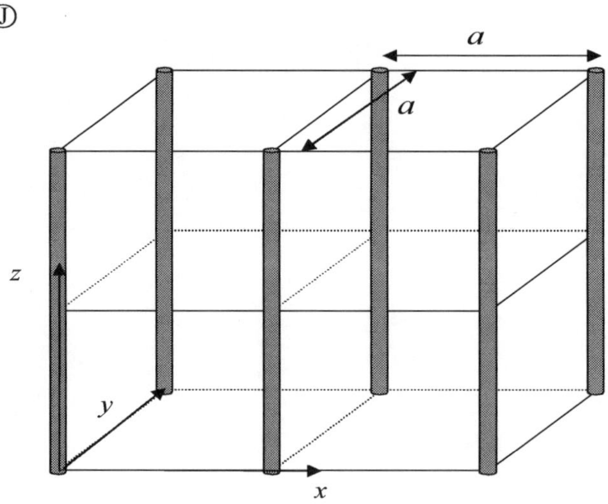

Figure 4.3: A simple square lattice.

In this experiment, you are given a ***simple*** square lattice made of metal rods, as illustrated in Fig. 4.3. The lattice is sealed in a box. You are asked to derive the lattice constant a of the lattice from the experiment. DO NOT open the box. No marks will be given to the experimental results if the seal is found broken after the experiment.

4.3. Tasks: Determination of lattice constant of given simple square lattice [6 Marks]

Task 1

Draw a top-view diagram of the simple square lattice shown in Fig. 4.3. In the diagram, indicate the lattice constant a of the given lattice and the interplanar spacing d of the diagonal planes. With the help of this diagram, derive Bragg's Law.

[1 Mark]

Task 2

Using Bragg's law and the apparatus provided, design an experiment to perform Bragg diffraction experiment to determine the lattice constant a of the lattice.

(a) Sketch the experimental set up. Mark all components using the labels in page 2 and indicate clearly the angle between the axis of the transmitter and lattice planes, θ, and the angle between the axis of the transmitter and the axis of the receiver, ζ. In your experiment, measure the diffraction on the diagonal planes the direction of which is indicated by the red line on the box.

[1.5 Marks]

(b) Carry out the diffraction experiment for $20° \leq \theta \leq 50°$. In this range, you will only observe the first order diffraction. In the answer sheet, tabulate your results and record both the θ and ζ.

[1.4 Marks]

(c) Plot the quantity proportional to the intensity of diffracted wave as a function of θ.

[1.3 Marks]

(d) Determine the lattice constant a using the graph and estimate the experimental error.

[0.8 Marks]

Note:

1. ***For best results, the transmitter should remain fixed during the experiment. The separation between the transmitter and the lattice, as well as that between lattice and receiver should be about 50 cm.***
2. ***Use only the diagonal planes in this experiment. Your result will not be correct if you try to use any other planes.***
3. ***The face of the lattice box with the red diagonal line must be at the top.***
4. ***To determine the position of the diffraction peak with better accuracy, use a number of data points around the peak position.***

Part 1

a. A sketch of the experimental setup (not required)

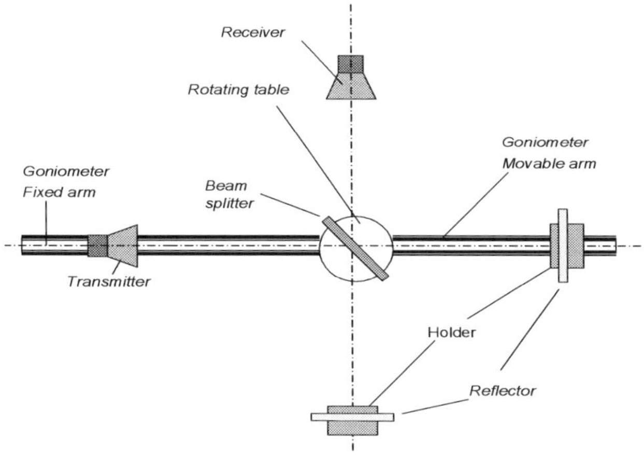

b. Data sheet (not required)

Position (cm)	Meter reading (mA)	Position (cm)	Meter reading (mA)	Position (cm)	Meter reading (mA)	Position (cm)	Meter reading (mA)
104.0	0.609	100.9	1.016	96.0	0.514	91.0	0.925
103.9	0.817	100.85	1.060	95.8	0.098	90.9	1.094
103.8	0.933	100.8	1.090	95.6	0.192	90.8	1.245
103.7	1.016	100.7	0.994	95.4	0.669	90.7	1.291
103.6	1.030	100.6	0.940	95.3	0.870	90.6	1.253
103.5	0.977	100.4	0.673	95.2	1.009	90.4	0.978
103.4	0.890	100.2	0.249	95.1	1.119	90.2	0.462
103.3	0.738	100.0	0.074	95.0	1.138	90.0	0.045
103.2	0.548	99.8	0.457	94.9	1.080	89.8	0.278
103.1	0.310	99.6	0.883	94.7	0.781	89.6	0.809
103.0	0.145	99.4	1.095	94.5	0.403	89.5	1.031
102.9	0.076	99.3	1.111	94.3	0.044	89.4	1.235
102.8	0.179	99.2	1.022	94.1	0.364	89.3	1.277
102.7	0.392	99.0	0.787	93.9	0.860	89.2	1.298
102.6	0.623	98.8	0.359	93.7	1.103	89.1	1.252
102.5	0.786	98.6	0.079	93.6	1.160	89.0	1.133
102.4	0.918	98.4	0.414	93.5	1.159	88.8	0.684
102.3	0.988	98.2	0.864	93.4	1.083	88.6	0.123
102.2	1.026	98..0	1.128	93.2	0.753	88.5	-0.020
102.1	1.006	97.9	1.183	93.0	0.331	88.4	0.123
102.0	0.945	97.8	1.132	92.8	0.073	88.2	0.679
101.9	0.747	97.7	1.015	92.6	0.515	88.0	1.116
101.8	0.597	97.5	0.713	92.4	0.968	87.9	1.265
101.7	0.363	97.2	0.090	92.2	1.217	87.8	1.339
101.6	0.161	97.0	0.342	92.15	1.234	87.7	1.313
101.5	0.055	96.8	0.714	92.1	1.230	87.6	1.190
101.4	0.139	96.6	1.007	92.0	1.165	87.4	0.867
101.3	0.357	96.5	1.087	91.8	0.871	87.2	0.316

101.2	0.589	96.4	1.070	91.6	0.353	87.1	0.034
101.1	0.781	96.3	1.018	91.4	0.018	87.0	-0.018
101.0	0.954	96.2	0.865	91.2	0.394	86.9	0.178
104.0	0.609	100.9	1.016	96.0	0.514	91.0	0.925
103.9	0.817	100.8	1.060	95.8	0.098	90.9	1.094
103.8	0.933	100.8	1.090	95.6	0.192	90.8	1.245
103.7	1.016	100.7	0.994	95.4	0.669	90.7	1.291
103.6	1.030	100.6	0.940	95.3	0.870	90.6	1.253
103.5	0.977	100.4	0.673	95.2	1.009	90.4	0.978
103.4	0.890	100.2	0.249	95.1	1.119	90.2	0.462
103.3	0.738	100.0	0.074	95.0	1.138	90.0	0.045
103.2	0.548	99.8	0.457	94.9	1.080	89.8	0.278
103.1	0.310	99.6	0.883	94.7	0.781	89.6	0.809
103.0	0.145	99.4	1.095	94.5	0.403	89.5	1.031
102.9	0.076	99.3	1.111	94.3	0.044	89.4	1.235
102.8	0.179	99.2	1.022	94.1	0.364	89.3	1.277
102.7	0.392	99.0	0.787	93.9	0.860	89.2	1.298
102.6	0.623	98.8	0.359	93.7	1.103	89.1	1.252
102.5	0.786	98.6	0.079	93.6	1.160	89.0	1.133
102.4	0.918	98.4	0.414	93.5	1.159	88.8	0.684
102.3	0.988	98.2	0.864	93.4	1.083	88.6	0.123
102.2	1.026	98.0	1.128	93.2	0.753	88.5	-0.020
102.1	1.006	97.9	1.183	93.0	0.331	88.4	0.123
102.0	0.945	97.8	1.132	92.8	0.073	88.2	0.679
101.9	0.747	97.7	1.015	92.6	0.515	88.0	1.116
101.8	0.597	97.5	0.713	92.4	0.968	87.9	1.265
101.7	0.363	97.2	0.090	92.2	1.217	87.8	1.339
101.6	0.161	97.0	0.342	92.15	1.234	87.7	1.313
101.5	0.055	96.8	0.714	92.1	1.230	87.6	1.190
101.4	0.139	96.6	1.007	92.0	1.165	87.4	0.867
101.3	0.357	96.5	1.087	91.8	0.871	87.2	0.316
101.2	0.589	96.4	1.070	91.6	0.353	87.1	0.034
101.1	0.781	96.3	1.018	91.4	0.018	87.0	-0.018
101.0	0.954	96.2	0.865	91.2	0.394	86.9	0.178

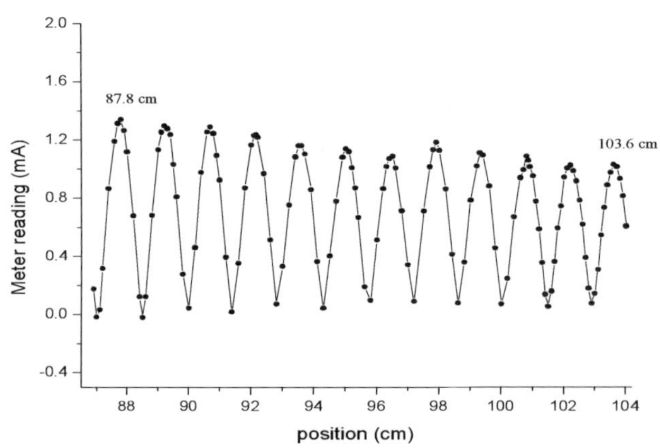

From the graph (not required) or otherwise, the positions of the first maximum point and 12th maximum point are measured at 87.8 cm and 103.6 cm.

The wavelength is calculated by

$$\frac{\lambda}{2} = \frac{103.6 - 87.8}{11}\text{ cm}$$

1.8 marks

Thus, $\boxed{\lambda = 2.87 \text{ cm}}$.

Error analysis

$$\lambda = \frac{2}{11}d, \qquad \Delta d = 0.05 \text{ x2 cm} = 0.1 \text{ cm}.$$

$$|\Delta\lambda| = \left|\frac{2}{11}\Delta d\right| = \frac{2}{11}\times 0.10 = 0.018 \quad cm < 0.02 \quad cm$$

0.2 marks

Part 2

(a) Deduction of interference conditions

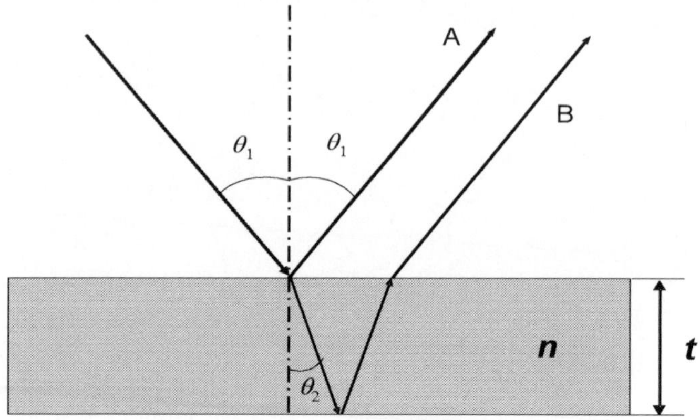

Assume that the thickness of the film is t and refractive index n. Let θ_1 be the incident angle and θ_2 the refracted angle. The difference of the optical paths ΔL is:

$$\Delta L = 2(nt/\cos\theta_2 - t\tan\theta_2 \sin\theta_1)$$

Law of refraction:

$$\sin\theta_1 = n\sin\theta_2$$

Thus

$$\Delta L = 2t\sqrt{n^2 - \sin^2\theta_1}$$

Considering the 180 deg (π) phase shift at the air- thin film interface for the reflected beam, we have interference conditions:

$$2t\sqrt{n^2 - \sin^2\theta_{\min}} = m\lambda \quad (m = 1, 2, 3, \ldots)$$ for the destructive peak

and $$2t\sqrt{n^2 - \sin^2\theta_{\max}} = (m \pm \frac{1}{2})\lambda$$ for the constructive peak

1 mark

If thickness t and wave length λ are known, one can determine the refractive index of the thin film from I - $\theta_1$spectrum (I is the intensity of the interfered beam).

(b) A sketch of the experimental setup

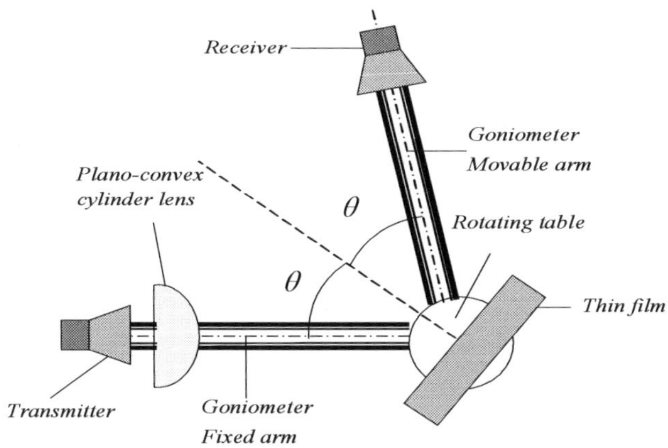

1 mark

Students should use the labeling on Page 2.

(c) Data Set

X: θ_l / degree	Y: Meter reading S/mA
40.0	0.309
41.0	0.270
42.0	0.226
43.0	0.196
44.0	0.164
45.0	0.114
46.0	0.063
47.0	0.036
48.0	0.022
49.0	0.039
50.0	0.066
51.0	0.135
52.0	0.215
53.0	0.262
54.0	0.321
55.0	0.391
56.0	0.454
57.0	0.511

58.0	0.566
59.0	0.622
60.0	0.664
61.0	0.691
62.0	0.722
63.0	0.754
64.0	0.796
65.0	0.831
66.0	0.836
67.0	0.860
68.0	0.904
69.0	0.970
70.0	1.022
71.0	1.018
72.0	0.926
73.0	0.800
74.0	0.770
75.0	0.915

Uncertainty: angle $\Delta\theta_1 = \pm 0.5^o$, current: ± 0.001 mA

0.5 marks

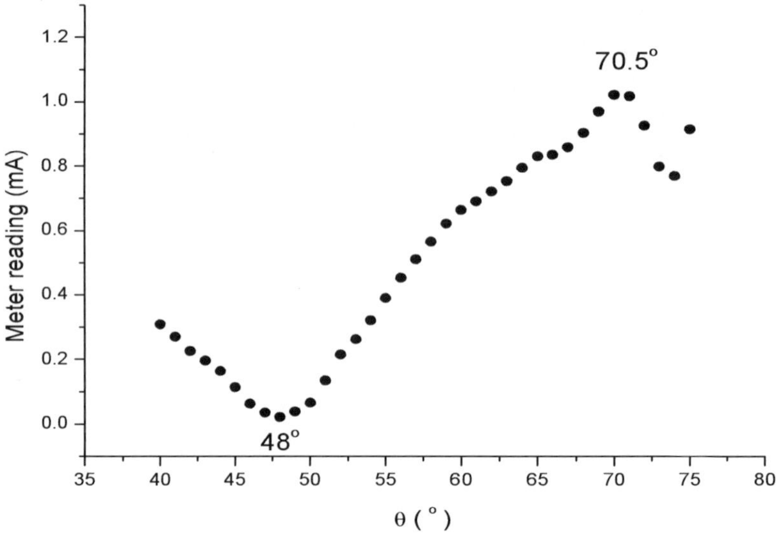

From the data, θ_{min} and θ_{max} can be found at 48° and 70.5° respectively.

0.9 marks

To calculate the refractive index, the following equations are used:

0.6 marks

$$2t\sqrt{n^2 - \sin^2 48^o} = m\lambda \quad (m = 1, 2, 3, \ldots) \qquad (1)$$

and
$$2t\sqrt{n^2 - \sin^2 70.5^o} = (m - \frac{1}{2})\lambda \qquad (2)$$

In this experiment, $t = 5.28$ cm, $\lambda = 2.85$cm (measured using other method).

Solving the simultaneous equations (1) and (2), we get

$$m = \frac{\sin^2 70.5^o - \sin^2 48^o}{(\frac{\lambda}{2t})^2} + 0.25$$

$$m = 4.83 \longrightarrow m = 5$$

1 mark

Substituting $m = 5$ in (1), we get $n = 1.54$

Substituting $m = 5$ in (2), we also get $n = 1.54$

0.5 marks

Error analysis:

$$n = \sqrt{\sin^2 \theta + (\frac{m\lambda}{2t})^2}$$

$$\Delta n = \frac{1}{\sqrt{\sin^2 \theta + (\frac{m\lambda}{2t})^2}}(\sin 2\theta \bullet \Delta\theta + \frac{m^2\lambda}{2t^2}\Delta\lambda - \frac{m^2\lambda^2}{2t^3}\Delta t)$$

$$= \frac{1}{n}(\sin 2\theta \bullet \Delta\theta + \frac{m^2\lambda}{2t^2}\Delta\lambda - \frac{m^2\lambda^2}{2t^3}\Delta t)$$

If we take $\Delta\theta = \pm 0.5^\circ = \pm 0.0087$ rad, $\Delta t = \pm 0.05$ cm, $\Delta\lambda = \pm 0.02$ cm, and $\theta = 48^\circ$

$$\Delta n = \frac{1}{1.54}(0.0087 \sin 96^o + \frac{5^2 \times 2.85}{2 \times 5.28^2} \times 0.01 + \frac{5^2 \times 2.85^2}{2 \times 5.28^3} \times 0.05) \approx 0.02$$

Thus, $\boxed{n + \Delta n = 1.54 \pm 0.02}$

0.5 marks

Part 3

Sample Solution

Task 1
Sketch your final experimental setup and mark all components using the labels given at page 2. In your sketch, write down the distance z (see Figure 3.2), where z is the distance from the tip of the prism to the central axis of the transmitter.

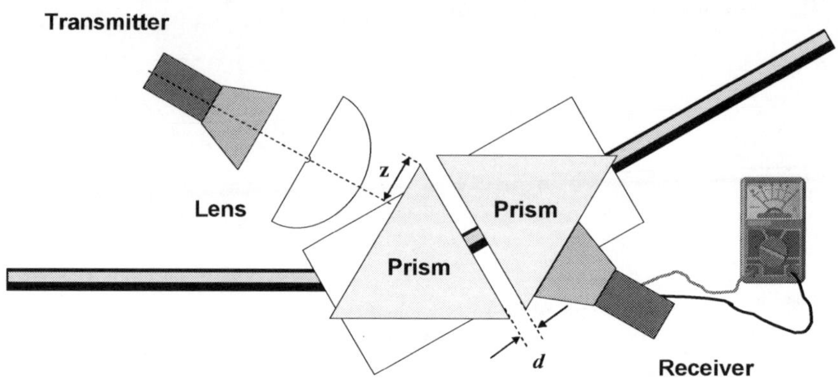

(Students should use labels on page 2.)

Task 2
Tabulate your data. Perform the experiment twice.

Data Set

X: d(cm)	ΔX(cm)	*Set 1* S_1 (mA)	*Set 2* S_2(mA)	$S_{average}$ (mA)	ΔS(mA)[#]	I_t (mA)[2*]	$\Delta(I_t)$[$]	Y: $\ln(I_t\,(\text{mA})^2)$	ΔY[&]
0.60	0.05	0.78	0.78	0.780	0.01	0.6080	0.016	-0.50	0.03
0.70	0.05	0.68	0.69	0.685	0.01	0.4690	0.014	-0.76	0.03
0.80	0.05	0.58	0.59	0.585	0.01	0.3420	0.012	-1.07	0.03
0.90	0.05	0.50	0.51	0.505	0.01	0.2550	0.010	-1.37	0.04
1.00	0.05	0.42	0.42	0.420	0.01	0.1760	0.008	-1.74	0.05
1.10	0.05	0.36	0.35	0.355	0.01	0.1260	0.007	-2.07	0.06
1.20	0.05	0.31	0.31	0.310	0.01	0.0961	0.006	-2.34	0.06
1.30	0.05	0.26	0.25	0.255	0.01	0.0650	0.005	-2.73	0.08
1.40	0.05	0.21	0.22	0.215	0.01	0.0462	0.004	-3.07	0.09

[#] $\Delta S = 0.01$ mA (for each set of current measurements)
[*] S^2 proportional to the intensity, I_t
[$] $\Delta(S^2) = \Delta I_t = 2\,S \times \Delta S$
[&] $\Delta Y = \Delta(\ln I_t) = \Delta(I_t)/I_t$

Task 3

By plotting appropriate graphs, determine the refractive index, n_1, of the prism with error analysis. Write the refractive index n_1, and its uncertainty Δn_1, of the prism in the answer sheet provided.

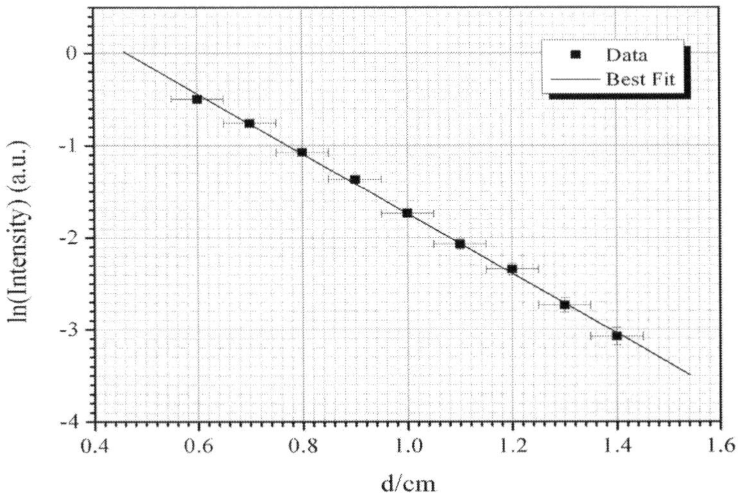

Least Square Fitting

X = d(cm)	ΔX(cm)	Y = ln(I_t)	ΔY	ΔY^2	XY	X^2	Y^2
0.60	0.05	-0.50	0.03	0.001	-0.298	0.360	0.247
0.70	0.05	-0.76	0.03	0.001	-0.530	0.490	0.573
0.80	0.05	-1.07	0.03	0.001	-0.858	0.640	1.150
0.90	0.05	-1.37	0.04	0.002	-1.230	0.810	1.867
1.00	0.05	-1.74	0.05	0.002	-1.735	1.000	3.010
1.10	0.05	-2.07	0.06	0.003	-2.278	1.210	4.290
1.20	0.05	-2.34	0.06	0.004	-2.811	1.440	5.487
1.30	0.05	-2.73	0.08	0.006	-3.553	1.690	7.469
1.40	0.05	-3.07	0.09	0.009	-4.304	1.960	9.451
ΣX =		ΣY =	ΣΔY =	$\Sigma(\Delta Y)^2$ =	ΣXY =	ΣX^2 =	ΣY^2 =
9.00		-15.648	0.469	0.029	-17.596	9.600	33.544

From $I_t = I_0 \exp(-2\gamma d)$, taking natural log on both sides, we obtain:

$$\ln(I_t) = -2\gamma d + \ln(I_0)$$

which is of the form $y = mx + c$.

To calculate the gradient, the following equation was used, where N is the number of data points:

$$m = \frac{N\sum(XY) - (\sum X)(\sum Y)}{N\sum X^2 - (\sum X)^2} = -3.247$$

To calculate the standard deviation σ_Y of the individual Y data values, the following equation was used:

$$\sigma_Y = \sqrt{\frac{\sum(\Delta Y)^2}{N-2}} = 0.064$$

Hence the standard deviation in the slope can be calculated:

$$\sigma_m = \sigma_Y \sqrt{\frac{N}{N\sum X^2 - (\sum X)^2}} = 0.082$$

From the gradient:

$$\begin{aligned} 2\gamma &= 3.247 \pm 0.082 \\ &\approx 3.25 \pm 0.08 \end{aligned}$$

Using:

$$n_1 = \frac{\sqrt{k_2^2 + \gamma^2}}{k_2 \sin\theta_1}$$

where $\theta_1 = 60^\circ$, $k_2 = 2\pi/\lambda \approx 2.20$ (using the wavelength determined from earlier part (using $\lambda = (2.85 \pm 0.02)$cm), we obtain:

$$\boxed{\begin{aligned} n_1 \pm \Delta n_1 &= 1.434 \pm 0.016 \\ &\approx 1.43 \pm 0.02 \end{aligned}}$$

Error Analysis for refractive index of n_1

$$\Delta n_1 = \frac{d}{dk_2}\left[\frac{\left(k_2^2+\gamma^2\right)^{\frac{1}{2}}}{k_2\sin\theta_1}\right]\Delta k_2 + \frac{d}{d\gamma}\left[\frac{\left(k_2^2+\gamma^2\right)^{\frac{1}{2}}}{k_2\sin\theta_1}\right]\Delta\gamma$$

$$\Delta n_1 = \left[\frac{\left(k_2^2+\gamma^2\right)^{-\frac{1}{2}}}{\sin\theta_1} - \frac{\left(k_2^2+\gamma^2\right)^{\frac{1}{2}}}{k_2^2\sin\theta_1}\right]\Delta k_2 + \left[\frac{\gamma\left(k_2^2+\gamma^2\right)^{-\frac{1}{2}}}{k_2\sin\theta_1}\right]\Delta\gamma$$

$$= 0.016$$

$$\approx 0.02$$

where:

$$\Delta k_2 = -\frac{2\pi}{\lambda^2}\Delta\lambda = -0.015$$

Note: Other methods of error analysis are also accepted.

Part 4

Task 1

Top-view of a simple square lattice.

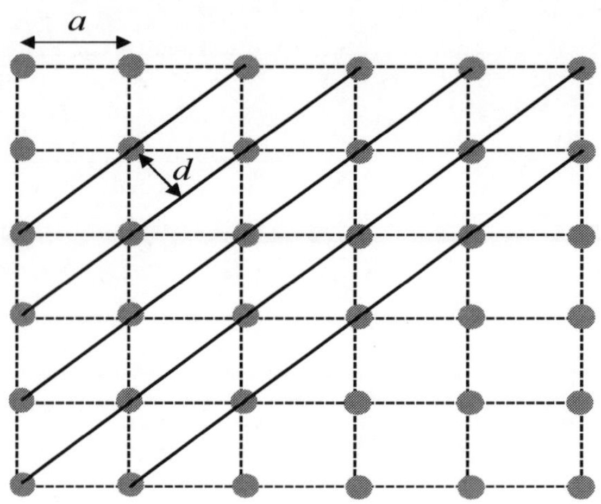

0.5 marks

Figure 4.1: *Schematic diagram of a simple square lattice with lattice constant a and interplaner d of the diagonal planes indicated.*

Deriving Bragg's Law

Conditions necessary for the observation of diffraction peaks:

1. The angle of incidence = angle of scattering.
2. The pathlength difference is equal to an integer number of wavelengths.

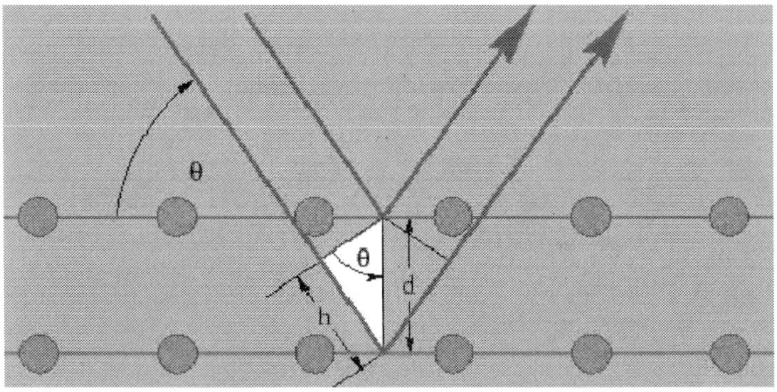

Figure 4.2: *Schematic diagram for deriving Bragg's law.*

$h = d\,sin\theta$ (1).

The path length difference is given by,

$2h = 2d\,sin\theta$ (2).

For diffraction to occur, the path difference must satisfy,

$\boxed{2\,d\,sin\theta = m\lambda,}$ $m = 1, 2, 3...$ (3).

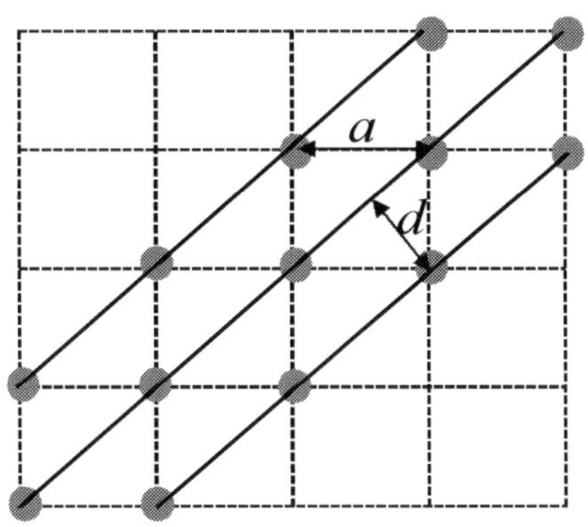

0.5 marks

Figure 4.3 *Illustration of the lattice used in the experiment (this Figure is not required)*

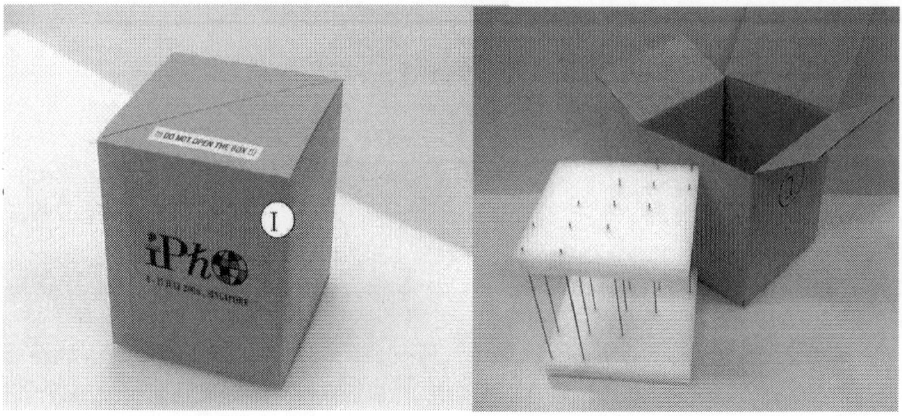

Fig. 4.4 *The actual lattice used in the experiment (not required)*

Task 2 (a)

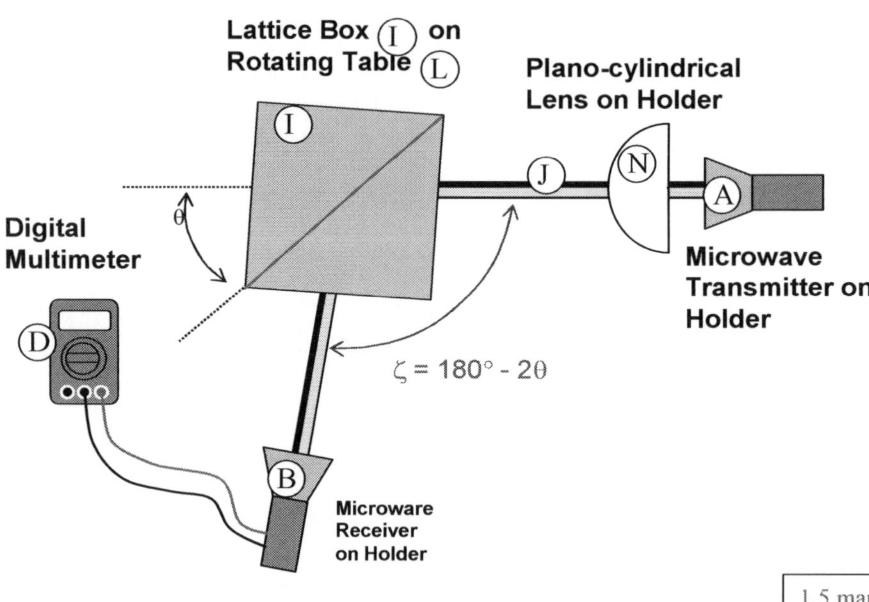

1.5 marks

Fig. 4.5 Sketch of the experimental set up

Task 2(b) & 2(c)

Data Set

θ/°	ζ/°	Output current S (mA)	Intensity $I=S^2$ $(mA)^2$
20.0	140.0	0.023	0.000529
21.0	138.0	0.038	0.001444
22.0	136.0	0.070	0.0049
23.0	134.0	0.109	0.011881
24.0	132.0	0.163	0.026569
25.0	130.0	0.201	0.040401
26.0	128.0	0.233	0.054289
27.0	126.0	0.275	0.075625
28.0	124.0	0.320	0.1024
29.0	122.0	0.350	0.1225
30.0	120.0	0.353	0.124609
31.0	118.0	0.358	0.128164
32.0	116.0	0.354	0.125316
33.0	114.0	0.342	0.116964
34.0	112.0	0.321	0.103041
35.0	110.0	0.303	0.091809
36.0	108.0	0.280	0.0784
37.0	106.0	0.241	0.058081
38.0	104.0	0.200	0.04
39.0	102.0	0.183	0.033489
40.0	100.0	0.162	0.026244
41.0	98.0	0.139	0.019321
42.0	96.0	0.120	0.0144
43.0	94.0	0.109	0.011881
44.0	92.0	0.086	0.007396
45.0	90.0	0.066	0.004356
46.0	88.0	0.067	0.004489
47.0	86.0	0.066	0.004356
48.0	84.0	0.070	0.0049
49.0	82.0	0.084	0.007056
50.0	80.0	0.080	0.0064

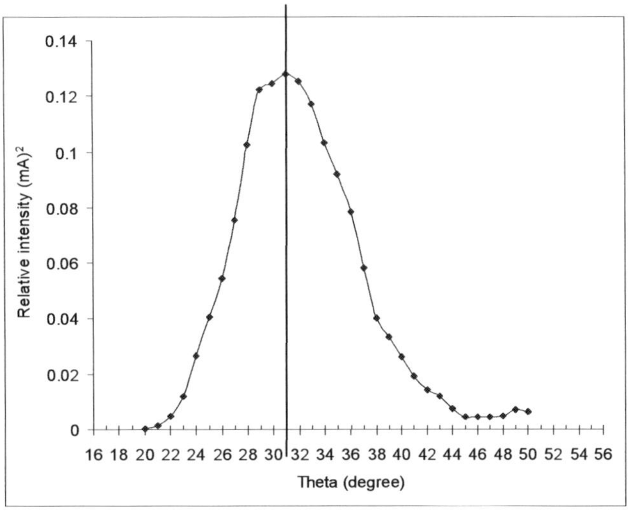

2.7 marks

Task 2(d)

From eq 3 and let $m = 1$,

$$2d \sin \theta_{max} = \lambda \qquad (4)$$

From Fig. 4.3,

$a = \sqrt{2}d$ (5)

Combine eqs (4) and (5), we obtain,

$$\boxed{a = \frac{\lambda}{\sqrt{2}\sin\theta_{\max}}}$$

From the symmetry of the data, the peak position is determined to be:

$\theta_{max} = 31°$ (The theoretical value is $\theta_{max} = 32°$)

$$\boxed{a = \frac{\lambda}{\sqrt{2}\sin\theta_{\max}} = \frac{2.85cm}{\sqrt{2}\sin 31^o} = 3.913cm}$$

(<u>Actual value a = 3.80 cm</u>)

[*The value 3.55 in the marking scheme is derived from:*

$$a = \frac{\lambda}{\sqrt{2}\sin\theta_{\max}} = \frac{2.83cm}{\sqrt{2}\sin 34^o} = 3.58cm$$

where 2.83 cm and 34 deg are the min and max allowed values for wavelength and peak position.

Similarly:

The value 4.10 is derived from: $a = \frac{\lambda}{\sqrt{2}\sin\theta_{\max}} = \frac{2.87cm}{\sqrt{2}\sin 30^o} = 4.06cm$

The value 3.55 is derived from: $a = \frac{\lambda}{\sqrt{2}\sin\theta_{\max}} = \frac{2.83cm}{\sqrt{2}\sin 34^o} = 3.58cm$

The value 3.40 is derived from: $a = \frac{\lambda}{\sqrt{2}\sin\theta_{\max}} = \frac{2.83cm}{\sqrt{2}\sin 35^o} = 3.49cm$

The value 4.20 is derived from: $a = \frac{\lambda}{\sqrt{2}\sin\theta_{\max}} = \frac{2.87cm}{\sqrt{2}\sin 29^o} = 4.18cm$]

Error analysis:

Known uncertainties:
$\Delta\lambda = 0.02$ cm;

$\Delta\theta = 0.5$ deg $= 0.014$ rad. (uncertainty in determining θ from graph).

From: $a = \dfrac{\lambda}{\sqrt{2}\sin\theta_{\max}}$

$$\Delta a = \frac{\Delta\lambda}{\sqrt{2}\sin\theta_{\max}} - \frac{\lambda}{\sqrt{2}(\sin\theta_{\max})^2}\frac{d}{d\theta}(\sin\theta_{\max})\Delta\theta$$

$$= a(\frac{\Delta\lambda}{\lambda} - \frac{1}{\sin\theta_{\max}}\frac{d}{d\theta}(\sin\theta_{\max})\Delta\theta)$$

$$= a(\frac{\Delta\lambda}{\lambda} - \cot\theta_{\max}\Delta\theta)$$

$$= 3.80(\frac{0.02}{2.85} - \cot(32°)\times(-0.014))cm$$

$$= 0.112cm \approx 0.1$$

Hence:

$$a \pm \Delta a = 3.913 \pm 0.112$$
$$\approx 3.9 \pm 0.1 \text{ cm}$$

0.8 marks

38TH

INTERNATIONAL PHYSICS OLYMPIAD

ISFAHAN, IRAN

2007

In this problem we deal with a simplified model of accelerometers designed to activate the safety air bags of automobiles during a collision. We would like to build an electromechanical system in such a way that when the acceleration exceeds a certain limit, one of the electrical parameters of the system such as the voltage at a certain point of the circuit will exceed a threshold and the air bag will be activated as a result.

Note: Ignore gravity in this problem.

1 Consider a capacitor with parallel plates as in Figure 1. The area of each plate in the capacitor is A and the distance between the two plates is d. The distance between the two plates is much smaller than the dimensions of the plates. One of these plates is in contact with a wall through a spring with a spring constant k, and the other plate is fixed. When the distance between the plates is d the spring is neither compressed nor stretched, in other words no force is exerted on the spring in this state. Assume that the permittivity of the air between the plates is that of free vacuum ε_0. The capacitance corresponding to this distance between the plates of the capacitor is $C_0 = \varepsilon_0 A/d$. We put charges $+Q$ and $-Q$ on the plates and let the system achieve mechanical equilibrium.

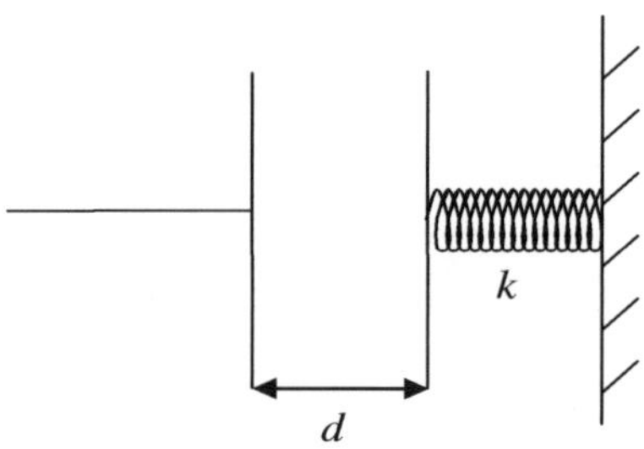

Figure 1

1.1	Calculate the electrical force, F_E, exerted by the plates on each other.	0.8
1.2	Let x be the displacement of the plate connected to the spring. Find x.	0.6
1.3	In this state, what is the electrical potential difference V between the plates of the capacitor in terms of Q, A, d, k?	0.4
1.4	Let C be the capacitance of the capacitor, defined as the ratio of charge to potential difference. Find C/C_0 as a function of Q, A, d and k.	0.3
1.5	What is the total energy, U, stored in the system in terms of Q, A, d and k?	0.6

Figure 2, shows a mass M which is attached to a conducting plate with negligible mass and also to two springs having identical spring constants k. The conducting plate can move back and forth in the space between two fixed conducting plates. All these plates are similar and have the same area A. Thus these three plates constitute two capacitors. As shown in Figure 2, the fixed plates are connected to the given potentials V and $-V$, and the middle plate is connected through a two-state switch to the ground. The wire connected to the movable plate does not disturb the motion of the plate and the three plates will always remain parallel. When the whole complex is not being accelerated, the distance from each fixed plate to the movable plate is d which is much smaller than the dimensions of the plates. The thickness of the movable plate can be ignored.

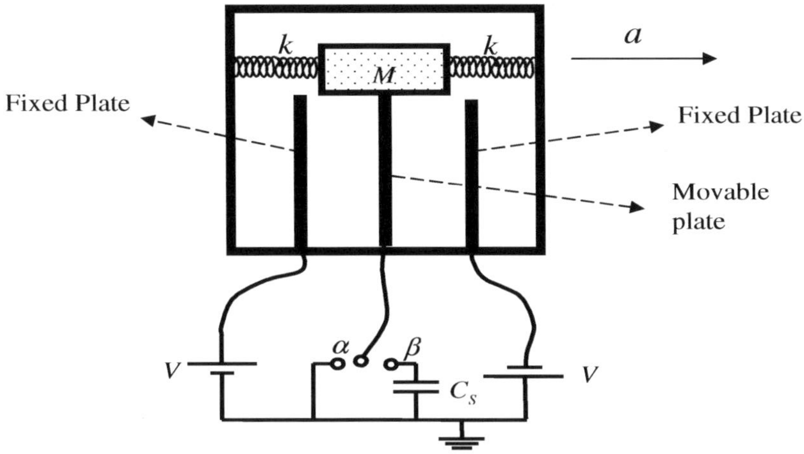

Figure 2

The switch can be in either one of the two states α and β. Assume that the capacitor complex is being accelerated along with the automobile, and the acceleration is constant. Assume that during this constant acceleration the spring does not oscillate and all components of this complex capacitor are in their equilibrium positions, i.e. they do not move with respect to each other, and hence with respect to the automobile.

Due to the acceleration, the movable plate will be displaced a certain amount x from the middle of the two fixed plates.

2 Consider the case where the switch is in state α i.e. the movable plate is connected to the ground through a wire, then

2.1	Find the charge on each capacitor as a function of x.	0.4
2.2	Find the net electrical force on the movable plate, F_E, as a function of x.	0.4
2.3	Assume $d >> x$ and terms of order x^2 can be ignored compared to terms of order d^2. Simplify the answer to the previous part.	0.2
2.4	Write the total force on the movable plate (the sum of the electrical and the spring forces) as $-k_{eff}x$ and give the form of k_{eff}.	0.7
2.5	Express the constant acceleration a as a function of x.	0.4

3 Now assume that the switch is in state β i.e. the movable plate is connected to the ground through a capacitor, the capacitance of which is C_S (there is no initial charge on the capacitors). If the movable plate is displaced by an amount x from its central position,

3.1	Find V_S the electrical potential difference across the capacitor C_S as a function of x.	1.5

3.2	Again assume that $d >> x$ and ignore terms of order x^2 compared to terms of order d^2. Simplify your answer to the previous part.	0.2

4 We would like to adjust the parameters in the problem such that the air bag will not be activated in normal braking but opens fast enough during a collision to prevent the driver's head from colliding with the windshield or the steering wheel. As you have seen in Part 2, the force exerted on the movable plate by the springs and the electrical charges can be represented as that of a spring with an effective spring constant k_{eff}. The whole capacitor complex is similar to a *mass and spring* system of mass M and spring constant k_{eff} under the influence of a constant acceleration a, which in this problem is the acceleration of the automobile.

Note: In this part of the problem, the assumption that the mass and spring are in equilibrium under a constant acceleration and hence are fixed relative to the automobile, no longer holds.

Ignore friction and consider the following numerical values for the parameters of the problem:

$d = 1.0\,\text{cm}$, $A = 2.5\times10^{-2}\,\text{m}^2$, $k = 4.2\times10^{3}\,\text{N/m}$, $\varepsilon_0 = 8.85\times10^{-12}\,\text{C}^2/\text{Nm}^2$, $V = 12\,\text{V}$, $M = 0.15\,\text{kg}$.

4.1	Using this data, find the ratio of the electrical force you calculated in section 2.3 to the force of the springs and show that one can ignore the electrical forces compared to the spring forces.	0.6

Although we did not calculate the electrical forces for the case when the switch is in the state β, it can be shown that in this situation, quite similarly, the electrical forces are as small and can be ignored.

4.2	If the automobile while traveling with a constant velocity, suddenly brakes with a constant acceleration a, what is the maximum displacement of the movable plate? Give your answer in parameter.	0.6

Assume that the switch is in state β and the system has been designed such that when the electrical voltage across the capacitor reaches $V_S = 0.15\text{V}$, the air bag is activated. We would like the air bag not to be activated during normal braking when the automobile's acceleration is less than the acceleration of gravity $g = 9.8\,m/s^2$, but be activated otherwise.

4.3	How much should C_S be for this purpose?	0.6

We would like to find out if the air bag will be activated fast enough to prevent the driver's head from hitting the windshield or the steering wheel. Assume that as a result of collision, the automobile experiences a deceleration equal to g but the driver's head keeps moving at a constant speed.

4.4	By estimating the distance between the driver's head and the steering wheel, find the time t_1 it takes before the driver's head hits the steering wheel.	0.8

4.5	Find the time t_2 before the air bag is activated and compare it to t_1. Is the air bag activated in time? Assume that airbag opens instantaneously.	0.9

Question "Orange"

1.1)
First of all, we use the Gauss's law for a single plate to obtain the electric field,

$$E = \frac{\sigma}{\varepsilon_0}. \qquad (0.2)$$

The density of surface charge for a plate with charge, Q and area, A is

$$\sigma = \frac{Q}{A}. \qquad (0.2)$$

Note that the electric field is resulted by two equivalent parallel plates. Hence the contribution of each plate to the electric field is $\frac{1}{2}E$. Force is defined by the electric filed times the charge, then we have

$$\text{Force} = \frac{1}{2}E\,Q = \frac{Q^2}{2\varepsilon_0 A}$$ (0.2)+ (0.2) (The ½ coefficient + the final result)

1.2)
The Hook's law for a spring is
$F_m = -k\,x$. (0.2)
In 1.2 we derived the electric force between two plates is

$$F_e = \frac{Q^2}{2\varepsilon_0 A}.$$

The system is stable. The equilibrium condition yields
$F_m = F_e$, (0.2)

$$\Rightarrow \qquad x = \frac{Q^2}{2\varepsilon_0 A\,k} \qquad (0.2)$$

1.3)
The electric field is constant thus the potential difference, V is given by
$V = E(d - x)$ (0.2)
(Other reasonable approaches are acceptable. For example one may use the definition of capacity to obtain V.)
By substituting the electric field obtained from previous section to the above equation, we get,

$$V = \frac{Q\,d}{\varepsilon_0 A}\left(1 - \frac{Q^2}{2\varepsilon_0 A\,k\,d}\right) \qquad (0.2)$$

1.4)
C is defined by the ratio of charge to potential difference, then

$$C = \frac{Q}{V}. \qquad (0.1)$$

Using the answer to 1.3, we get $\dfrac{C}{C_0}=\left(1-\dfrac{Q^2}{2\varepsilon_0 A k d}\right)^{-1}$ (0.2)

1.5)
Note that we have both the mechanical energy due to the spring

$$U_m=\frac{1}{2}kx^2, \qquad (0.2)$$

and the electrical energy stored in the capacitor.

$$U_E=\frac{Q^2}{2C}. \qquad (0.2)$$

Therefore the total energy stored in the system is

$$U=\frac{Q^2 d}{2\varepsilon_0 A}\left(1-\frac{Q^2}{4\varepsilon_0 A k d}\right) \qquad (0.2)$$

2.1)
For the given value of x, the amount of charge on each capacitor is

$$Q_1=V\,C_1=\frac{\varepsilon_0 A V}{d-x}, \qquad (0.2)$$

$$Q_2=V\,C_2=\frac{\varepsilon_0 A V}{d+x}. \qquad (0.2)$$

2.2)
Note that we have two capacitors. By using the answer to 1.1 for each capacitor, we get

$$F_1=\frac{Q_1^{\;2}}{2\varepsilon_0 A},$$

$$F_2=\frac{Q_2^{\;2}}{2\varepsilon_0 A}.$$

As these two forces are in the opposite directions, the net electric force is

$$F_E=F_1-F_2,\ (0.2) \quad\Rightarrow\quad F_E=\frac{\varepsilon_0 A V^2}{2}\left(\frac{1}{(d-x)^2}-\frac{1}{(d+x)^2}\right) \quad (0.2)$$

2.3)
Ignoring terms of order x^2 in the answer to 2.2., we get

$$F_E=\frac{2\varepsilon_0 A V^2}{d^3}x \qquad (0.2)$$

2.4)
There are two springs placed in series with the same spring constant, k, then the mechanical force is

$F_m = -2kx$. (The coefficient (2) has (0.2))

Combining this result with the answer to 2.4 and noticing that these two forces are in the opposite directions, we get

$F = F_m + F_E$, $\Rightarrow$ $F = -2\left(k - \dfrac{\varepsilon_0 A V^2}{d^3}\right)x$, (Opposite signs of the two forces have (0.3))

$$\Rightarrow \quad k_{eff} = 2\left(k - \frac{\varepsilon_0 A V^2}{d^3}\right) \quad (0.2)$$

2.5)

By using the Newtown's second law,

$F = ma$ (0.2)

and the answer to 2.4, we get

$$a = -\frac{2}{m}\left(k - \frac{\varepsilon_0 A V^2}{d^3}\right)x \quad (0.2)$$

3.1)

Starting with Kirchhoff's laws, for two electrical circuits, we have

$$\begin{cases} \dfrac{Q_S}{C_S} + V - \dfrac{Q_2}{C_2} = 0 \\ -\dfrac{Q_S}{C_S} + V - \dfrac{Q_1}{C_1} = 0 \\ Q_2 - Q_1 + Q_S = 0 \end{cases}$$

(Each has (0.3), Note: the sings may depend on the specific choice made)

Noting that $V_s = \dfrac{Q_S}{C_S}$ one obtains

$$\Rightarrow V_S = V \frac{\dfrac{2\varepsilon_0 A x}{d^2 - x^2}}{C_S + \dfrac{2\varepsilon_0 A d}{d^2 - x^2}}.$$

((0.4) + (0.2): (0.4) for solving the above equations and (0.2) for final result)

Note: Students may simplify the above relation using the approximation $d^2 >> x^2$. It does not matter in this section.

3.2)
Ignoring terms of order x^2 in the answer to 3.1., we get

$$V_S = V \frac{2\varepsilon_0 A x}{d^2 C_S + 2\varepsilon_0 A d} . \qquad (0.2)$$

4.1)
The ratio of the electrical force to the mechanical (spring) force is

$$\frac{F_E}{F_m} = \frac{\varepsilon_0 A V^2}{k d^3} ,$$

Putting the numerical values:

$$\frac{F_E}{F_m} = 7.6\times10^{-9} .$$

((0.2) + (0.2) + (0.2): (0.2) for order of magnitude, (0.2) for two significant digits and (0.2) for correct answer (7.6 or 7.5)).

As it is clear from this result, we can ignore the electrical forces compared to the electric force.

4.2)
As seen in the previous section, one may assume that the only force acting on the moving plate is due to springs:

$F = 2k x$. (The concept of equilibrium (0.2))

Hence in mechanical equilibrium, the displacement of the moving plate is

$$x = \frac{ma}{2k} .$$

The maximum displacement is twice this amount, like the mass spring system in a gravitational force field, when the mass is let to fall.

$x_{\max} = 2x$ (0.2)

$$x_{\max} = \frac{m a}{k} \qquad (0.2)$$

4.3)
At the acceleration

$a = g$, (0.2)

The maximum displacement is

$$x_{\max} = \frac{m g}{k} .$$

Moreover, from the result obtained in 3.2, we have

$$V_S = V\frac{2\varepsilon_0 A\,x_{\max}}{d^2 C_S + 2\varepsilon_0 A\,d}$$

This should be the same value given in the problem, $0.15\,\text{V}$.

$$\Rightarrow \qquad C_S = \frac{2\varepsilon_0 A}{d}\left(\frac{V\,x_{\max}}{V_S\,d} - 1\right) \qquad (0.2)$$

$$\Rightarrow \qquad C_S = 8.0\times10^{-11}\,\text{F} \qquad (0.2)$$

4.4)

Let ℓ be the distance between the driver's head and the steering wheel. It can be estimated to be about

$\ell = 0.4m - 1m$. (0.2)

Just at the time the acceleration begins, the relative velocity of the driver's head with respect to the automobile is zero.

$\Delta v(t=0) = 0$, (0.2)

then

$$\ell = \frac{1}{2}g\,t_1^{\,2} \qquad \Rightarrow \qquad t_1 = \sqrt{\frac{2\ell}{g}} \qquad (0.2)$$

$t_1 = 0.3 - 0.5\ s$ (0.2)

4.5)

The time t_2 is half of period of the harmonic oscillator, hence

$$t_2 = \frac{T}{2}, \qquad (0.3)$$

The period of harmonic oscillator is simply given by

$$T = 2\pi\sqrt{\frac{m}{2k}}, \qquad (0.2)$$

therefore,

$t_2 = 0.013\ s$. (0.2)

As $t_1 > t_2$, the airbag activates in time. (0.2)

In physics, whenever we have an equality relation, both sides of the equation should be of the same type i.e. they must have the same dimensions. For example you cannot have a situation where the quantity on the right-hand side of the equation represents a length and the quantity on the left-hand side represents a time interval. Using this fact, sometimes one can nearly deduce the form of a physical relation without solving the problem analytically. For example if we were asked to find the time it takes for an object to fall from a height of h under the influence of a constant gravitational acceleration g, we could argue that one only needs to build a quantity representing a time interval, using the quantities g and h and the only possible way of doing this is $T = a(h/g)^{1/2}$. Notice that this solution includes an as yet undetermined coefficient a which is *dimensionless* and thus cannot be determined, using this method. This coefficient can be a number such as 1, $1/2$, $\sqrt{3}$, π, or any other real number. This method of deducing physical relations is called *dimensional analysis*. In dimensional analysis the dimensionless coefficients are not important and we do not need to write them. Fortunately in most physical problems these coefficients are of the order of 1 and eliminating them does not change the order of magnitude of the physical quantities. Therefore, by applying the dimensional analysis to the above problem, one obtains $T = (h/g)^{1/2}$.

Generally, the dimensions of a physical quantity are written in terms of the dimensions of four fundamental quantities: M (mass), L (length), T (time), and K (temperature). The dimensions of an arbitrary quantity, x is denoted by $[x]$. As an example, to express the dimensions of velocity v, kinetic energy E_k, and heat capacity C_V we write: $[v] = LT^{-1}$, $[E_k] = ML^2T^{-2}$, $[C_V] = ML^2T^{-2}K^{-1}$.

1 Fundamental Constants and Dimensional Analysis

1.1	Find the dimensions of *the fundamental constants,* i.e. the Planck's constant, h, the speed of light, c, the universal constant of gravitation, G, and the Boltzmann constant, k_B, in terms of the dimensions of length, mass, time, and temperature.	0.8

The Stefan-Boltzmann law states that the black body emissive power which is the total energy radiated per unit surface area of a black body in unit time is equal to $\sigma\theta^4$ where σ is the Stefan-Boltzmann's constant and θ is the absolute temperature of the black body.

1.2	Determine the dimensions of the Stefan-Boltzmann's constant in terms of the dimensions of length, mass, time, and temperature.	0.5

The Stefan-Boltzmann's constant is not a fundamental constant and one can write it in terms of fundamental constants i.e. one can write $\sigma = a\,h^{\alpha}c^{\beta}G^{\gamma}k_B^{\ \delta}$. In this relation a is a dimensionless parameter of the order of 1. As mentioned before, the exact value of a is not significant from our viewpoint, so we will set it equal to 1.

1.3	Find α, β, γ, and δ using dimensional analysis.	1.0

2 Physics of Black Holes

In this part of the problem, we would like to find out some properties of black holes using dimensional analysis. According to a certain theorem in physics known as the *no hair theorem*, all the characteristics of the black hole which we are considering in this problem depend only on the mass of the black hole. One characteristic of a black hole is the area of its *event horizon*. Roughly speaking, the event horizon is the boundary of the black hole. Inside this boundary, the gravity is so strong that even light cannot emerge from the region enclosed by the boundary.

We would like to find a relation between the mass of a black hole, m, and the area of its event horizon, A. This area depends on the mass of the black hole, the speed of light, and the universal constant of gravitation. As in 1.3 we shall write $A = G^{\alpha} c^{\beta} m^{\gamma}$.

2.1	Use dimensional analysis to find α, β, and γ.	0.8

From the result of 2.1 it becomes clear that the area of the event horizon of a black hole increases with its mass. From a classical point of view, nothing comes out of a black hole and therefore in all physical processes the area of the event horizon can only increase. In analogy with the second law of thermodynamics, Bekenstein proposed to assign entropy, S, to a black hole, proportional to the area of its event horizon i.e. $S = \eta A$. This conjecture has been made more plausible using other arguments.

2.2	Use the thermodynamic definition of entropy $dS = dQ/\theta$ to find the dimensions of entropy. dQ is the exchanged heat and θ is the absolute temperature of the system.	0.2

2.3	As in 1.3, express the dimensioned constant η as a function of the fundamental constants h, c, G, and k_B.	1.1

*Do **not** use dimensional analysis for the rest of problem, but you may use the results you have obtained in previous sections.*

3 Hawking Radiation

With a semi-quantum mechanical approach, Hawking argued that contrary to the classical point of view, black holes emit radiation similar to the radiation of a black body at a temperature which is called the *Hawking temperature*.

3.1	Use $E = mc^2$, which gives the energy of the black hole in terms of its mass, and the laws of thermodynamics to express the Hawking temperature θ_H of a black hole in terms of its mass and the fundamental constants. Assume that the black hole does no work on its surroundings.	0.8

3.2	The mass of an isolated black hole will thus change because of the Hawking radiation. Use Stefan-Boltzmann's law to find the dependence of this rate of change on the Hawking temperature of the black hole, θ_H and express it in terms of mass of the black hole and the fundamental constants.	0.7

3.3	Find the time t^*, that it takes an isolated black hole of mass m to evaporate completely i.e. to lose all its mass.	1.1

From the viewpoint of thermodynamics, black holes exhibit certain exotic behaviors. For example the heat capacity of a black hole is negative.

3.4	Find the heat capacity of a black hole of mass m.	0.6

4 Black Holes and the Cosmic Background Radiation

Consider a black hole exposed to the cosmic background radiation. The cosmic background radiation is a black body radiation with a temperature θ_B which fills the entire universe. An object with a total area A will thus receive an energy equal to $\sigma\theta_B^4 \times A$ per unit time. A black hole, therefore, loses energy through Hawking radiation and gains energy from the cosmic background radiation.

4.1	Find the rate of change of a black hole's mass, in terms of the mass of the black hole, the temperature of the cosmic background radiation, and the fundamental constants.	0.8

4.2	At a certain mass, m^*, this rate of change will vanish. Find m^* and express it in terms of θ_B and the fundamental constants.	0.4

4.3	Use your answer to 4.2 to substitute for θ_B in your answer to part 4.1 and express the rate of change of the mass of a black hole in terms of m, m^*, and the fundamental constants.	0.2

4.4	Find the Hawking temperature of a black hole at thermal equilibrium with cosmic background radiation.	0.4

4.5	Is the equilibrium stable or unstable? Why? (Express your answer mathematically)	0.6

1.1) One may use any reasonable equation to obtain the dimension of the questioned quantities.

I) The Planck relation is $h\nu = E \quad \Rightarrow \quad [h][\nu] = [E] \quad \Rightarrow \quad [h] = [E][\nu]^{-1} = ML^2T^{-1}$
(0.2)

II) $[c] = LT^{-1}$ (0.2)

III) $F = \dfrac{Gmm}{r^2} \quad \Rightarrow \quad [G] = [F][r^2][m]^{-2} = M^{-1}L^3T^{-2}$ (0.2)

IV) $E = K_B\theta \quad \Rightarrow \quad [K_B] = [\theta]^{-1}[E] = ML^2T^{-2}K^{-1}$ (0.2)

1.2) Using the Stefan-Boltzmann's law,

$\dfrac{Power}{Area} = \sigma\theta^4$, or any equivalent relation, one obtains:
(0.3)

$[\sigma]K^4 = [E]L^{-2}T^{-1} \Rightarrow [\sigma] = MT^{-3}K^{-4}$. (0.2)

1.3) The Stefan-Boltzmann's constant, up to a numerical coefficient, equals $\sigma = h^\alpha c^\beta G^\gamma {k_B}^\delta$, where $\alpha, \beta, \gamma, \delta$ can be determined by dimensional analysis. Indeed, $[\sigma] = [h]^\alpha [c]^\beta [G]^\gamma [k_B]^\delta$, where e.g. $[\sigma] = MT^{-3}K^{-4}$.

$$MT^{-3}K^{-4} = \left(ML^2T^{-1}\right)^\alpha \left(LT^{-1}\right)^\beta \left(M^{-1}L^3T^{-2}\right)^\gamma \left(ML^2T^{-2}K^{-1}\right)^\delta = M^{\alpha-\gamma+\delta}L^{2\alpha+\beta+3\gamma+2\delta}T^{-\alpha-\beta-2\gamma-2\delta}K^{-\delta},$$
(0.2)

The above equality is satisfied if,

$$\Rightarrow \begin{cases} \alpha-\gamma+\delta = 1, \\ 2\alpha+\beta+3\gamma+2\delta = 0, \\ -\alpha-\beta-2\gamma-2\delta = -3, \\ -\delta = -4, \end{cases} \text{(Each one (0.1))} \quad \Rightarrow \begin{cases} \alpha = -3, \\ \beta = -2, \\ \gamma = 0, \\ \delta = 4. \end{cases} \text{(Each one (0.1))}$$

$$\Rightarrow \quad \sigma = \frac{{k_B}^4}{c^2h^3}.$$

2.1) Since A, the area of the event horizon, is to be calculated in terms of m from a classical theory of relativistic gravity, e.g. the General Relativity, it is a combination of c, characteristic of special relativity, and G characteristic of gravity. Especially, it is independent of the Planck constant h which is characteristic of quantum mechanical phenomena.

$A = G^{\alpha} c^{\beta} m^{\gamma}$

Exploiting dimensional analysis,

$$\Rightarrow \ [A]=[G]^{\alpha}[c]^{\beta}[m]^{\gamma} \ \Rightarrow \ L^2 = \left(M^{-1}L^3T^{-2}\right)^{\alpha}\left(LT^{-1}\right)^{\beta} M^{\gamma} = M^{-\alpha+\gamma}L^{3\alpha+\beta}T^{-2\alpha-\beta} \quad (0.2)$$

The above equality is satisfied if,

$$\Rightarrow \begin{cases} -\alpha+\gamma=0, \\ 3\alpha+\beta=2, \\ -2\alpha-\beta=0, \end{cases} \text{(Each one (0.1))} \quad \Rightarrow \quad \begin{cases} \alpha=2, \\ \beta=-4, \\ \gamma=2, \end{cases} \text{(Each one (0.1))} \Rightarrow$$

$$A = \frac{m^2 G^2}{c^4}.$$

2.2)

From the definition of entropy $dS = \frac{dQ}{\theta}$, one obtains $[S]=[E][\theta]^{-1} = ML^2T^{-2}K^{-1}$ (0.2)

2.3) Noting $\eta = S/A$, one verifies that,

$$\begin{cases} [\eta]=[S][A]^{-1} = MT^{-2}K^{-1}, \\ [\eta]=[G]^{\alpha}[h]^{\beta}[c]^{\gamma}[k_B]^{\delta} = M^{-\alpha+\beta+\delta}L^{3\alpha+2\beta+\gamma+2\delta}T^{-2\alpha-\beta-\gamma-2\delta}K^{-\delta}, \end{cases} \quad (0.2)$$

Using the same scheme as above,

$$\Rightarrow \begin{cases} -\alpha+\beta+\delta=1, \\ 3\alpha+2\beta+\gamma+2\delta=0, \\ -2\alpha-\beta-\gamma-2\delta=-2, \\ \delta=1, \end{cases} \text{(Each one (0.1))} \quad \Rightarrow \begin{cases} \alpha=-1, \\ \beta=-1, \\ \gamma=3, \\ \delta=1, \end{cases} \text{(Each one (0.1))}$$

thus, $\eta = \frac{c^3 k_B}{G h}$. (0.1)

3.1)

The first law of thermodynamics is $dE = dQ + dW$. By assumption, $dW = 0$. Using the definition of entropy, $dS = \frac{dQ}{\theta}$, one obtains,

$$dE = \theta_H\, dS + 0, \qquad \text{(0.2) + (0.1), for setting } dW = 0.$$

Using, $\begin{cases} S = \dfrac{Gk_B}{ch} m^2, & \text{[(0.1) for } S\text{]} \\ E = mc^2, \end{cases}$

one obtains, $\theta_H = \dfrac{dE}{dS} = \left(\dfrac{dS}{dE}\right)^{-1} = c^2\left(\dfrac{dS}{dm}\right)^{-1}$ (0.2)

Therefore, $\theta_H = \left(\dfrac{1}{2}\right)\dfrac{c^3 h}{Gk_B}\dfrac{1}{m}$. (0.1)+(0.1) (for the coefficient)

3.2) The Stefan-Boltzmann's law gives the rate of energy radiation per unit area. Noting that $E = mc^2$ we have:

$$\begin{cases} dE/dt = -\sigma\theta_H{}^4 A, \quad (0.2) \\ \sigma = \dfrac{k_B{}^4}{c^2 h^3}, \\ A = \dfrac{m^2 G^2}{c^4} \\ E = mc^2, \end{cases} \Rightarrow c^2 \frac{dm}{dt} = -\frac{k_B{}^4}{c^2 h^3}\left(\frac{c^3 h}{2Gk_B}\frac{1}{m}\right)^4 \frac{m^2 G^2}{c^4}, \quad (0.2)$$

$$\Rightarrow \quad \frac{dm}{dt} = -\frac{1}{16}\frac{c^4 h}{G^2}\frac{1}{m^2}. \quad \text{(0.1) (for simplification) + (0.2) (for the minus sign)}$$

3.3)
By integration:

$$\frac{dm}{dt} = -\frac{1}{16}\frac{c^4 h}{G^2}\frac{1}{m^2}. \quad \Rightarrow \int m^2 dm = -\int \frac{c^4 h}{16G^2} dt \quad (0.3)$$

$$\Rightarrow m^3(t) - m^3(0) = -\frac{3c^4 h}{16G^2} t, \quad \text{(0.2) + (0.2) (Integration and correct boundary values)}$$

At $t = t^*$ the black hole evaporates completely:

$$m(t^*) = 0 \quad (0.1) \qquad \Rightarrow t^* = \frac{16G^2}{3c^4 h} m^3 \quad \text{(0.2)+(0.1) (for the coefficient)}$$

3.4) C_V measures the change in E with respect to variation of θ.

$$\begin{cases} C_V = \dfrac{dE}{d\theta}, & (0.2) \\ E = mc^2, & (0.2) \\ \theta = \dfrac{c^3 h}{2Gk_B}\dfrac{1}{m} \end{cases} \Rightarrow \quad C_V = -\frac{2Gk_B}{ch}m^2.$$ 0.1)+(0.1) (for the coefficient)

4.1) Again the Stefan-Boltzmann's law gives the rate of energy loss per unit area of the black hole. A similar relation can be used to obtain the energy gained by the black hole due to the background radiation. To justify it, note that in the thermal equilibrium, the total change in the energy is vanishing. The blackbody radiation is given by the Stefan-Boltzmann's law. Therefore the rate of energy gain is given by the same formula.

(0.1) + (0.4) (For the first and the second terms respectively)

$$\begin{cases} \dfrac{dE}{dt} = -\sigma\theta^4 A + \sigma\theta_B{}^4 A \\ E = mc^2, \end{cases} \Rightarrow \quad \frac{dm}{dt} = -\frac{hc^4}{16G^2}\frac{1}{m^2} + \frac{G^2}{c^8 h^3}(k_B\,\theta_B)^4 m^2 \qquad (0.3)$$

4.2)

Setting $\dfrac{dm}{dt} = 0$, we have:

$$-\frac{hc^4}{16G^2}\frac{1}{m^{*2}} + \frac{G^2}{c^8 h^3}(k_B\,\theta_B)^4 m^{*2} = 0 \qquad (0.2)$$

and consequently,

$$m^* = \frac{c^3 h}{2G\,k_B}\frac{1}{\theta_B} \qquad (0.2)$$

4.3)

$$\theta_B = \frac{c^3 h}{2G\,k_B}\frac{1}{m^*} \quad \Rightarrow \quad \frac{dm}{dt} = -\frac{hc^4}{16G^2}\frac{1}{m^2}\left(1 - \frac{m^4}{m^{*4}}\right) \qquad (0.2)$$

4.4) Use the solution to 4.2,

$m^* = \dfrac{c^3 h}{2G\,k_B}\dfrac{1}{\theta_B}$ (0.2) and 3.1 to obtain, $\theta^* = \dfrac{c^3 h}{2Gk_B}\dfrac{1}{m^*} = \theta_B$ (0.2)

One may also argue that m^* corresponds to thermal equilibrium. Thus for $m = m^*$ the black hole temperature equals θ_B.

Or one may set $\dfrac{dE}{dt} = -\sigma\left(\theta^{*4} - \theta_B{}^4\right)A = 0$ to get $\theta^* = \theta_B$.

4.5) Considering the solution to 4.3, one verifies that it will go away from the equilibrium. (0.6)

$$\frac{dm}{dt}=-\frac{hc^4}{G^2}\frac{1}{m^2}\left(1-\frac{m^4}{m^{*4}}\right)\Rightarrow\begin{cases}m>m^* & \Rightarrow & \dfrac{dm}{dt}>0\\ m<m^* & \Rightarrow & \dfrac{dm}{dt}<0\end{cases}$$

Two stars rotating around their center of mass form a binary star system. Almost half of the stars in our galaxy are binary star systems. It is not easy to realize the binary nature of most of these star systems from Earth, since the distance between the two stars is much less than their distance from us and thus the stars cannot be resolved with telescopes. Therefore, we have to use either photometry or spectrometry to observe the variations in the intensity or the spectrum of a particular star to find out whether it is a binary system or not.

Photometry of Binary Stars

If we are exactly on the plane of motion of the two stars, then one star will occult (pass in front of) the other star at certain times and the intensity of the whole system will vary with time from our observation point. These binary systems are called ecliptic binaries.

1 Assume that two stars are moving on circular orbits around their common center of mass with a constant angular speed ω and we are exactly on the plane of motion of the binary system. Also assume that the surface temperatures of the stars are T_1 and T_2 $(T_1 > T_2)$, and the corresponding radii are R_1 and R_2 $(R_1 > R_2)$, respectively. The total intensity of light, measured on Earth, is plotted in Figure 1 as a function of time. Careful measurements indicate that the intensities of the incident light from the stars corresponding to the minima are respectively 90 and 63 percent of the maximum intensity, I_0, received from both stars $(I_0 = 4.8\times10^{-9}\ \text{W/m}^2)$. The vertical axis in Figure 1 shows the ratio I/I_0 and the horizontal axis is marked in days.

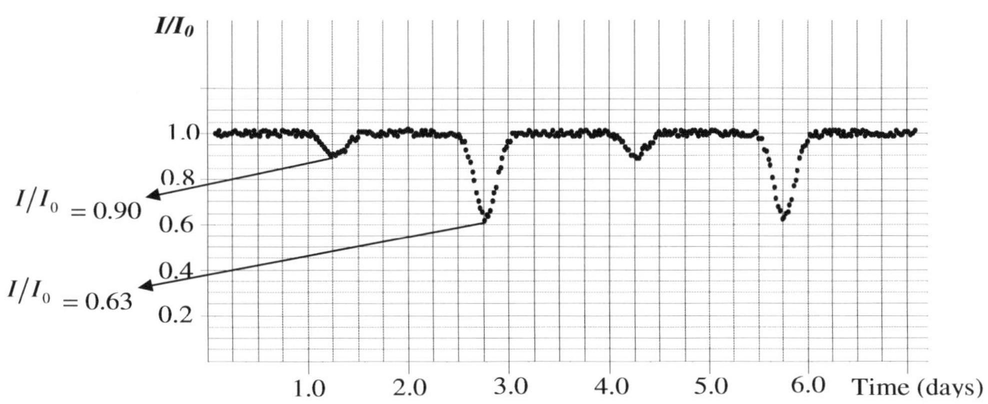

Figure 1. The relative intensity received from the binary star system as a function of time. The vertical axis has been scaled by $I_0 = 4.8\times10^{-9}\ \text{W/m}^2$. Time is given in days.

1.1	Find the period of the orbital motion. Give your answer in **seconds** up to two significant digits. What is the angular frequency of the system in **rad/sec**?	0.8

To a good approximation, the receiving radiation from a star is a uniform black body radiation from a flat disc with a radius equal to the radius of the star. Therefore, the power received from the star is proportional to AT^4 where A is area of the disc and T is the surface temperature of the star.

1.2	Use the diagram in Figure 1 to find the ratios T_1/T_2 and R_1/R_2.	1.6

Spectrometry of Binary Systems

In this section, we are going to calculate the astronomical properties of a binary star by using experimental spectrometric data of the binary system.

Atoms absorb or emit radiation at their certain characteristic wavelengths. Consequently, the observed spectrum of a star contains *absorption lines* due to the atoms in the star's atmosphere. Sodium has a characteristic yellow line spectrum (D_1 line) with a wavelength 5895.9Å (10 Å = 1 nm). We examine the absorption spectrum of atomic Sodium at this wavelength for the binary system of the previous section. The spectrum of the light that we receive from the binary star is Doppler-shifted, because the stars are moving with respect to us. Each star has a different speed. Accordingly the absorption wavelength for each star will be shifted by a different amount. Highly accurate wavelength measurements are required to observe the Doppler shift since the speed of the stars is much less than the speed of light. The speed of the center of mass of the binary system we consider in this problem is much smaller than the orbital velocities of the stars. Hence all the Doppler shifts can be attributed to the orbital velocity of the stars. Table 1 shows the measured spectrum of the stars in the binary system we have observed.

Table 1: Absorption spectrum of the binary star system for the Sodium D_1 line

t/days	0.3	0.6	0.9	1.2	1.5	1.8	2.1	2.4
λ_1 (Å)	5897.5	5897.7	5897.2	5896.2	5895.1	5894.3	5894.1	5894.6
λ_2 (Å)	5893.1	5892.8	5893.7	5896.2	5897.3	5898.7	5899.0	5898.1

t/days	2.7	3.0	3.3	3.6	3.9	4.2	4.5	4.8
λ_1 (Å)	5895.6	5896.7	5897.3	5897.7	5897.2	5896.2	5895.0	5894.3
λ_2 (Å)	5896.4	5894.5	5893.1	5892.8	5893.7	5896.2	5897.4	5898.7

(*Note*: There is no need to make a graph of the data in this table)

2 Using Table 1,

2.1	Let v_1 and v_2 be the orbital velocity of each star. Find v_1 and v_2. The speed of light $c = 3.0\times10^8$ m/s. Ignore all relativistic effects.	1.8

2.2	Find the mass ratio of the stars (m_1/m_2).	0.7

2.3	Let r_1 and r_2 be the distances of each star from their center of mass. Find r_1 and r_2.	0.8

2.4	Let r be the distance between the stars. Find r.	0.2

3 The gravitational force is the only force acting between the stars.

3.1	Find the mass of each star up to one significant digit. The universal gravitational constant $G = 6.7\times10^{-11}\ \mathrm{m^3kg^{-1}s^{-2}}$.	1.2

General Characteristics of Stars

4 Most of the stars generate energy through the same mechanism. Because of this, there is an empirical relation between their mass, M, and their luminosity, L, which is the total radiant power of the star. This relation could be written in the form $L/L_{Sun} = (M/M_{Sun})^{\alpha}$. Here, $M_{Sun} = 2.0\times10^{30}$ kg is the solar mass and, $L_{Sun} = 3.9\times10^{26}$ W is the solar luminosity. This relation is shown in a log-log diagram in Figure 2.

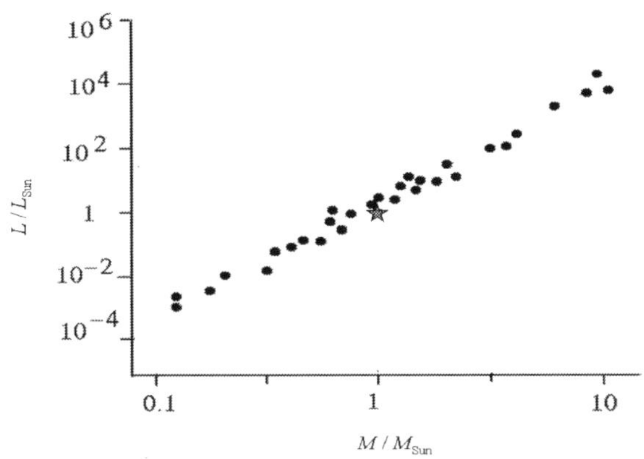

Figure 2. The luminosity of a star versus its mass varies as a power law. The diagram is log-log. The star-symbol represents Sun with a mass of $2.0\times10^{30}\ kg$ and luminosity of 3.9×10^{26} W.

4.1	Find α up to one significant digit.	0.6

4.2	Let L_1 and L_2 be the luminosity of the stars in the binary system studied in the previous sections. Find L_1 and L_2.	0.6

4.3	What is the distance, d, of the star system from us in light years? To find the distance you can use the diagram of Figure 1. One light year is the distance light travels in one year.	0.9

4.4	What is the maximum angular distance, θ, between the stars from our observation point?	0.4

4.5	What is the smallest aperture size for an optical telescope, D, that can resolve these two stars?	0.4

Question "Pink"

1.1

Period = 3.0 days = 2.6×10^5 s. (0.4)

Period = $\dfrac{2\pi}{\omega}$ (0.2) $\Rightarrow$ $\omega = 2.4\times10^{-5}$ rad s^{-1}. (0.2)

1.2

Calling the minima in the diagram 1, $I_1/I_0 = \alpha = 0.90$ and $I_2/I_0 = \beta = 0.63$, we have:

$$\frac{I_0}{I_1} = 1 + \left(\frac{R_2}{R_1}\right)^2\left(\frac{T_2}{T_1}\right)^4 = \frac{1}{\alpha} \quad (0.4)$$

$$\frac{I_2}{I_1} = 1 - \left(\frac{R_2}{R_1}\right)^2\left(1 - \left(\frac{T_2}{T_1}\right)^4\right) = \frac{\beta}{\alpha} \quad (0.4) \text{ (or equivalent relations)}$$

From above, one finds:

$$\frac{R_1}{R_2} = \sqrt{\frac{\alpha}{1-\beta}} \Rightarrow \frac{R_1}{R_2} = 1.6 \quad (0.2+0.2) \text{ and } \frac{T_1}{T_2} = \sqrt[4]{\frac{1-\beta}{1-\alpha}} \Rightarrow \frac{T_1}{T_2} = 1.4 \ (0.2+0.2)$$

2.1)

Doppler-Shift formula:

$\dfrac{\Delta\lambda}{\lambda_0} \cong \dfrac{v}{c}$ (or equivalent relation) (0.4)

Maximum and minimum wavelengths: $\lambda_{1,\max} = 5897.7$ Å, $\lambda_{1,\min} = 5894.1$ Å

$\lambda_{2,\max} = 5899.0$ Å, $\lambda_{2,\min} = 5892.8$ Å

Difference between maximum and minimum wavelengths:

$\Delta\lambda_1 = 3.6$ Å , $\Delta\lambda_2 = 6.2$ Å (All 0.6)

Using the Doppler relation and noting that the shift is due to twice the orbital speed: (Factor of two 0.4)

$$v_1 = c\frac{\Delta\lambda_1}{2\lambda_0} = 9.2\times10^4 \text{ m/s} \quad (0.2)$$

$$v_2 = c\frac{\Delta\lambda_2}{2\lambda_0} = 1.6\times10^5 \text{ m/s} \quad (0.2)$$

The student can use the wavelength of central line and maximum (or minimum) wavelengths. Marking scheme is given in the Excel file.

2.2) As the center of mass is not moving with respect to us: (0.5)

$$\frac{m_1}{m_2} = \frac{v_2}{v_1} = 1.7 \quad (0.2)$$

2.3)

Writing $r_i = \frac{v_i}{\omega}$ for $i = 1,2$, we have (0.4)

$r_1 = 3.8 \times 10^9$ m, (0.2) $r_2 = 6.5 \times 10^9$ m (0.2)

2.4)

$r = r_1 + r_2 = 1.0 \times 10^{10}$ m (0.2)

3.1)

The gravitational force is equal to mass times the centrifugal acceleration

$$G\frac{m_1 m_2}{r^2} = m_1 \frac{v_1^2}{r_1} = m_2 \frac{v_2^2}{r_2} \quad (0.7)$$

Therefore,

$$\begin{cases} m_1 = \dfrac{r^2 v_2^2}{G r_2} \\ m_2 = \dfrac{r^2 v_1^2}{G r_1} \end{cases} \quad (0.1) \quad \Rightarrow \quad \begin{cases} m_1 = 6 \times 10^{30}\,\text{kg} \\ m_2 = 3 \times 10^{30}\,\text{kg} \end{cases} \quad (0.2 + 0.2)$$

4.1) As it is clear from the diagram, with one significant digit, $\alpha = 4$. (0.6)

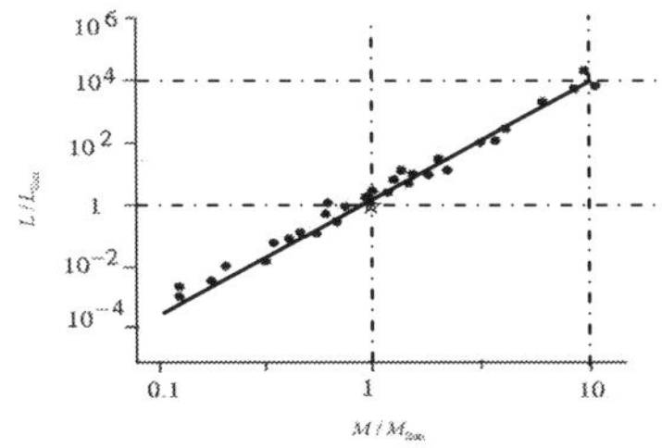

4.2)

As we have found in the previous section: $L_i = L_{Sun}\left(\frac{M_i}{M_{Sun}}\right)^4$ (0.2)

So,

$$L_1 = 3\times10^{28}\,\text{Watt (0.2)}$$
$$L_2 = 4\times10^{27}\,\text{Watt (0.2)}$$

4.3) The total power of the system is distributed on a sphere with radius d to produce I_0, that is:

$$I_0 = \frac{L_1 + L_2}{4\pi d^2} \quad (0.5) \qquad \Rightarrow d = \sqrt{\frac{L_1 + L_2}{4\pi I_0}} = 1\times10^{18}\,\text{m} \quad (0.2)$$
$$= 100\ \text{ly.}\ (0.2)$$

4.4) $$\theta \cong \tan\theta = \frac{r}{d} = 1\times10^{-8}\,\text{rad.} \quad (0.2 + 0.2)$$

4.5)
A typical optical wavelength is λ_0. Using uncertainty relation:

$$D = \frac{d\,\lambda_0}{r} \cong 50\ \text{m.} \quad (0.2 + 0.2)$$

Experimental Problem

Determination of energy band gap of semiconductor thin films

I. Introduction

Semiconductors can be roughly characterized as materials whose electronic properties fall somewhere between those of conductors and insulators. To understand semiconductor electronic properties, one can start with the *photoelectric effect* as a well-known phenomenon. The photoelectric effect is a quantum electronic phenomenon, in which photoelectrons are emitted from the matter through the absorption of sufficient energy from electromagnetic radiation (i.e. photons). The minimum energy which is required for the emission of an electron from a metal by light irradiation (*photoelectron*) is defined as "*work function*". Thus, only photons with a frequency ν higher than a characteristic threshold, i.e. with an energy $h\nu$ (h is the Planck's constant) more than the material's work function, are able to knock out the photoelectrons.

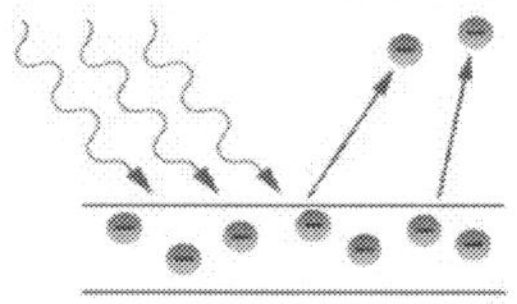

Figure 1. An illustration of photoelectron emission from a metal plate: The incoming photon should have an energy which is more than the work function of the material.

In fact, the concept of work function in the photoelectric process is similar to the concept of the energy band gap of a semiconducting material. In solid state physics, the band gap E_g is the energy difference between the top of the valence band and the bottom of the conduction band of insulators and semiconductors. The valence band is completely filled with electrons, while the conduction band is empty however electrons can go from the valence band to the conduction band if they acquire sufficient energy (at least equal to the band gap energy).The semiconductor's conductivity strongly depends on its energy band gap.

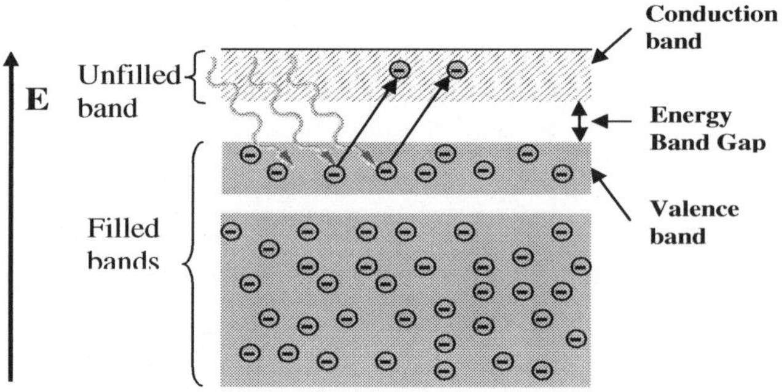

Figure 2. Energy band scheme for a semiconductor.

Band gap engineering is the process of controlling or altering the band gap of a material by controlling the composition of certain semiconductor alloys. Recently, it has been shown that by changing the nanostructure of a semiconductor it is possible to manipulate its band gap.

In this experiment, we are going to obtain the energy band gap of a thin-film semiconductor containing nano-particle chains of iron oxide (Fe_2O_3) by using an optical method. To measure the band gap, we study the optical absorption properties of the transparent film using its optical transmission spectrum. As a rough statement, the absorption spectra shows a sharp increase when the energy of the incident photons equals to the energy band gap.

II. Experimental Setup

You will find the following items on your desk:

1. A large white box containing a spectrometer with a halogen lamp.
2. A small box containing a sample, a glass substrate, a sample-holder, a grating, and a photoresistor.
3. A multimeter.
4. A calculator.
5. A ruler.
6. A card with a hole punched in its center.
7. A set of blank labels.

The spectrometer contains a goniometer with a precision of $5'$. The Halogen lamp acts as the source of radiation and is installed onto the fixed arm of the spectrometer (for detailed information see the enclosed "Description of Apparatus").

The small box contains the following items:

1. A sample-holder with two windows: a glass substrate coated with Fe_2O_3 film mounted on one window and an uncoated glass substrate mounted on the other.
2. A photoresistor mounted on its holder, which acts as a light detector.
3. A transparent diffraction grating (600 line/mm).

Note: *Avoid touching the surface of any component in the small box!*

A schematic diagram of the setup is shown in Figure 3:

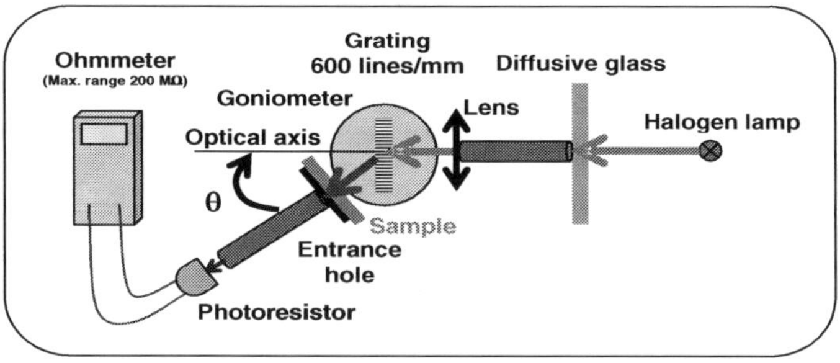

Figure 3. Schematic diagram of the experimental setup.

III. Methods

To obtain the transmission of a film at each wavelength, $T_{film}(\lambda)$, one can use the following formula:

$$T_{film}(\lambda) = I_{film}(\lambda)/I_{glass}(\lambda) \qquad (1)$$

where I_{film} and I_{glass} are respectively the intensity of the light transmitted from the coated glass substrate, and the intensity of the light transmitted from the uncoated glass slide. The value of I can be measured using a light detector such as a photoresistor. In a photoresistor, the electrical resistance decreases when the intensity of the incident light increases. Here, the value of I can be determined from the following relation:

$$I(\lambda) = C(\lambda)R^{-1} \qquad \textbf{(2)}$$

where R is the electrical resistance of the photoresistor, C is a λ-dependent coefficient.

The transparent grating on the spectrometer diffracts different wavelengths of light into different angles. Therefore, to study the variations of T as a function of λ, it is enough to change the angle of the photoresistor (θ') with respect to the optical axis (defined as the direction of the incident light beam on the grating), as shown in Figure 4.

From the principal equation of a diffraction grating:

$$n\lambda = d[\sin(\theta' - \theta_0) + \sin\theta_0] \qquad (3)$$

one can obtain the angle θ' corresponding to a particular λ: n is an integer number representing the order of diffraction, d is the period of the grating, and θ_o is the angle the normal vector to the surface of grating makes with the optical axis (see Fig. 4). (In this experiment we shall try to place the grating perpendicular to the optical axis making $\theta_o = 0$, but since this cannot be achieved with perfect precision the error associated with this adjustment will be measured in task 1-e.)

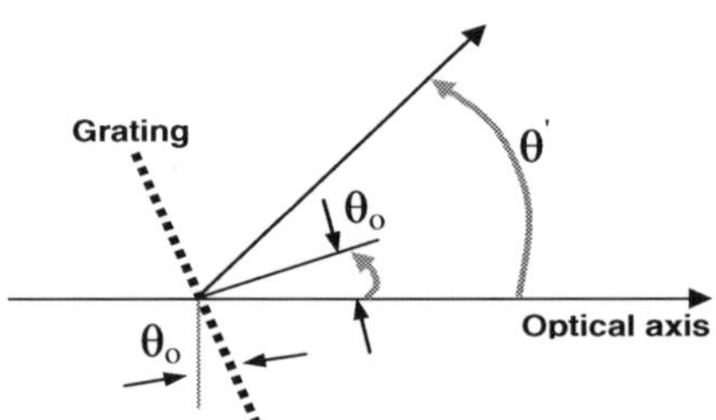

Figure 4. Definition of the angles involved in Equation 3.

Experimentally it has been shown that for photon energies slightly larger than the band gap energy, the following relation holds:

$$\alpha h\nu = A(h\nu - E_g)^{\eta} \qquad (4)$$

where α is the absorption coefficient of the film, A is a constant that depends on the film's material, and η is the constant determined by the absorption mechanism of the film's material and structure. Transmission is related to the value of α through the well-known absorption relation:

$$T_{film} = \exp(-\alpha t) \qquad (5)$$

where t is thickness of the film.

IV. Tasks:

0. *Your apparatus and sample box (small box containing the sample holder) are marked with numbers. Write down the* ***Apparatus number*** *and* ***Sample number*** *in their appropriate boxes, in the answer sheet.*

1. Adjustments and Measurements:

1-a	Check the vernier scale and report the maximum precision ($\Delta\theta$).	*0.1 pt*

Note: Magnifying glasses are available on request.

Step1:
To start the experiment, turn on the Halogen lamp to warm up. It would be better not to turn off the lamp during the experiment. Since the halogen lamp heats up during the experiment, please be careful not to touch it.

Place the lamp as far from the lens as possible, this will give you a parallel light beam.

We are going to make a rough zero-adjustment of the goniometer without utilizing the photoresistor. Unlock the rotatable arm with screw 18 (underneath the arm), and visually align the rotatable arm with the optical axis. Now, firmly lock the rotatable arm with screw 18. Unlock the vernier with screw 9 and rotate the stage to 0 on the vernier scale. Now firmly lock the vernier with screw 9 and use the vernier fine-adjustment screw (screw 10) to set the zero of the vernier scale. Place the grating inside its holder. Rotate the grating's stage until the diffraction grating is roughly perpendicular to the optical axis. Place the card with a hole in front of the light source and position the hole such that a beam of light is incident on the grating. Carefully rotate the grating so that the spot of reflected light falls onto the hole. Then the reflected light beam coincides with the incident beam. Now lock the grating's stage by tightening screw 12.

1-b	By measuring the distance between the hole and the grating, estimate the precision of this adjustment ($\Delta\theta_o$).	*0.3 pt*
	Now, by rotating the rotatable arm, determine and report the range of angles for which the first-order diffraction of visible light (from blue to red) is observed.	*0.2 pt*

Step 2:
Now, install the photoresistor at the end of the rotatable arm. To align the system optically, by using the photoresistor, loosen the screw 18, and slightly turn the rotatable arm so that the photoresistor shows a minimum resistance. For fine positioning, firmly lock screw 18, and use the fine adjustment screw of the rotatable arm.

Use the vernier fine-adjustment screw to set the zero of the vernier scale.

<table>
<tr><td rowspan="2">1-c</td><td>Report the measured minimum resistance value ($R_{\min}^{(0)}$).</td><td>0.1 pt</td></tr>
<tr><td>Your zero-adjustment is more accurate now, report the precision of this new adjustment ($\Delta\varphi_o$) .
Note: $\Delta\varphi_o$ is the error in this alignment i.e. it is a measure of misalignment of the rotatable arm and the optical axis.</td><td>0.1 pt</td></tr>
</table>

➢ **Hint:** After this task you should tighten the fixing screws of the vernier. Moreover, tighten the screw of the photoresistor holder to fix it and do not remove it during the experiment.

Step 3:
Move the rotatable arm to the region of the first-order diffraction. Find the angle at which the resistance of the photoresistor is minimum (maximum light intensity). Using the balancing screws, you can slightly change the *tilt* of the grating's stage, to achieve an even lower resistance value.

1-c	Report the minimum value of the observed resistance ($R_{\min}^{(1)}$) in its appropriate box.	*0.1 pt*

It is now necessary to check the perpendicularity of the grating for zero adjustment, again. For this you must use the reflection-coincidence method of Step 1.

Important: From here onwards carry out the experiment in dark (close the cover).

Measurements: Screw the sample-holder onto the rotatable arm. Before you start the measurements, examine the appearance of your semiconductor film (sample). Place the sample in front of the entrance hole S_1 on the rotatable arm such that a uniformly coated part of the sample covers the hole. To make sure that every time you will be working with the same part of the sample make proper markings on the sample holder and the rotatable arm with blank labels.

Attention: At higher resistance measurements it is necessary to allow the photoresistor to relax, therefore for each measurement in this range wait 3 to 4 minutes before recording your measurement.

<table>
<tr><td rowspan="2">1-d</td><td>Measure the resistance of the photoresistor for the uncoated glass substrate and the glass substrate coated with semiconductor layer as a function of the angle θ (the value read by the goniometer for the angle between the photoresistor and your specified optical axis). Then fill in Table 1d. Note that you need at least 20 data points in the range you found in Step 1b. Carry out your measurement using the appropriate range of your ohmmeter.</td><td>2.0 pt</td></tr>
<tr><td>Consider the error associated with each data point. Base your</td><td>1.0 pt</td></tr>
</table>

	answer only on your direct readings of the ohmmeter.	

Step 4:
The precision obtained so far is still limited since it is impossible to align the rotatable arm with the optical axis and/or position the grating perpendicular to the optical axis with 100% precision. So we still need to find the asymmetry of the measured transmission at both sides of the optical axis (resulting from the deviation of the normal to the grating surface from the optical axis (θ_o)).

To measure this asymmetry, follow these steps:

1-e	First, measure T_{film} at $\theta = -20°$. Then, obtain values for T_{film} at some other angles around $+20°$. Complete Table 1e (you can use the values obtained in Table 1d).	*0.6 pt*
	Draw T_{film} versus θ and visually draw a curve.	*0.6 pt*

On your curve find the angle γ for which the value of T_{film} is equal to the T_{film} that you measured at $\theta = -20^o$ ($\gamma \equiv \theta|_{T_{film}=T_{film}(-20°)}$). Denote the difference of this angle with $+20°$ as δ, in other words:

$$\delta = \gamma - 20° \qquad (6)$$

1-e	Report the value of δ in the specified box.	*0.2 pt*

Then for the first-order diffraction, Eq. (3) can be simplified as follows:

$$\lambda = d\sin(\theta - \delta/2), \quad (7)$$

where θ is the angle read on the goniometer.

2. Calculations:

2-a	Use Eq. (7) to express $\Delta\lambda$ in terms of the errors of the other parameters (assume d is exact and there is no error is associated with it). Also using Eqs. (1), (2), and (5), express ΔT_{film} in terms of R and ΔR.	*0.6 pt*

2-b	Report the range of values of $\Delta\lambda$ over the region of first-order diffraction.	*0.3 pt*

2-c	Based on the measured parameters in Task 1, complete Table 2c for each θ. Note that the wavelength should be calculated using Eq. (7).	*2.4 pt*

2-d	Plot R_{glass}^{-1} and R_{film}^{-1} as a function of wavelength together on the same diagram. Note that on the basis of Eq. (2) behaviors of R_{glass}^{-1} and R_{film}^{-1} can reasonably give us an indication of the way I_{glass} and I_{film} behave, respectively.	*1.5 pt*
	In Table 2d, report the wavelengths at which R_{glass} and R_{film} attain their minimum values.	*0.4 pt*

2-e	For the semiconductor layer (sample) plot T_{film} as a function of wavelength. This quantity also represents the variation of the film transmission in terms of wavelength.	*1.0 pt*

3. **Data analysis:**

By substituting $\eta = 1/2$ and A=0.071 ((eV)$^{1/2}$/nm) in Eq. (4) one can find values for E_g and t in units of eV and nm, respectively. This will be accomplished by plotting a suitable diagram in an $x-y$ coordinate system and doing an extrapolation in the region satisfying this equation.

3-a	By assuming $x = h\nu$ and $y = (\alpha t h\nu)^2$ and by using your measurements in Task 1, fill in Table 3a for wavelengths around 530 nm and higher. Express your results (x and y) with the correct number of significant figures (digits), based on the estimation of the error on one single data point. Note that $h\nu$ should be calculated in units of eV and wavelength in units of nm. Write the unit of each variable between the parentheses in the top row of the table.	*2.4 pt*

3-b	Plot y versus x. Note that the y parameter corresponds to the absorption of the film. Fit a line to the points in the linear region around 530 nm. Specify the region where Eq. (4) is satisfied, by reporting the values of the smallest and the largest x-coordinates for the data points to which you fit the line.	*2.6 pt*

3-c	Call the slope of this line m, and find an expression for the film thickness (t) and its error (Δt) in terms of m and A (consider A to have no error).	*0.5 pt*

3-d	Obtain the values of E_g and t and their associated errors in units of eV and nm, respectively. Fill in Table 3d.	*3.0 pt*

❖ Some useful physical constants required for your analysis:

- Speed of the light: $c = 3.00 \times 10^8$ m/s
- Plank's constant: $h = 6.63 \times 10^{-34}$ J.s
- Electron charge: $e = 1.60 \times 10^{-19}$ C

Solution (The Experimental Question):

Task 1

1a.

$\Delta\theta_{\text{nominal}}$=5′=0.08°

$\Delta\theta_{\text{nominal}}$ (degree)	0.08

1b.

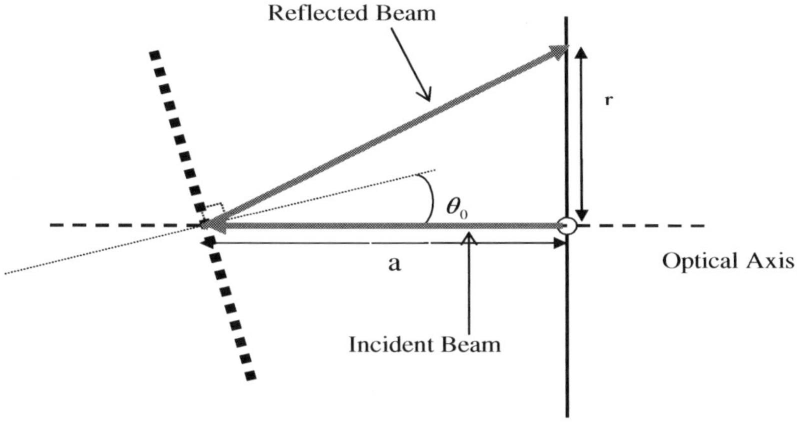

If "a" is the distance between card and the grating and "r" is the distance between the hole and the light spot so we have

$$\Delta f(x_1, x_2, \ldots) = \sqrt{\left(\frac{\partial f}{\partial x_1}\Delta x_1\right)^2 + \left(\frac{\partial f}{\partial x_2}\Delta x_2\right)^2 + \ldots}$$

$$\tan(2\theta_0) = \frac{r}{a},\ If\ \theta_0 << 1 \Rightarrow \theta_0 = \frac{r}{2a} \Rightarrow \Delta\theta_0 = \sqrt{\left(\frac{\Delta r}{2a}\right)^2 + \left(\frac{r\,\Delta \mathrm{a}}{2a^2}\right)^2}$$

We want θ_0 to be zero i.e. $r = 0 \Rightarrow \Delta\theta_0 = \dfrac{\Delta r}{2a}$

$$\Delta r = 1\,mm,\ a = (70 \pm 1)mm \Rightarrow \theta_0 = \frac{\Delta r}{2a}\ rad = 0.007\ rad = 0.4^\circ$$

$\Delta\theta_0$	0.4°
θ range of visible light (degree)	13°≤ θ ≤26°

1c.

$R_{\min}^{(0)}$	(21.6±0.1) kΩ
$\Delta\varphi_0$	5´ = 0.08°
$R_{\min}^{(1)}$	R=(192±1) kΩ

$\Delta\varphi_0$=5´ because

θ= 5´ => R= (21.9±0.1) kΩ

θ=-5´ => R= (21.9±0.1) kΩ

1d.

Table 1d. The measured parameters

θ (degree)	R_{glass}(MΩ)	ΔR_{glass}(MΩ)	R_{film}(MΩ)	ΔR_{film}(MΩ)
15.00	3.77	0.03	183	3
15.50	2.58	0.02	132	2
16.00	1.88	0.01	87	1
16.50	1.19	0.01	51.5	0.5
17.00	0.89	0.01	33.4	0.3
17.50	0.68	0.01	19.4	0.1
18.00	0.486	0.005	10.4	0.1
18.50	0.365	0.005	5.40	0.03
19.00	0.274	0.003	2.66	0.02
19.50	0.225	0.002	1.42	0.01
20.00	0.200	0.002	0.880	0.005
20.50	0.227	0.002	0.822	0.005
21.00	0.368	0.003	1.123	0.007
21.50	0.600	0.005	1.61	0.01
22.00	0.775	0.005	1.85	0.01
22.50	0.83	0.01	1.87	0.01
23.00	0.88	0.01	1.93	0.02
23.50	1.01	0.01	2.14	0.02
24.00	1.21	0.01	2.58	0.02
24.50	1.54	0.01	3.27	0.02
25.00	1.91	0.01	4.13	0.02
16.25	1.38	0.01	66.5	0.5
16.75	1.00	0.01	40.0	0.3
17.25	0.72	0.01	23.4	0.2
17.75	0.535	0.005	12.8	0.1
18.25	0.391	0.003	6.83	0.05
18.75	0.293	0.003	3.46	0.02
19.25	0.235	0.003	1.76	0.01
19.75	0.195	0.002	0.988	0.005
20.25	0.201	0.002	0.776	0.005
20.75	0.273	0.003	0.89	0.01

1e.

In θ=-20° => R_{glass}= (132± 2) kΩ , R_{film}= (518±5) kΩ

θ	T_{film}	θ	T_{film}
θ = -20°	0.255	19.25	0.134
		19.50	0.158
		19.75	0.197
		20.00	0.227
		20.25	0.259
		20.50	0.276
		20.75	0.307

Graphics

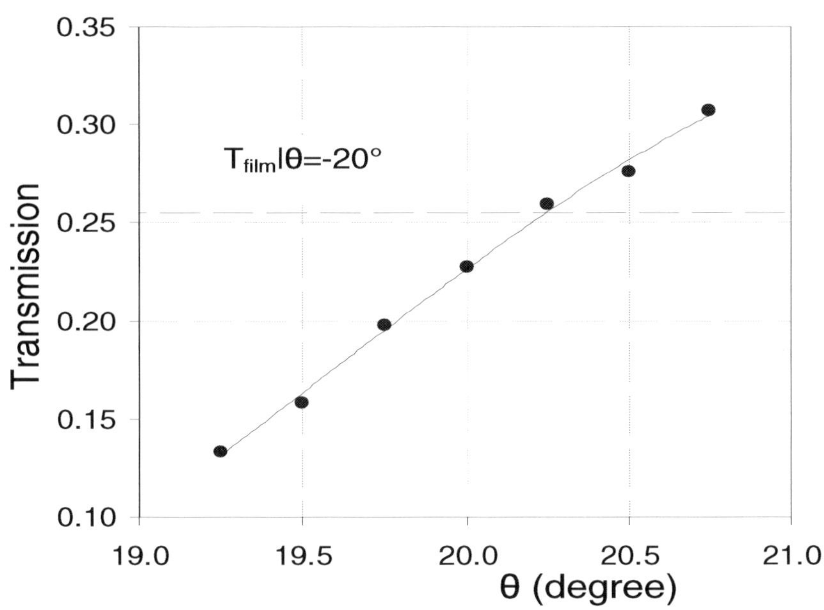

We see that: T(θ= 20.25°) = T(θ= -20°)

δ (degree)	0.25±0.08

Task 2.

2a.

$$\lambda = d\sin\left(\theta - \frac{\delta}{2}\right) \Rightarrow \Delta\lambda = \lambda\sqrt{\left(\frac{\Delta d}{d}\right)^2 + \cot^2\left(\theta - \frac{\delta}{2}\right)\left(\Delta\theta^2 + \frac{\Delta\delta^2}{4}\right)} \approx d\ \cos(\theta)\left(\frac{0.1\pi}{180}\right)$$

where $\Delta\theta = \Delta\delta = 5' = 0.08$ degree

and $d = \frac{1}{600}$ mm

$$\boxed{\Delta\lambda = 2.9\cos(\theta)\ (\text{nm})}$$

$$T_{film} = \frac{R_{glass}}{R_{film}} \Rightarrow \Delta T = T_{film}\sqrt{\left(\frac{\Delta R_{film}}{R_{film}}\right)^2 + \left(\frac{\Delta R_{glass}}{R_{glass}}\right)^2}$$

$$\boxed{\Delta T = \frac{R_{glass}}{R_{film}}\sqrt{\left(\frac{\Delta R_{film}}{R_{film}}\right)^2 + \left(\frac{\Delta R_{glass}}{R_{glass}}\right)^2}}$$

2b.

$$13 \le \theta \le 26$$

$$\boxed{2.6 \le \Delta\lambda \le 2.8\ \text{nm}}$$

2c.

Table 2c. The calculated parameters using the measured parameters

θ (degree)	λ (nm)	$I_g/C(\lambda)$ ($M\Omega^{-1}$)	$I_s/C(\lambda)$ ($M\Omega^{-1}$)	T_{film}	αt
15.0	428	0.265	0.00546	0.0206	3.88
15.5	442	0.388	0.00758	0.0195	3.94
16.0	456	0.532	0.0115	0.0216	3.83
16.25	463	0.725	0.0150	0.0208	3.88
16.5	470	0.840	0.0194	0.0231	3.77
16.75	477	1.00	0.0250	0.0250	3.69
17.0	484	1.12	0.0299	0.0266	3.63
17.25	491	1.39	0.0427	0.0308	3.48
17.5	498	1.47	0.0515	0.0351	3.35
17.75	505	1.87	0.0781	0.0418	3.17
18.0	512	2.06	0.096	0.0467	3.06
18.25	518	2.56	0.146	0.0572	2.86
18.5	525	2.74	0.185	0.0676	2.69
18.75	532	3.41	0.289	0.0847	2.47
19.0	539	3.65	0.376	0.103	2.27
19.25	546	4.26	0.568	0.134	2.01
19.5	553	4.44	0.704	0.158	1.84
19.75	560	5.13	1.01	0.197	1.62
20.0	567	5.00	1.14	0.227	1.48
20.25	573	4.98	1.29	0.259	1.35
20.5	580	4.41	1.22	0.276	1.29
20.75	587	3.66	1.12	0.307	1.18
21.0	594	2.72	0.890	0.328	1.12
21.5	607	1.67	0.621	0.373	0.99
22.0	621	1.29	0.541	0.419	0.87
22.5	634	1.20	0.535	0.444	0.81
23.0	648	1.14	0.518	0.456	0.79
23.5	661	0.99	0.467	0.472	0.75
24.0	675	0.826	0.388	0.469	0.76
24.5	688	0.649	0.306	0.471	0.75
25.0	701	0.524	0.242	0.462	0.77

2d.

Graphics

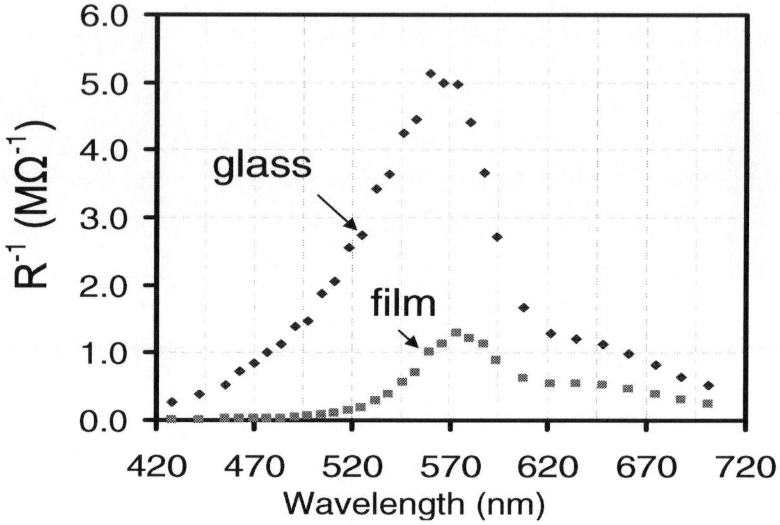

$\lambda_{max}(I_{glass})$	564±5 (nm)
$\lambda_{max}(I_{film})$	573±5 (nm)

2e. Graphics

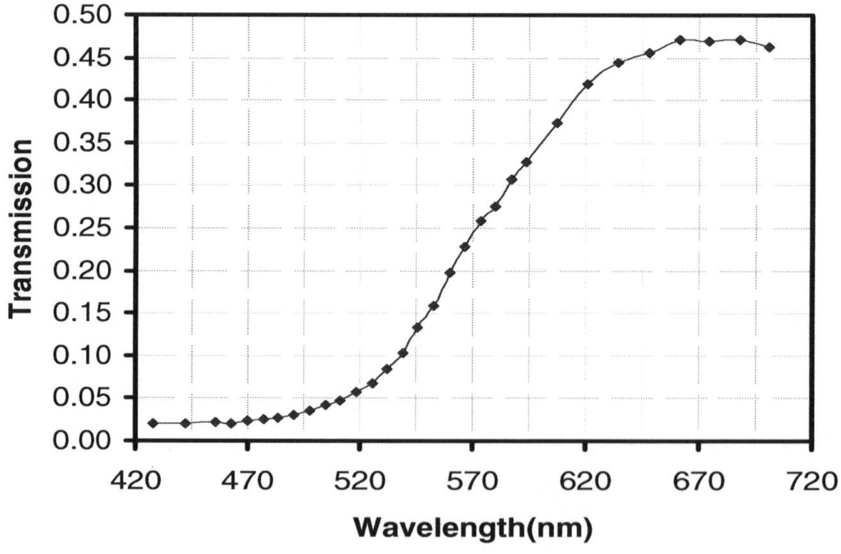

Task 3.

3a.

Table 3a. The calculated parameters for each measured data point

θ (degree)	x (eV)	y (eV^2)
15.00	2.898	126.6
15.50	2.806	121.9
16.00	2.720	108.8
16.25	2.679	107.8
16.50	2.639	98.9
16.75	2.600	92.0
17.00	2.563	86.3
17.25	2.527	77.4
17.50	2.491	69.7
17.75	2.457	60.9
18.00	2.424	55.1
18.25	2.392	46.8
18.50	2.360	40.4
18.75	2.330	33.1
19.00	2.300	27.3
19.25	2.271	20.91
19.50	2.243	17.07
19.75	2.215	12.92
20.00	2.188	10.51
20.25	2.162	8.53
20.50	2.137	7.56
20.75	2.112	6.23
21.00	2.088	5.43
21.50	2.041	4.06
22.00	1.997	3.02
22.50	1.954	2.52
23.00	1.914	2.26
23.50	1.875	1.98
24.00	1.838	1.94
24.50	1.803	1.84
25.00	1.769	1.86

3b.

Graphics

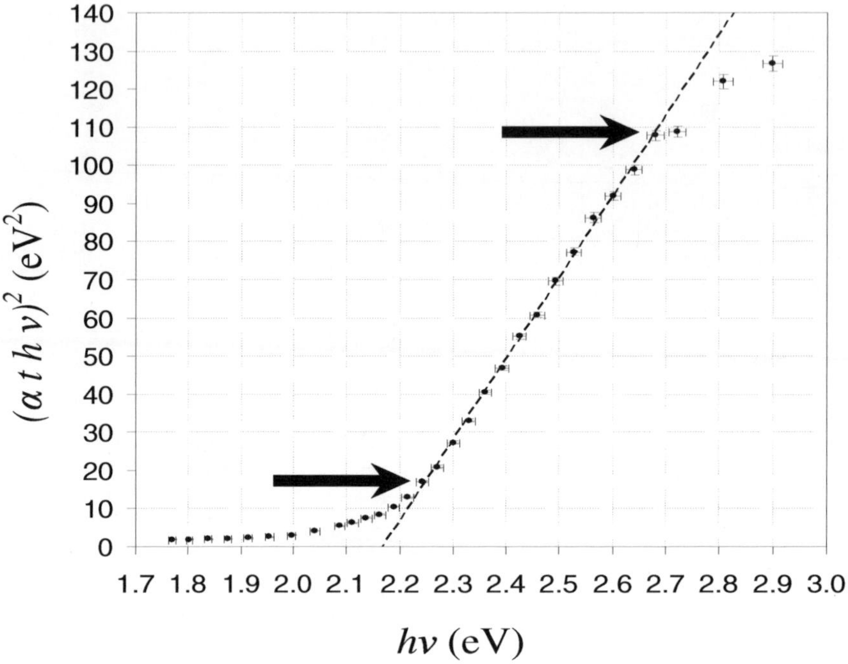

$x_{min} = 2.24(\text{eV})$	$x_{max} = 2.68(\text{eV})$

3c.

$$\alpha h\nu = A(h\nu - E_g)^{\frac{1}{2}} \Rightarrow (\alpha t h\nu)^2 = (At)^2(h\nu - E_g)$$

$$\Rightarrow y = (At)^2(x - E_g) \Rightarrow m = (At)^2 \Rightarrow t = \frac{\sqrt{m}}{A}$$

$$\Rightarrow \frac{\Delta t}{t} = \frac{\Delta m}{2m}$$

$t = \dfrac{\sqrt{m}}{A}$
$\Delta t = \dfrac{\Delta m}{2A\sqrt{m}}$

In linear range we have, m=213 (eV), r^2= 0.9986, E_g=2.17 (eV)
and we have $A = 0.071\left(\mathrm{eV}^{1/2}/\mathrm{nm}\right)$ so we find t= 206 (nm)

$$\Delta m = \sqrt{\frac{(\delta\, y)^2 + \frac{m^2}{R^2}(\delta\, x)^2}{\sum_i x_i^2 - N\bar{x}^2}} \approx \sqrt{\frac{(\delta\, y)^2 + (m\,\delta\, x)^2}{\sum_i x_i^2 - N\bar{x}^2}} = \sqrt{\frac{(\delta\, xy)^2}{\sum_i x_i^2 - N\bar{x}^2}},\ (\delta\, xy)^2 = (\delta\, y)^2 + (m\,\delta\, x)^2$$

where δx & δy are the mean of error range of x & y

$$\delta\, x \approx \sqrt{\frac{\sum_i \delta x_i^{\,2}}{N}}\ \&\ \delta\, y = \sqrt{\frac{\sum_i \delta y_i^{\,2}}{N}} \quad \text{So} \qquad \delta x \approx 0.014\,(\mathrm{eV}),\ \delta y \approx 0.9(\mathrm{eV})^2$$

→ Δm ≈ 10 (eV) → Δt = t × Δm/(2 m) ≈ 5 (nm)

$$\Delta E_g = \frac{1}{m}\sqrt{\left(\left(\frac{m^2\,\delta x^2 + \delta y^2}{N}\right) + \left(\frac{\bar{y}}{m}\right)^2 \Delta m^2\right)} = \frac{1}{m}\sqrt{\left(\left(\frac{\delta xy^{\,2}}{N}\right) + \left(\frac{\bar{y}}{m}\right)^2 \Delta m^2\right)}$$

$$\Delta E_g \approx 0.02(\mathrm{eV})$$

Table 3d. The calculated values of E_g and t using Fig. 3

E_g (eV)	ΔE_g (eV)	t (nm)	Δt (nm)
2.17	0.02	206	5

Description of the Apparatus

In Fig.1 you can see the general view of the apparatus set up on your desk, which will be used in the experiment. The instrument is a spectroscope to be equipped with a detector to act as a simple spectrometer.

To start adjusting the apparatus, you should first pull up the white cover of the box (Fig.1). The cover pivots on one side of the base of the apparatus. In order to establish a dark environment for the detector, the cover should be returned to its initial position and kept tightly closed during the measurement of the spectra. The power cord has a switch that turns the halogen lamp on and off. There are four screws to level the apparatus (a magnified view of which you can see in right inset of Fig.1)

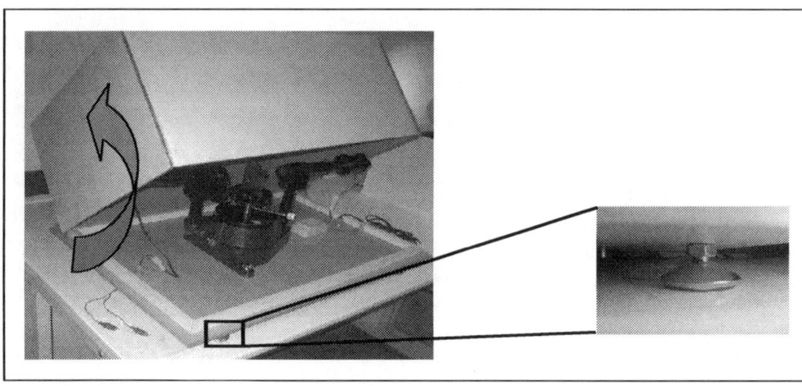

Figure 1. Apparatus of the experiment. One of the level adjusting screws is enlarged in the right inset.

Warning 1: Avoid touching the halogen lamp and its holder which will be ***hot*** after the lamp is turned on!

Warning 2: Do not manipulate the adaptor and its connections. Power is supplied to the apparatus through 220 V outlets!

The top view of the apparatus is shown in Fig.2 . The details are introduced in the figure.

Figure 2.

1. Power cord
2. Halogen lamp and its cooling fan
3. On/Off switch
4. Arm of adjustable length
5. Adjusting screw
6. Adaptor: 220V – less than 12 V
7. Lens
8. Vernier
9. Vernier's lock
10. Fine adjustment screw for the vernier
11. Grating's stage
12. Grating's stage's fixing screw
13. Adjustment screw for leveling the grating's stage (shown in Fig. 4)
14. Grating holder
15. Sample holder
16. Fixing and adjusting screw for the sample and glass holder (Fig. 6)
17. Rotatable arm
18. Rotatable arm's lock (Fig.4)
19. Fine adjustment for the rotatable arm
20. Detector position
21. Fixing screw for the detector
22. Connecting socket for the detector
23. Connection to the multimeter
24. Fixing screw to the base

The number mentioned on the top-left corner, is the **apparatus number**.

The angle, which the rotatable arm makes with the direction of the fixed arm of the apparatus, could be measured by a protractor equipped by a vernier. In this vernier resolution scale is 30' (minutes of arc). This instrument is able to measure an angle with accuracy of 5'.

In addition to the apparatus you should find a box (Figure 3), containing the following elements:

1: a detector in its holder; 2: a 600 line/mm grating; 3: the sample and a glass substrate mounted in a frame.

Figure 3. The small box, containing the glass and sample holder, a diffraction grating and a photoresistor.

First, you should take the grating out of its cover and put it into its frame (the grating holder, Fig. 4), carefully.

CAUTION: Touching the surface of the grating could reduce its diffraction efficiency seriously, or even damage it!

There are three adjustment screws (Fig. 4) for making the grating stand vertically in its position.

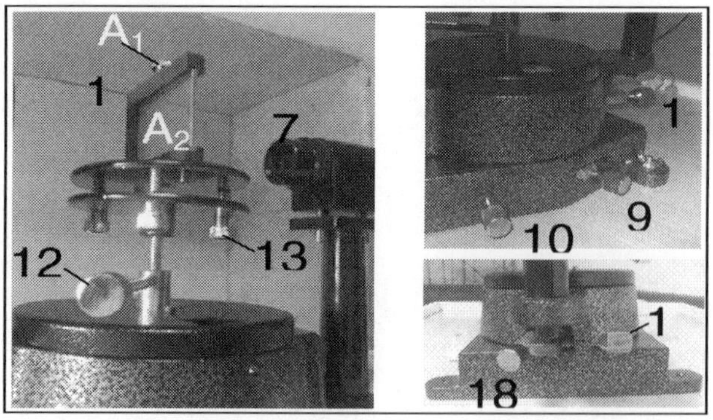

Figure 4. Locking, fixing and adjusting screws of the apparatus. A_1: Fixing screw for the grating; A_2: The grating. 7, 9, 10, 12-14, 18 and 19 are explained in Figure 2.

The detector should be tight to its position, in the end of the rotatable arm, (Figure 5):

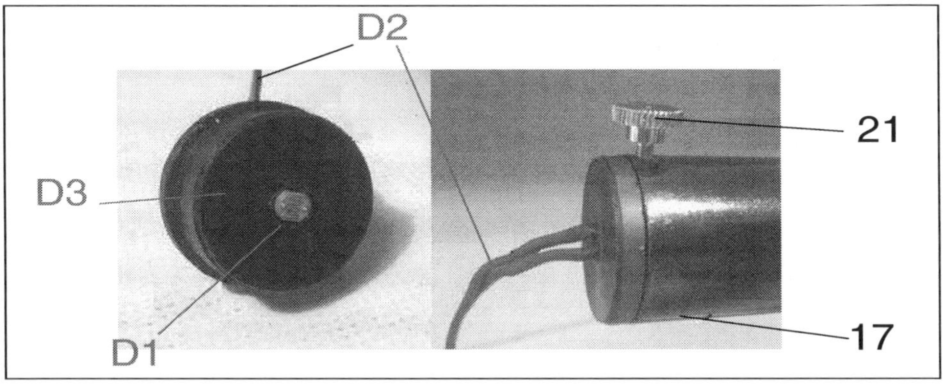

Figure 5 . The detector and its holder. D1: The photoresistor; D2: connecting wire. D3: The detector holder. 17 and 21 are explained in Fig. 2.

The sample and the glass substrate are fixed to a frame (holder) (Fig. 6c), which would be attached to the instrument by a fixing screw (Fig. 6a, item 16). This frame is rotatable and one can put the sample or the glass substrate in front of the entrance hole, by turning the frame around the fixing screw (Fig. 6a).

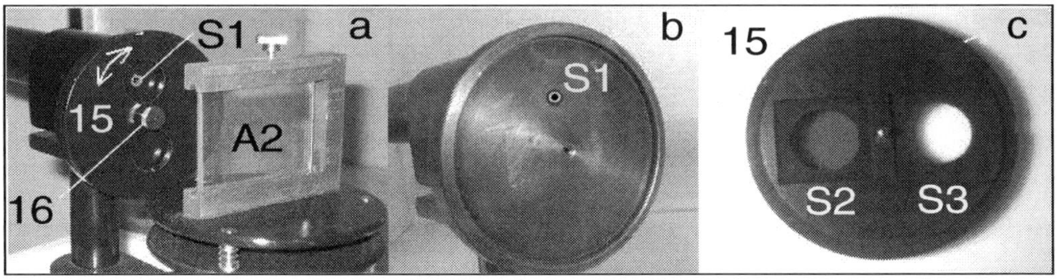

Figure 6 . The Sample and the glass holder. S1: Entrance hole; S2: Sample; S3: Glass substrate. 15 and 16 are explained in Fig. 2.

The Multimeter which you should use for recording the signal detected by the photoresistor is shown in the Fig. 7. This multimeter can measure up to 200 MΩ. The red and black probe wires should be connected to the instrument as is shown in the Fig. 7. The on/off button is placed on the left hand side of the multimeter (Fig. 7, item M1).

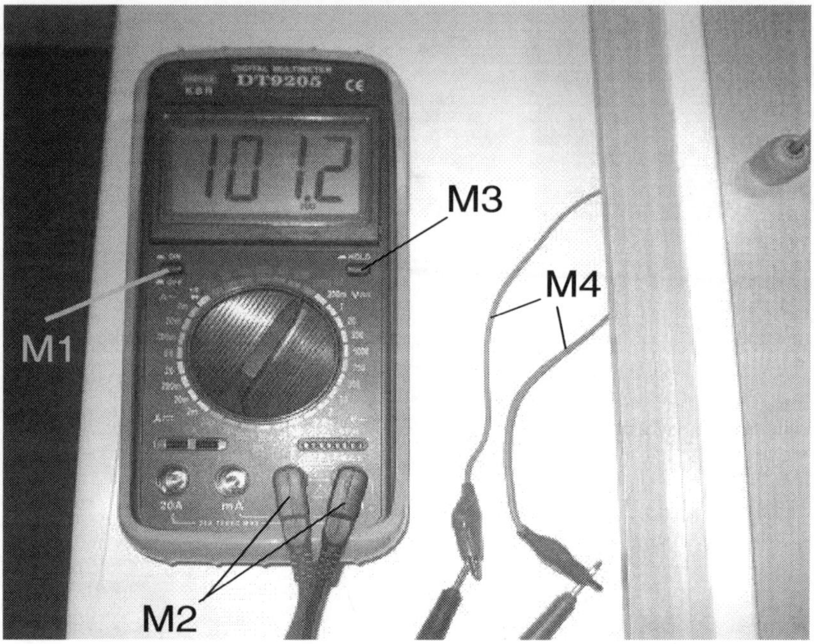

Figure 7. The Multimeter for measuring the resistance of the photoresistor. M1: on/off switch; M2: probe wires; M3: Hold button; M4: connections to the apparatus.

Note: The multimeter has auto-off feature. In the case of auto-off, you should push on/off button (M1) twice, successively.

❖ Hold button should not be active during the experiment.

39TH

INTERNATIONAL PHYSICS OLYMPIAD

HANOI, VIETNAM

2008

WATER-POWERED RICE-POUNDING MORTAR

A. Introduction

Rice is the main staple food of most people in Vietnam. To make white rice from paddy rice, one needs separate of the husk (a process called "hulling") and separate the bran layer ("milling"). The hilly parts of northern Vietnam are abundant with water streams, and people living there use *water-powered rice-pounding mortar* for bran layer separation. Figure 1 shows one of such mortars., Figure 2 shows how it works.

B. Design and operation

***1. Design*.**

The rice-pounding mortar shown in Figure 1 has the following parts:

The mortar, basically a wooden container for rice.

The lever, which is a tree trunk with one larger end and one smaller end. It can rotate around a horizontal axis. A *pestle* is attached perpendicularly to the lever at the smaller end. The length of the pestle is such that it touches the rice in the mortar when the lever lies horizontally. The larger end of the lever is carved hollow to form a bucket. The shape of the bucket is crucial for the mortar's operation.

2. Modes of operation

The mortar has two modes.

Working mode. In this mode, the mortar goes through an operation cycle illustrated in Figure 2.

The rice-pounding function comes from the work that is transferred from the pestle to the rice during stage f) of Figure 2. If, for some reason, the pestle never touches the rice, we say that the mortar is not working.

*Rest mode with the lever lifted up***.** During stage c) of the operation cycle (Figure 2), as the tilt angle α increases, the amount of water in the bucket decreases. At one particular moment in time, the amount of water is just enough to counterbalance the weight of the lever. Denote the tilting angle at this instant by β. If the lever is kept at angle β and the initial angular velocity is zero, then the lever will remain at this position forever. This is the rest mode with the lever lifted up. The stability of this position depends on the flow rate of water into the bucket, Φ. If Φ exceeds some value Φ_2, then this rest mode is stable, and the mortar cannot be in the working mode.

In other words, Φ_2 is the minimal flow rate for the mortar not to work.

OPERATION CYCLE OF A WATER-POWERED RICE-POUNDING MORTAR

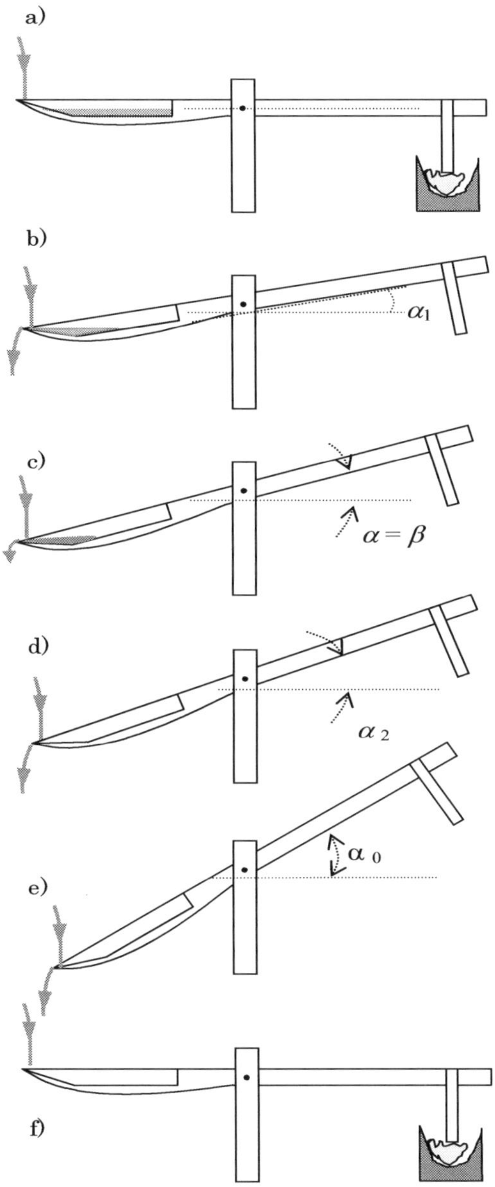

Figure 2

a) At the beginning there is no water in the bucket, the pestle rests on the mortar. Water flows into the bucket with a small rate, but for some time the lever remains in the horizontal position.

b) At some moment the amount of water is enough to lift the lever up. Due to the tilt, water rushes to the farther side of the bucket, tilting the lever more quickly. Water starts to flow out at $\alpha = \alpha_1$.

c) As the angle α increases, water starts to flow out. At some particular tilt angle, $\alpha = \beta$, the total torque is zero.

d) α continues increasing, water continues to flow out until no water remains in the bucket.

e) α keeps increasing because of inertia. Due to the shape of the bucket, water falls into the bucket but immediately flows out. The inertial motion of the lever continues until α reaches the maximal value α_0.

f) With no water in the bucket, the weight of the lever pulls it back to the initial horizontal position. The pestle gives the mortar (with rice inside) a pound and a new cycle begins.

C. The problem

Consider a water-powered rice-pounding mortar with the following parameters (Figure 3)

The mass of the lever (including the pestle but without water) is $M = 30$ kg,

The center of mass of the lever is G. The lever rotates around the axis T (projected onto the point T on the figure).

The moment of inertia of the lever around T is $I = 12\ \text{kg}\cdot\text{m}^2$.

When there is water in the bucket, the mass of water is denoted as m, the center of mass of the water body is denoted as N.

The tilt angle of the lever with respect to the horizontal axis is α.

The main length measurements of the mortar and the bucket are as in Figure 3.

Neglect friction at the rotation axis and the force due to water falling onto the bucket. In this problem, we make an approximation that the water surface is always horizontal.

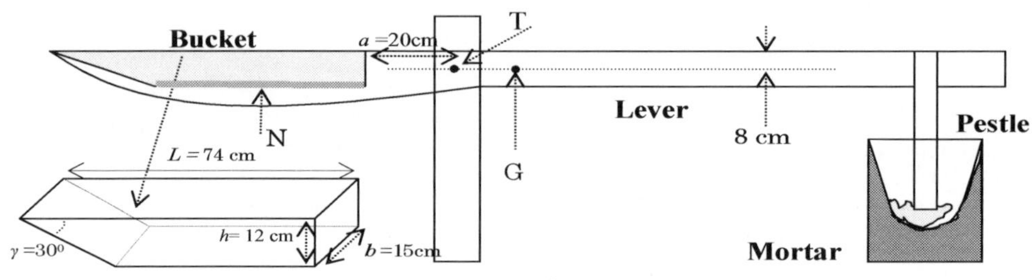

Figure 3 Design and dimensions of the rice-pounding mortar

1. The structure of the mortar

At the beginning, the bucket is empty, and the lever lies horizontally. Then water flows into the bucket until the lever starts rotating. The amount of water in the bucket at this moment is $m = 1.0$ kg.

1.1. Determine the distance from the center of mass G of the lever to the rotation axis T. It is known that GT is horizontal when the bucket is empty.

1.2. Water starts flowing out of the bucket when the angle between the lever and the horizontal axis reaches α_1. The bucket is completely empty when this angle is α_2. Determine α_1 and α_2.

1.3. Let $\mu(\alpha)$ be the total torque (relative to the axis T) which comes from the weight of the lever and the water in the bucket. $\mu(\alpha)$ is zero when $\alpha=\beta$. Determine β and the mass m_1 of water in the bucket at this instant.

2. Parameters of the working mode

Let water flow into the bucket with a flow rate Φ which is constant and small. The amount of water flowing into the bucket when the lever is in motion is negligible. In this part, neglect the change of the moment of inertia during the working cycle.

2.1. Sketch a graph of the torque μ as a function of the angle α, $\mu(\alpha)$, during one operation cycle. Write down explicitly the values of $\mu(\alpha)$ at angle α_1, α_2, and $\alpha = 0$.

2.2. From the graph found in section 2.1., discuss and give the geometric interpretation of the value of the total energy W_{total} produced by $\mu(\alpha)$ and the work W_{pounding} that is transferred from the pestle to the rice.

2.3. From the graph representing μ versus α, estimate α_0 and W_{pounding} (assume the kinetic energy of water flowing into the bucket and out of the bucket is negligible.) You may replace curve lines by zigzag lines, if it simplifies the calculation.

3. The rest mode

Let water flow into the bucket with a constant rate Φ, but one cannot neglect the amount of water flowing into the bucket during the motion of the lever.

3.1. Assuming the bucket is always overflown with water,

3.1.1. Sketch a graph of the torque μ as a function of the angle α in the vicinity of $\alpha=\beta$. To which kind of equilibrium does the position $\alpha=\beta$ of the lever belong?

3.1.2. Find the analytic form of the torque $\mu(\alpha)$ as a function of $\Delta\alpha$ when $\alpha=\beta+\Delta\alpha$, and $\Delta\alpha$ is small.

3.1.3. Write down the equation of motion of the lever, which moves with zero initial velocity from the position $\alpha=\beta+\Delta\alpha$ ($\Delta\alpha$ is small). Show that the motion is, with good accuracy, harmonic oscillation. Compute the period τ.

3.2. At a given Φ, the bucket is overflown with water at all times only if the lever moves sufficiently slowly. There is an upper limit on the amplitude of harmonic oscillation, which depends on Φ. Determine the minimal value Φ_1 of Φ (in kg/s) so that the lever can make a harmonic oscillator motion with amplitude 1°.

3.3. Assume that Φ is sufficiently large so that during the free motion of the lever when the tilting angle decreases from α_2 to α_1 the bucket is always overflown with water. However, if Φ is too large the mortar cannot operate. Assuming that the motion of the lever is that of a harmonic oscillator, estimate the minimal flow rate Φ_2 for the rice-pounding mortar to not work.

Solution

1. The structure of the mortar

1.1. Calculating the distance TG

The volume of water in the bucket is $V = 1000\,\text{cm}^3 = 10^{-3}\,\text{m}^3$. The length of the bottom of the bucket is $d = L - h\tan 60^0 = (0.74 - 0.12\tan 60^0)\,\text{m} = 0.5322\,\text{m}$.

(as the initial data are given with two significant digits, we shall keep only two significant digits in the final answer, but we keep more digits in the intermediate steps). The height c of the water layer in the bucket is calculated from the formula:

$$V = bcd + b\frac{c}{2}c\tan 60^0 \Rightarrow c = \frac{(d^2 + 2\sqrt{3}V/b)^{1/2} - d}{\sqrt{3}}$$

Inserting numerical values for V, b and d, we find $c = 0.01228\,\text{m}$.
When the lever lies horizontally, the distance, on the horizontal axis, between the rotation axis and the center of mass of water N, is $\text{TH} \approx a + \frac{d}{2} + \frac{c}{4}\tan 60^\text{o} = 0.4714\,\text{m}$, and $\text{TG} = (m/M)\text{TH} = 0.01571\,\text{m}$ (see the figure below).

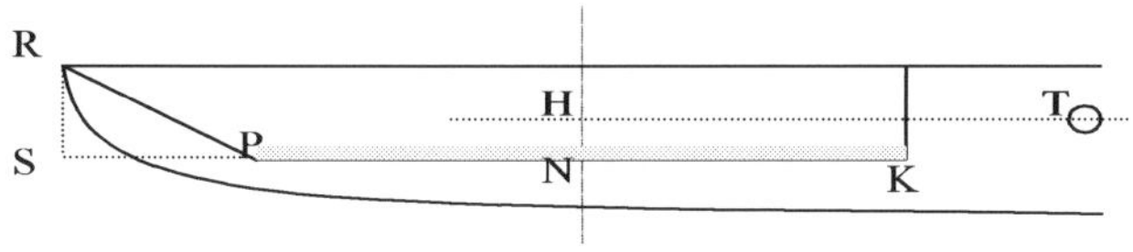

Answer: $\text{TG} = 0.016\,\text{m}$.

1.2. Calculating the values of α_1 and α_2.

When the lever tilts with angle α_1, water level is at the edge of the bucket. At that point the water volume is $10^{-3}\,\text{m}^3$. Assume $\text{PQ} < d$. From geometry $V = hb \times \text{PQ}/2$, from which $\text{PQ} = 0.1111\,\text{m}$. The assumption $\text{PQ} < d$ is obviously satisfied ($d = 0.5322\,\text{m}$).

To compute the angle α_1, we note that $\tan\alpha_1 = h/\text{QS} = h/(\text{PQ} + \sqrt{3}h)$. From this we find $\alpha_1 = 20.6^\text{o}$.

When the tilt angle is 30°, the bucket is empty: $\alpha_2 = 30^{\circ}$.

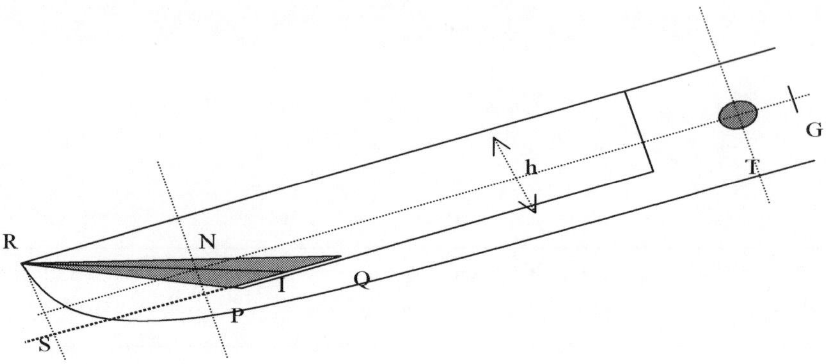

1.3. Determining the tilt angle β of the lever and the amount of water in the bucket m when the total torque μ on the lever is equal to zero

Denote $PQ = x(m)$. The amount of water in the bucket is $m = \rho_{\text{water}} \dfrac{xhb}{2} = 9x$ (kg).

$\mu = 0$ when the torque coming from the water in the bucket cancels out the torque coming from the weight of the lever. The cross section of the water in the bucket is the triangle PQR in the figure. The center of mass N of water is located at 2/3 of the meridian RI, therefore NTG lies on a straight line. Then: $mg \times TN = Mg \times TG$ or

$$m \times TN = M \times TG = 30 \times 0.1571 = 0.4714 \qquad (1)$$

Calculating TN from x then substitute (1):

$$TN = L + a - \frac{2}{3}(h\sqrt{3} + \frac{x}{2}) = 0.94 - 0.08\sqrt{3} - \frac{x}{3} = 0.8014 - \frac{x}{3}$$

which implies $m \times TN = 9x(0.8014 - x/3) = -3x^2 + 7.213x$ (2)

So we find an equation for x:

$$-3x^2 + 7.213x = 0.4714 \qquad (3)$$

The solutions to (3) are $x = 2.337$ and $x = 0.06723$. Since x has to be smaller than 0.5322, we have to take $x = x_0 = 0.06723$ and $m = 9x_0 = 0.6051\,\text{kg}$.

$$\tan\beta = \frac{h}{x + h\sqrt{3}} = 0.4362, \text{ or } \beta = 23.57^{\circ}.$$

Answer: $m = 0.61\,\text{kg}$ and $\beta = 23.6^{\circ}$.

2. Parameters of the working mode

2.1.Graphs of $\mu(\alpha)$, $\alpha(t)$, and $\mu(t)$ during one operation cycle.

Initially when there is no water in the bucket, $\alpha = 0$, μ has the largest magnitude equal to $gM \times \text{TG} = 30 \times 9.81 \times 0.01571 = 4.624\,\text{N} \cdot \text{m}$. Our convention will be that the sign of this torque is negative as it tends to decrease α.

As water flows into the bucket, the torque coming from the water (which carries positive sign) makes μ increase until μ is slightly positive, when the lever starts to lift up. From that moment, by assumption, the amount of water in the bucket is constant.

The lever tilts so the center of mass of water moves away from the rotation axis, leading to an increase of μ, which reaches maximum when water is just about to overflow the edge of the bucket. At this moment $\alpha = \alpha_1 = 20.6^\circ$.

A simple calculation shows that

$$\text{SI} = \text{SP} + \text{PQ}/2 = 0.12 \times 1.732 + 0.1111/2 = 0.2634\,\text{m}.$$

$$\text{TN} = 0.20 + 0.74 - \frac{2}{3}\text{SI} = 0.7644\,\text{m}.$$

$$\mu_{max} = (1.0 \times \text{TN} - 30 \times \text{TG})g\cos 20.6^\circ$$

$$=(1.0 \times 0.7644 - 30 \times 0.01571) \times 9.81 \times \cos 20.6^\circ = 2.690\ \text{N} \cdot \text{m}.$$

Therefore $\mu_{max} = 2.7\ \text{N} \cdot \text{m}$.

As the bucket tilts further, the amount of water in the bucket decreases, and when $\alpha = \beta$, $\mu = 0$. Due to inertia, α keeps increasing and μ keeps decreasing. The bucket is empty when $\alpha = 30^\circ$, when μ equals $-30 \times g \times \text{TG} \times \cos 30^\circ = -4.0\,\text{N} \cdot \text{m}$. After that α keeps increasing due to inertia to α_0 ($\mu = -gM\,\text{TG}\cos\alpha_0 = -4.62\cos\alpha_0\,\text{N} \cdot \text{m}$), then quickly decreases to 0 ($\mu = -4.62\,\text{N} \cdot \text{m}$).

On this basis we can sketch the graphs of $\alpha(t)$, $\mu(t)$, and $\mu(\alpha)$ as in the figure below

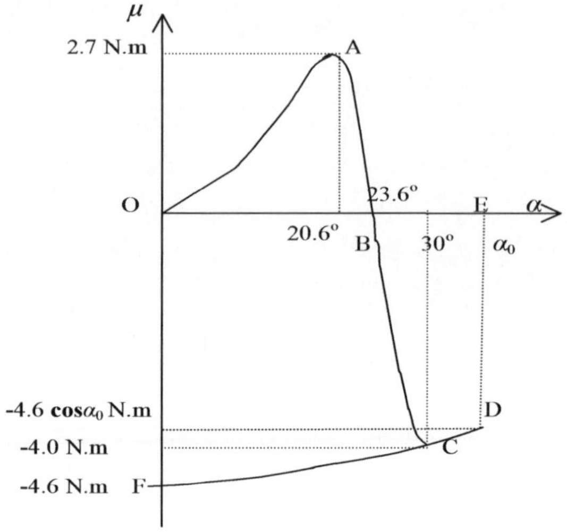

2.2. The infinitesimal work produced by the torque $\mu(\alpha)$ is $dW = \mu(\alpha)d\alpha$. The energy obtained by the lever during one cycle due to the action of $\mu(\alpha)$ is $W = \oint \mu(\alpha)d\alpha$, which is the area limited by the line $\mu(\alpha)$. Therefore W_{total} is equal to the area enclosed by the curve (OABCDFO) on the graph $\mu(\alpha)$.

The work that the lever transfers to the mortar is the energy the lever receives as it moves from the position $\alpha = \alpha_o$ to the horizontal position $\alpha = 0$. We have W_{pounding} equals to the area of (OEDFO) on the graph $\mu(\alpha)$. It is equal to $gM \times \text{TG} \times \mathbf{sin}\,\alpha_0 = \mathbf{4.6 sin}\,\alpha_0$ (J).

2.3. The magnitudes of α_0 can be estimated from the fact that at point D the energy of the lever is zero. We have

area (OABO) = area (BEDCB)

Approximating OABO by a triangle, and BEDCB by a trapezoid, we obtain:

$$23.6 \times 2.7 \times (1/2) = 4.0 \times [(\alpha_0 - 23.6) + (\alpha_0 - 30)] \times (1/2),$$

which implies $\alpha_0 = 34.7^{\circ}$. From this we find

$$W_{\text{pounding}} = \text{area (OEDFO)} = \int_{34.76}^{0} -Mg \times \text{TG} \times \mathbf{cos}\,\alpha\, d\alpha = 4.62 \times \mathbf{sin}\,34.7^{\circ} = 2.63$$

Thus we find $W_{\text{pounding}} \approx 2.6$ J.

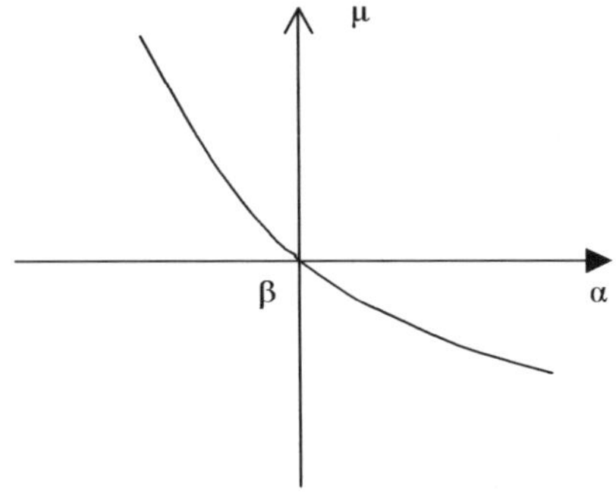

3. The rest mode

3.1.

3.1.1. The bucket is always overflown with water. The two branches of $\mu(\alpha)$ in the vicinity of $\alpha = \beta$ corresponding to increasing and decreasing α coincide with each other.

The graph implies that $\alpha = \beta$ is a stable equilibrium of the mortar.

3.1.2. Find the expression for the torque μ when the tilt angle is $\alpha = \beta + \Delta\alpha$ ($\Delta\alpha$ is small **).**

The mass of water in bucket when the lever tilts with angle α is $m = (1/2)\rho bh\text{PQ}$, where $\text{PQ} = h\left(\frac{1}{\tan\alpha} - \frac{1}{\tan 30^0}\right)$**.** A simple calculation shows that when α increases from β to $\beta + \Delta\alpha$, the mass of water increases by $\Delta m = -\frac{bh^2\rho}{2\sin^2\alpha}\Delta\alpha \approx -\frac{bh^2\rho}{2\sin^2\beta}\Delta\alpha$. The torque μ acting on the lever when the tilt is $\beta + \Delta\alpha$ equals the torque due to Δm.

We have $\mu = \Delta m \times g \times \text{TN} \times \cos(\beta + \Delta\alpha)$. TN is found from the equilibrium condition of the lever at tilting angle β:

$\text{TN} = M \times \text{TG} / m = 30 \times 0.01571 / 0.605 = 0.779\,\text{m}$.

We find at the end $\mu = -47.2 \times \Delta\alpha\ \text{N}\cdot\text{m} \approx -47 \times \Delta\alpha\ \text{N}\cdot\text{m}$.

3.1.3. Equation of motion of the lever

$\mu = I\frac{d^2\alpha}{dt^2}$ where $\mu = -47 \times \Delta\alpha$, $\alpha = \beta + \Delta\alpha$, and I is the sum of moments of inertia of the lever and of the water in bucket relative to the axis T. Here I is not constant the amount of water in the bucket depends on α. When $\Delta\alpha$ is small, one can consider the amount and the shape of water in the bucket to be constant, so I is approximatey a constant. Consider water in bucket as a material point with mass 0.6 kg, a simple calculation gives $I = 12 + 0.6 \times 0.78^2 = 12.36 \approx 12.4\ \text{kg}\,\text{m}^2$. We have $-47 \times \Delta\alpha = 12.4 \times \frac{d^2\Delta\alpha}{dt^2}$. That is the equation for a harmonic oscillator with period

$\tau = 2\pi\sqrt{\dfrac{12.4}{47}} = 3.227$. The answer is therefore $\tau = 3.2\,\text{s}$.

3.2. Harmonic oscillation of lever (around $\alpha = \beta$) when bucket is always overflown. Assume the lever oscillate harmonically with amplitude $\Delta\alpha_0$ around $\alpha = \beta$. At time $t = 0$, $\Delta\alpha = 0$, the bucket is overflown. At time dt the tilt changes by $d\alpha$. We are interested in the case $d\alpha < 0$, i.e., the motion of lever is in the direction of decreasing α, and one needs to add more water to overflow the bucket. The equation of motion is:

$\Delta\alpha = -\Delta\alpha_0 \sin(2\pi t/\tau)$, therefore $d(\Delta\alpha) = d\alpha = -\Delta\alpha_0(2\pi/\tau)\cos(2\pi t/\tau)dt$.

For the bucket to be overflown, during this time the amount of water falling to the bucket should be at least $dm = -\dfrac{bh^2\rho}{2\sin^2\beta}d\alpha = \dfrac{2\Delta\alpha_0\pi bh^2\rho dt}{2\tau\sin^2\beta}\cos\left(\dfrac{2\pi t}{\tau}\right)$; dm is maximum at $t = 0$, $dm_0 = \dfrac{\pi bh^2\rho\Delta\alpha_0}{\tau\sin^2\beta}dt$.

The amount of water falling to the bucket is related to flow rate Φ; $dm_0 = \Phi dt$, therefore $\Phi = \dfrac{\pi bh^2\rho\Delta\alpha_0}{\tau\sin^2\beta}$.

An overflown bucket is the necessary condition for harmonic oscillations of the lever, therefore the condition for the lever to have harmonic oscillations with ampltude 1° or $2\pi/360$ rad is $\Phi \geq \Phi_1$ with

$$\Phi_1 = \frac{\pi bh^2\rho 2\pi}{360\tau\sin^2\beta} = 0.2309\,\text{kg/s}$$

So $\Phi_1 = 0.23\,\text{kg/s}$.

3.3 Determination of Φ_2

If the bucket remains overflown when the tilt decreases to 20.6°, then the amount of water in bucket should reach 1 kg at this time, and the lever oscillate harmonically with amplitude equal $23.6^\circ - 20.6^\circ = 3^\circ$. The flow should exceed $3\Phi_1$, therefore

$\Phi_2 = 3 \times 0.23 \approx 0.7\,\text{kg/s}$.

This is the minimal flow rate for the rice-pounding mortar not to work.

CHERENKOV LIGHT AND RING IMAGING COUNTER

Light propagates in vacuum with the speed c. There is no particle which moves with a speed higher than c. However, it is possible that in a transparent medium a particle moves with a speed v higher than the speed of the light in the same medium $\frac{c}{n}$, where n is the refraction index of the medium. Experiment (Cherenkov, 1934) and theory (Tamm and Frank, 1937) showed that a charged particle, moving with a speed v in a transparent medium with refractive index n such that $v > \frac{c}{n}$, radiates light, called *Cherenkov light*, in directions forming with the trajectory an angle

$$\theta = \arccos \frac{1}{\beta n} \qquad (1)$$

where $\beta = \frac{v}{c}$.

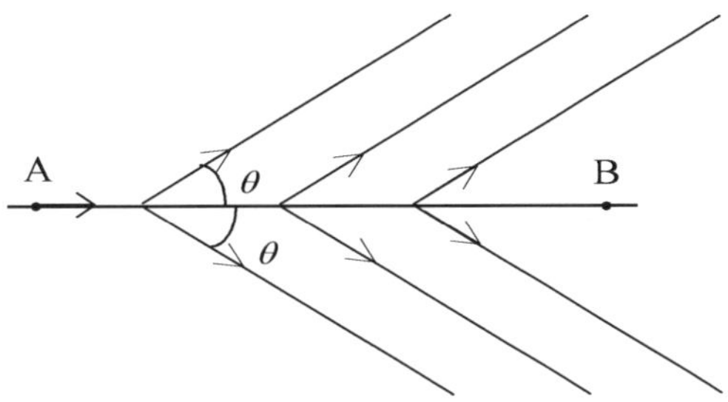

1. To establish this fact, consider a particle moving at constant velocity $v > \frac{c}{n}$ on a straight line. It passes A at time 0 and B at time t_1. As the problem is symmetric with respect to rotations around AB, it is sufficient to consider light rays in a plane containing AB.

At any point C between A and B, the particle emits a spherical light wave, which propagates with velocity $\frac{c}{n}$. We define the wave front at a given time t as the envelope of all these spheres at this time.

1.1. Determine the wave front at time t_1 and draw its intersection with a plane containing the trajectory of the particle.

1.2. Express the angle φ between this intersection and the trajectory of the particle in terms of n and β.

2. Let us consider a beam of particles moving with velocity $v > \frac{c}{n}$, such that the angle θ is small, along a straight line IS. The beam crosses a concave spherical mirror of focal length f and center C, at point S. SC makes with SI a small angle α (see the figure in the Answer Sheet). The particle beam creates a ring image in the focal plane of the mirror.

Explain why with the help of a sketch illustrating this fact. Give the position of the center O and the radius r of the ring image.

This set up is used in *ring imaging Cherenkov counters* (RICH) and the medium which the particle traverses is called the *radiator*.

Note: in all questions of the present problem, terms of second order and higher in α and θ will be neglected.

3. A beam of particles of known momentum $p = 10.0$ GeV/c consists of three types of particles: protons, kaons and pions, with rest mass $M_{\text{p}} = 0.94$ GeV$/c^2$, $M_\kappa = 0.50$ GeV$/c^2$ and $M_\pi = 0.14$ GeV$/c^2$, respectively. Remember that pc and Mc^2 have the dimension of an energy, and 1 eV is the energy acquired by an electron after being accelerated by a voltage 1 V, and 1 GeV = 10^9 eV, 1 MeV = 10^6 eV.

The particle beam traverses an air medium (the radiator) under the pressure P. The refraction index of air depends on the air pressure P according to the relation $n = 1 + aP$ where $a = 2.7\times10^{-4}$ atm^{-1}

3.1. Calculate for each of the three particle types the minimal value $P_{\min}$ of the air pressure such that they emit Cherenkov light.

3.2. Calculate the pressure $P_{\frac{1}{2}}$ such that the ring image of kaons has a radius equal to one half of that corresponding to pions. Calculate the values of θ_κ and θ_π in this case.

Is it possible to observe the ring image of protons under this pressure?

4. Assume now that the beam is not perfectly monochromatic: the particles momenta are distributed over an interval centered at 10 GeV$/c$ having a half width at half height Δp. This makes the ring image broaden, correspondingly θ distribution has a half width at half height $\Delta\theta$. The pressure of the radiator is $P_{\frac{1}{2}}$ determined in 3.2.

4.1. Calculate $\dfrac{\Delta\theta_\kappa}{\Delta p}$ and $\dfrac{\Delta\theta_\pi}{\Delta p}$, the values taken by $\dfrac{\Delta\theta}{\Delta p}$ in the pions and kaons cases.

4.2. When the separation between the two ring images, $\theta_\pi - \theta_\kappa$, is greater than 10 times the half-width sum $\Delta\theta = \Delta\theta_{\text{K}} + \Delta\theta_{\pi}$, that is $\theta_{\pi} - \theta_{\text{K}} > 10\ \Delta\theta$, it is possible to distinguish well the two ring images. Calculate the maximal value of Δp such that the two ring images can still be well distinguished.

5. Cherenkov first discovered the effect bearing his name when he was observing a bottle of water located near a radioactive source. He saw that the water in the bottle emitted light.

5.1. Find out the minimal kinetic energy $T_{\min}$ of a particle with a rest mass M moving in water, such that it emits Cherenkov light. The index of refraction of water is $n = 1.33$.

5.2. The radioactive source used by Cherenkov emits either α particles (i.e. helium nuclei) having a rest mass $M_{\alpha} = 3.8\ \text{GeV}/c^2$ or β particles (i.e. electrons) having a rest mass $M_{\text{e}} = 0.51\ \text{MeV}/c^2$. Calculate the numerical values of $T_{\min}$ for α particles and β particles.

Knowing that the kinetic energy of particles emitted by radioactive sources never exceeds a few MeV, find out which particles give rise to the radiation observed by Cherenkov.

6. In the previous sections of the problem, the dependence of the Cherenkov effect on wavelength λ has been ignored. We now take into account the fact that the Cherenkov radiation of a particle has a broad continuous spectrum including the visible range (wavelengths from 0.4 μm to 0.8 μm). We know also that the index of refraction n of the radiator decreases linearly by 2% of $n-1$ when λ increases over this range.

6.1. Consider a beam of pions with definite momentum of $10.0\ \text{GeV}/c$ moving in air at pressure 6 atm. Find out the angular difference $\delta\theta$ associated with the two ends of the visible range.

6.2. On this basis, study qualitatively the effect of the dispersion on the ring image of pions with momentum distributed over an interval centered at $p = 10\ \text{GeV}/c$ and having a half width at half height $\Delta p = 0.3\ \text{GeV}/c$.

6.2.1. Calculate the broadening due to dispersion (varying refraction index) and that due to achromaticity of the beam (varying momentum).

6.2.2. Describe how the color of the ring changes when going from its inner to outer edges by checking the appropriate boxes in the Answer Sheet.

Solution

1.

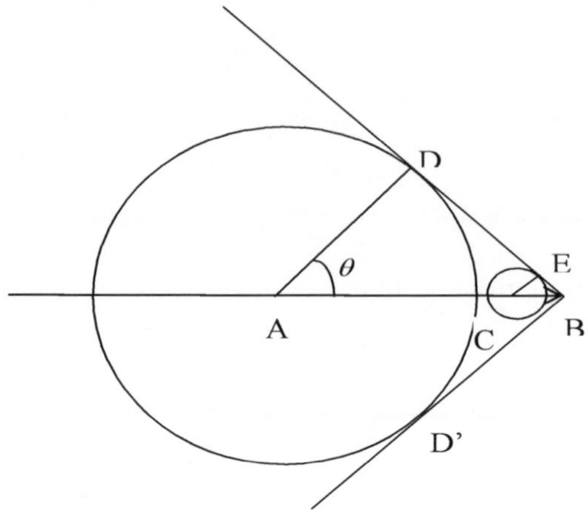

Figure 1

Let us consider a plane containing the particle trajectory. At $t=0$, the particle position is at point A. It reaches point B at $t=t_1$. According to the Huygens principle, at moment $0<t<t_1$, the radiation emitted at A reaches the circle with a radius equal to AD and the one emitted at C reaches the circle of radius CE. The radii of the spheres are proportional to the distance of their centre to B:

$$\frac{\text{CE}}{\text{CB}}=\frac{c(t_1-t)/n}{v(t_1-t)}=\frac{1}{\beta n}=\text{const}$$

The spheres are therefore transformed into each other by homothety of vertex B and their envelope is the cone of summit B and half aperture $\varphi=\textbf{Arcsin}\frac{1}{\beta n}=\frac{\pi}{2}-\theta$, where θ is the angle made by the light ray CE with the particle trajectory.

1.1. The intersection of the wave front with the plane is two straight lines, BD and BD'.

1.2. They make an angle $\varphi=\textbf{Arcsin}\frac{1}{\beta n}$ with the particle trajectory.

2. The construction for finding the ring image of the particles beam is taken in the plane containing the trajectory of the particle and the optical axis of the mirror.

We adopt the notations:

S – the point where the beam crosses the spherical mirror

F – the focus of the spherical mirror

C – the center of the spherical mirror

IS – the straight-line trajectory of the charged particle making a small angle α with the optical axis of the mirror.

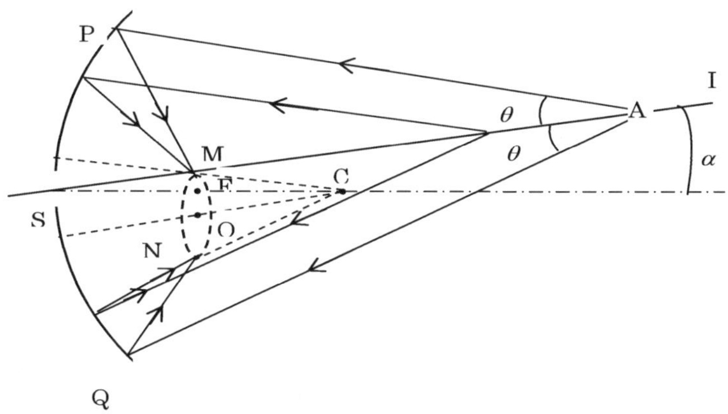

Figure 2

CF = FS = f

CO//IS

CM//AP

CN//AQ

$$\widehat{FCO} = \alpha \Rightarrow FO = f \times \alpha$$

$$\widehat{MCO} = \widehat{OCN} = \theta \Rightarrow MO = f \times \theta$$

We draw a straight line parallel to IS passing through the center C. The line intersects the focal plane at O. We have FO $\approx f \times \alpha$.

Starting from C, we draw two lines in both sides of the line CO making with it an angle θ. These two lines intersect the focal plane at M and N, respectively. All the rays of Cherenkov radiation in the plane of the sketch, striking the mirror and being reflected, intersect at M or N.

In three-dimension case, the Cherenkov radiation gives a ring in the focal plane with the center at O (FO $\approx f\times\alpha$) and with the radius MO $\approx f\times\theta$.

In the construction, all the lines are in the plane of the sketch. Exceptionally, the ring is illustrated spatially by a dash line.

3.

3.1. For the Cherenkov effect to occur it is necessary that $n>\frac{c}{v}$, that is $n_{\min}=\frac{c}{v}$.

Putting $\zeta=n-1=2.7\times10^{-4}P$, we get

$$\zeta_{\min}=2.7\times10^{-4}P_{\min}=\frac{c}{v}-1=\frac{1}{\beta}-1 \tag{1}$$

Because

$$\frac{Mc^2}{pc}=\frac{Mc}{p}=\frac{Mc}{\frac{Mv}{\sqrt{1-\beta^2}}}=\frac{\sqrt{1-\beta^2}}{\beta}=K \tag{2}$$

then $K=0.094$; 0.05 ; 0.014 for proton, kaon and pion, respectively.

From (2) we can express β through K as

$$\beta=\frac{1}{\sqrt{1+K^2}} \tag{3}$$

Since $K^2<<1$ for all three kinds of particles we can neglect the terms of order higher than 2 in K. We get

$$1-\beta=1-\frac{1}{\sqrt{1+K^2}}\approx\frac{1}{2}K^2=\frac{1}{2}\left(\frac{Mc}{p}\right)^2 \tag{3a}$$

$$\frac{1}{\beta}-1=\sqrt{1+K^2}-1\approx\frac{1}{2}K^2=\frac{1}{2}\left(\frac{Mc}{p}\right)^2 \tag{3b}$$

Putting (3b) into (1), we obtain

$$P_{\min} = \frac{1}{2.7 \times 10^{-4}} \times \frac{1}{2} K^2 \tag{4}$$

We get the following numerical values of the minimal pressure:

$P_{\min} = 16$ atm for protons,

$P_{\min} = 4.6$ atm for kaons,

$P_{\min} = 0.36$ atm for pions.

3.2. For $\theta_\pi = 2\theta_\kappa$ we have

$$\cos\theta_\pi = \cos 2\theta_\kappa = 2\cos^2\theta_\kappa - 1 \tag{5}$$

We denote

$$\varepsilon = 1 - \beta = 1 - \frac{1}{\sqrt{1+K^2}} \approx \frac{1}{2} K^2 \tag{6}$$

From (5) we obtain

$$\frac{1}{\beta_\pi n} = \frac{2}{\beta_\kappa^2 n^2} - 1 \tag{7}$$

Substituting $\beta = 1 - \varepsilon$ and $n = 1 + \zeta$ into (7), we get approximately:

$$\zeta_{\frac{1}{2}} = \frac{4\varepsilon_\kappa - \varepsilon_\pi}{3} = \frac{1}{6}\left(4K_\kappa^2 - K_\pi^2\right) = \frac{1}{6}\left[4.(0.05)^2 - (0.014)^2\right],$$

$$P_{\frac{1}{2}} = \frac{1}{2.7 \times 10^{-4}} \zeta_{\frac{1}{2}} = 6 \text{ atm}.$$

The corresponding value of refraction index is $n = 1.00162$. We get:

$\theta_\kappa = 1.6^\circ$; $\theta_\pi = 2\theta_\kappa = 3.2^\circ$.

We do not observe the ring image of protons since

$P_{\frac{1}{2}} = 6\,\text{atm} < 16\,\text{atm} = P_{\min}$ for protons.

4.

4.1. Taking logarithmic differentiation of both sides of the equation $\cos\theta = \frac{1}{\beta n}$, we obtain

$$\frac{\sin\theta \times \Delta\theta}{\cos\theta} = \frac{\Delta\beta}{\beta} \quad (8)$$

Logarithmically differentiating equation (3a) gives

$$\frac{\Delta\beta}{1-\beta} = 2\frac{\Delta p}{p} \quad (9)$$

Combining (8) and (9), taking into account (3b) and putting approximately $\tan\theta = \theta$, we derive

$$\frac{\Delta\theta}{\Delta p} = \frac{2}{\theta} \times \frac{1-\beta}{p\beta} = \frac{K^2}{\theta p} \quad (10)$$

We obtain

-for kaons $K_\kappa = 0.05$, $\theta_\kappa = 1.6^\circ = 1.6\frac{\pi}{180}$ rad, and so, $\frac{\Delta\theta_\kappa}{\Delta p} = 0.51\frac{1^\circ}{\text{GeV}/c}$,

-for pions $K_\pi = 0.014$, $\theta_\pi = 3.2^\circ$, and $\frac{\Delta\theta_\pi}{\Delta p} = 0.02\frac{1^\circ}{\text{GeV}/c}$.

4.2. $\frac{\Delta\theta_\kappa + \Delta\theta_\pi}{\Delta p} \equiv \frac{\Delta\theta}{\Delta p} = (0.51+0.02)\frac{1^\circ}{\text{GeV}/c} = 0.53\frac{1^\circ}{\text{GeV}/c}$.

The condition for two ring images to be distinguishable is $\Delta\theta < 0.1(\theta_\pi - \theta_\kappa) = 0.16^\circ$.

It follows $\Delta p < \frac{1}{10} \times \frac{1.6}{0.53} = 0.3\ \text{GeV}/c$.

5.

5.1. The lower limit of β giving rise to Cherenkov effect is

$$\beta = \frac{1}{n} = \frac{1}{1.33}. \tag{11}$$

The kinetic energy of a particle having rest mass M and energy E is given by the expression

$$T = E - Mc^2 = \frac{Mc^2}{\sqrt{1-\beta^2}} - Mc^2 = Mc^2\left[\frac{1}{\sqrt{1-\beta^2}} - 1\right]. \tag{12}$$

Substituting the limiting value (11) of β into (12), we get the minimal kinetic energy of the particle for Cherenkov effect to occur:

$$T_{\text{min}} = Mc^2\left[\frac{1}{\sqrt{1-\left(\frac{1}{1.33}\right)^2}} - 1\right] = 0.517\ Mc^2 \tag{13}$$

5.2.

For α particles, $T_{\text{min}} = 0.517 \times 3.8\ \text{GeV} = 1.96\ \text{GeV}$.

For electrons, $T_{\text{min}} = 0.517 \times 0.51\ \text{MeV} = 0.264\ \text{MeV}$.

Since the kinetic energy of the particles emitted by radioactive source does not exceed a few MeV, these are electrons which give rise to Cherenkov radiation in the considered experiment.

6. For a beam of particles having a definite momentum the dependence of the angle θ on the refraction index n of the medium is given by the expression

$$\cos\theta = \frac{1}{n\beta} \tag{14}$$

6.1. Let $\delta\theta$ be the difference of θ between two rings corresponding to two wavelengths limiting the visible range, i.e. to wavelengths of 0.4 μm (violet) and 0.8 μm (red), respectively. The difference in the refraction indexes at these wavelengths is $n_v - n_r = \delta n = 0.02(n-1)$.

Logarithmically differentiating both sides of equation (14) gives

$$\frac{\sin\theta \times \delta\theta}{\cos\theta} = \frac{\delta n}{n} \tag{15}$$

Corresponding to the pressure of the radiator $P = 6$ atm we have from 4.2. the values $\theta_\pi = 3.2^\circ$, $n = 1.00162$.

Putting approximately $\tan\theta = \theta$ and $n = 1$, we get $\delta\theta = \frac{\delta n}{\theta} = 0.033^\circ$.

6.2.

6.2.1. The broadening due to dispersion in terms of half width at half height is, according to (6.1), $\frac{1}{2}\delta\theta = 0.017^\circ$.

6.2.2. The broadening due to achromaticity is, from 4.1., $0.02 \frac{1^\circ}{\text{GeV/c}} \times 0.3\ \text{GeV/c} = 0.006^\circ$, that is three times smaller than above.

6.2.3. The color of the ring changes from red to white then blue from the inner edge to the outer one.

CHANGE OF AIR TEMPERATURE WITH ALTITUDE, ATMOSPHERIC STABILITY AND AIR POLLUTION

Vertical motion of air governs many atmospheric processes, such as the formation of clouds and precipitation and the dispersal of air pollutants. If the atmosphere is *stable*, vertical motion is restricted and air pollutants tend to be accumulated around the emission site rather than dispersed and diluted. Meanwhile, in an *unstable* atmosphere, vertical motion of air encourages the vertical dispersal of air pollutants. Therefore, the pollutants' concentrations depend not only on the strength of emission sources but also on the *stability* of the atmosphere.

We shall determine the atmospheric stability by using the concept of *air parcel* in meteorology and compare the temperature of the air parcel rising or sinking adiabatically in the atmosphere to that of the surrounding air. We will see that in many cases an air parcel containing air pollutants and rising from the ground will come to rest at a certain altitude, called a *mixing height*. The greater the mixing height, the lower the air pollutant concentration. We will evaluate the mixing height and the concentration of carbon monoxide emitted by motorbikes in the Hanoi metropolitan area for a morning rush hour scenario, in which the vertical mixing is restricted due to a temperature inversion (air temperature increases with altitude) at elevations above 119 m.

Let us consider the air as an ideal diatomic gas, with molar mass μ = 29 g/mol.

Quasi equilibrium adiabatic transformation obey the equation $pV^{\gamma} = \text{const}$, where

$\gamma = \dfrac{c_p}{c_V}$ is the ratio between isobaric and isochoric heat capacities of the gas.

The student may use the following data if necessary:

The universal gas constant is $R = 8.31$ J/(mol.K).

The atmospheric pressure on ground is $p_0 = 101.3$ kPa

The acceleration due to gravity is constant, $g = 9.81$ m/s^2

The molar isobaric heat capacity is $c_p = \dfrac{7}{2}R$ for air.

The molar isochoric heat capacity is $c_V = \dfrac{5}{2}R$ for air.

Mathematical hints

a. $\int \frac{dx}{A+Bx} = \frac{1}{B}\int \frac{d(A+Bx)}{A+Bx} = \frac{1}{B}\ln(A+Bx)$

b. The solution of the differential equation $\frac{dx}{dt} + Ax = B$ (with A and B constant) is $x(t) = x_1(t) + \frac{B}{A}$ where $x_1(t)$ is the solution of the differential equation $\frac{dx}{dt} + Ax = 0$.

c. $\lim_{x\to\infty}\left(1+\frac{1}{x}\right)^x = e$

1. Change of pressure with altitude.

1.1. Assume that the temperature of the atmosphere is uniform and equal to T_0. Write down the expression giving the atmospheric pressure p as a function of the altitude z.

1.2. Assume that the temperature of the atmosphere varies with the altitude according to the relation

$$T(z) = T(0) - \Lambda z$$

where Λ is a constant, called the *temperature lapse rate* of the atmosphere (the vertical gradient of temperature is $-\Lambda$).

1.2.1. Write down the expression giving the atmospheric pressure p as a function of the altitude z.

1.2.2. A process called free convection occurs when the air density increases with altitude. At which values of Λ does the free convection occur?

2. Change of the temperature of an air parcel in vertical motion

Consider an air parcel moving upward and downward in the atmosphere. An air parcel is a body of air of sufficient dimension, several meters across, to be treated as an independent thermodynamical entity, yet small enough for its temperature to be considered uniform. The vertical motion of an air parcel can be treated as a quasi adiabatic process, i.e. the exchange of heat with the surrounding air is negligible. If the air parcel rises in the atmosphere, it expands and cools. Conversely, if it moves downward, the increasing outside pressure will compress the air inside the parcel and its temperature will increase.

As the size of the parcel is not large, the atmospheric pressure at different points on the parcel boundary can be considered to have the same value $p(z)$, with z - the altitude of the parcel center. The temperature in the parcel is uniform and equals to $T_{\text{parcel}}(z)$, which is generally different from the temperature of the surrounding air $T(z)$. In parts 2.1 and 2.2, we do not make any assumption about the form of $T(z)$.

2.1. The change of the parcel temperature T_{parcel} with altitude is defined by $\frac{dT_{\text{parcel}}}{dz} = -G$. Derive the expression of $G(T, T_{\text{parcel}})$.

2.2. Consider a special atmospheric condition in which at any altitude z the temperature T of the atmosphere equals to that of the parcel T_{parcel}, $T(z) = T_{\text{parcel}}(z)$. We use Γ to denote the value of G when $T = T_{\text{parcel}}$, that is $\Gamma = -\frac{dT_{\text{parcel}}}{dz}$ (with $T = T_{\text{parcel}}$). Γ is called *dry adiabatic lapse rate*.

2.2.1. Derive the expression of Γ

2.2.2. Calculate the numerical value of Γ.

2.2.3. Derive the expression of the atmospheric temperature $T(z)$ as a function of the altitude.

2.3. Assume that the atmospheric temperature depends on altitude according to the relation $T(z) = T(0) - \Lambda z$, where Λ is a constant. Find the dependence of the parcel temperature $T_{\text{parcel}}(z)$ on altitude z.

2.4. Write down the approximate expression of $T_{\text{parcel}}(z)$ when $|\Lambda z| << T(0)$ and $T(0) \approx T_{\text{parcel}}(0)$.

3. The atmospheric stability.

In this part, we assume that T changes linearly with altitude.

3.1. Consider an air parcel initially in equilibrium with its surrounding air at altitude

z_0, i.e. it has the same temperature $T(z_0)$ as that of the surrounding air. If the parcel is moved slightly up and down (e.g. by atmospheric turbulence), one of the three following cases may occur:

- The air parcel finds its way back to the original altitude z_0, the equilibrium of the parcel is stable. The atmosphere is said to be stable.
- The parcel keeps moving in the original direction, the equilibrium of the parcel is unstable. The atmosphere is unstable.
- The air parcel remains at its new position, the equilibrium of the parcel is indifferent. The atmosphere is said to be neutral.

What is the condition on Λ for the atmosphere to be stable, unstable or neutral?

3.2. A parcel has its temperature on ground $T_{\text{parcel}}(0)$ higher than the temperature $T(0)$ of the surrounding air. The buoyancy force will make the parcel rise. Derive the expression for the maximal altitude the parcel can reach in the case of a stable atmosphere in terms of Λ and Γ.

4. The mixing height

4.1. Table 1 shows air temperatures recorded by a radio sounding balloon at 7: 00 am on a November day in Hanoi. The change of temperature with altitude can be approximately described by the formula $T(z)=T(0)-\Lambda z$ with different lapse rates Λ in the three layers $0 < z < 96$ m, 96 m $< z <$ 119 m and 119 m $< z <$ 215 m.

Consider an air parcel with temperature $T_{\text{parcel}}(0) = 22^{\circ}\text{C}$ ascending from ground. On the basis of the data given in Table 1 and using the above linear approximation, calculate the temperature of the parcel at the altitudes of 96 m and 119 m.

4.2. Determine the maximal elevation H the parcel can reach, and the temperature $T_{\text{parcel}}(H)$ of the parcel.

H is called the mixing height. Air pollutants emitted from ground can mix with the air in the atmosphere (e.g. by wind, turbulence and dispersion) and become diluted within this layer.

Table 1

Data recorded by a radio sounding balloon at 7:00 am on a November day in Hanoi.

Altitude, m	Temperature, °C
5	21.5
60	20.6
64	20.5
69	20.5
75	20.4
81	20.3
90	20.2
96	20.1
102	20.1
109	20.1
113	20.1
119	20.1
128	20.2
136	20.3
145	20.4
153	20.5
159	20.6
168	20.8
178	21.0
189	21.5
202	21.8
215	22.0
225	22.1
234	22.2
246	22.3
257	22.3

5. Estimation of carbon monoxide (CO) pollution during a morning motorbike rush hour in Hanoi.

Hanoi metropolitan area can be approximated by a rectangle with base dimensions L and W as shown in the figure, with one side taken along the south-west bank of the Red River.

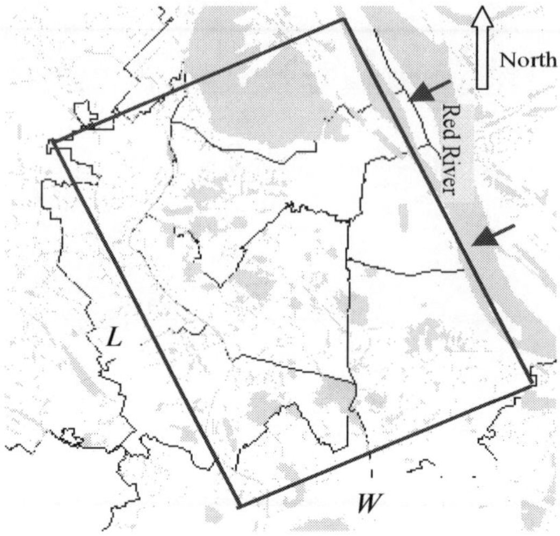

It is estimated that during the morning rush hour, from 7:00 am to 8:00 am, there are $8\text{x}10^5$ motorbikes on the road, each running on average 5 km and emitting 12 g of CO per kilometer. The amount of CO pollutant is approximately considered as emitted uniformly in time, at a constant rate *M* during the rush hour. At the same time, the clean north-east wind blows perpendicularly to the Red River (i.e. perpendicularly to the sides *L* of the rectangle) with velocity *u*, passes the city with the same velocity, and carries a part of the CO-polluted air out of the city atmosphere.

Also, we use the following rough approximate model:

- The CO spreads quickly throughout the entire volume of the mixing layer above the Hanoi metropolitan area, so that the concentration $C(t)$ of CO at time t can be assumed to be constant throughout that rectangular box of dimensions *L, W* and *H.*
- The upwind air entering the box is clean and no pollution is assumed to be lost from the box through the sides parallel to the wind.
- Before 7:00 am, the CO concentration in the atmosphere is negligible.

5.1. Derive the differential equation determining the CO pollutant concentration $C(t)$ as a function of time.

5.2. Write down the solution of that equation for $C(t)$.

5.3. Calculate the numerical value of the concentration $C(t)$ at 8:00 a.m.

Given $L = 15$ km, $W = 8$ km, $u = 1$ m/s.

Solution

1. For an altitude change dz , the atmospheric pressure change is :

$$dp = -\rho g dz \qquad (1)$$

where g is the acceleration of gravity, considered constant, ρ is the specific mass of air, which is considered as an ideal gas:

$$\rho = \frac{m}{V} = \frac{p\mu}{RT}$$

Put this expression in (1) :

$$\frac{dp}{p} = -\frac{\mu g}{RT} dz$$

1.1. If the air temperature is uniform and equals T_0, then

$$\frac{dp}{p} = -\frac{\mu g}{RT_0} dz$$

After integration, we have :

$$p(z) = p(0)\mathbf{e}^{-\frac{\mu g}{RT_0} z} \qquad (2)$$

1.2. If

$$T(z) = T(0) - \Lambda z \qquad (3)$$

then

$$\frac{dp}{p} = -\frac{\mu g}{R\left[T(0) - \Lambda z\right]} dz \qquad (4)$$

1.2.1. Knowing that :

$$\int \frac{dz}{T(0) - \Lambda z} = -\frac{1}{\Lambda} \int \frac{d\left[T(0) - \Lambda z\right]}{T(0) - \Lambda z} = -\frac{1}{\Lambda} \mathbf{ln}\left(T(0) - \Lambda z\right)$$

by integrating both members of (4), we obtain :

$$\mathbf{ln} \frac{p(z)}{p(0)} = \frac{\mu g}{R\Lambda} \mathbf{ln} \frac{T(0) - \Lambda z}{T(0)} = \frac{\mu g}{R\Lambda} \mathbf{ln}\left(1 - \frac{\Lambda z}{T(0)}\right)$$

$$p(z) = p(0)\left(1 - \frac{\Lambda z}{T(0)}\right)^{\frac{\mu g}{R\Lambda}} \qquad (5)$$

1.2.2. The free convection occurs if:

$$\frac{\rho(z)}{\rho(0)} > 1$$

The ratio of specific masses can be expressed as follows:

$$\frac{\rho(z)}{\rho(0)} = \frac{p(z)}{p(0)}\frac{T(0)}{T(z)} = \left(1 - \frac{\Lambda z}{T(0)}\right)^{\frac{\mu g}{R\Lambda}-1}$$

The last term is larger than unity if its exponent is negative:

$$\frac{\mu g}{R\Lambda} - 1 < 0$$

Then :

$$\Lambda > \frac{\mu g}{R} = \frac{0.029 \times 9.81}{8.31} = 0.034 \ \frac{\mathrm{K}}{\mathrm{m}}$$

2. In vertical motion, the pressure of the parcel always equals that of the surrounding air, the latter depends on the altitude. The parcel temperature T_{parcel} depends on the pressure.

2.1. We can write:

$$\frac{dT_{\text{parcel}}}{dz} = \frac{dT_{\text{parcel}}}{dp}\frac{dp}{dz}$$

p is simultaneously the pressure of air in the parcel and that of the surrounding air.

Expression for $\dfrac{dT_{\text{parcel}}}{dp}$

By using the equation for adiabatic processes $pV^{\gamma} = \text{const}$ and equation of state, we can deduce the equation giving the change of pressure and temperature in a quasi-equilibrium adiabatic process of an air parcel:

$$T_{\text{parcel}}\, p^{\frac{1-\gamma}{\gamma}} = \text{const} \qquad (6)$$

where $\gamma = \frac{c_p}{c_V}$ is the ratio of isobaric and isochoric thermal capacities of air. By logarithmic differentiation of the two members of (6), we have:

$$\frac{dT_{\text{parcel}}}{T_{\text{parcel}}} + \frac{1-\gamma}{\gamma}\frac{dp}{p} = 0$$

Or

$$\frac{dT_{\text{parcel}}}{dp} = \frac{T_{\text{parcel}}}{p}\frac{\gamma - 1}{\gamma} \qquad (7)$$

Note: we can use the first law of thermodynamic to calculate the heat received by the parcel in an elementary process: $dQ = \frac{m}{\mu} c_V dT_{\text{parcel}} + pdV$, this heat equals zero in an adiabatic process. Furthermore, using the equation of state for air in the parcel $pV = \frac{m}{\mu} RT_{\text{parcel}}$ we can derive (6)

Expression for $\frac{dp}{dz}$

From (1) we can deduce:

$$\frac{dp}{dz} = -\rho g = -\frac{pg\mu}{RT}$$

where T is the temperature of the surrounding air.

On the basis of these two expressions, we derive the expression for dT_{parcel} / dz :

$$\frac{dT_{\text{parcel}}}{dz} = -\frac{\gamma - 1}{\gamma}\frac{\mu g}{R}\frac{T_{\text{parcel}}}{T} = -G \qquad (8)$$

In general, G is not a constant.

2.2.

2.2.1. If at any altitude, $T = T_{\text{parcel}}$, then instead of G in (8), we have :

$$\Gamma = \frac{\gamma - 1}{\gamma}\frac{\mu g}{R} = \text{const} \qquad (9)$$

or

$$\Gamma = \frac{\mu g}{c_p} \tag{9'}$$

2.2.2. Numerical value:

$$\Gamma = \frac{1.4-1}{1.4}\frac{0.029 \times 9.81}{8.31} = 0.00978\,\frac{\text{K}}{\text{m}} \approx 10^{-2}\,\frac{\text{K}}{\text{m}}$$

2.2.3. Thus, the expression for the temperature at the altitude z in this special atmosphere (called adiabatic atmosphere) is :

$$T(z) = T(0) - \Gamma z \tag{10}$$

2.3. Search for the expression of $T_{\text{parcel}}(z)$

Substitute T in (7) by its expression given in (3), we have:

$$\frac{dT_{\text{parcel}}}{T_{\text{parcel}}} = -\frac{\gamma-1}{\gamma}\frac{\mu g}{R}\frac{dz}{T(0)-\Lambda z}$$

Integration gives:

$$\textbf{ln}\frac{T_{\text{parcel}}(z)}{T_{\text{parcel}}(0)} = -\frac{\gamma-1}{\gamma}\frac{\mu g}{R}\left(-\frac{1}{\Lambda}\right)\textbf{ln}\frac{T(0)-\Lambda z}{T(0)}$$

Finally, we obtain:

$$T_{\text{parcel}}(z) = T_{\text{parcel}}(0)\left(\frac{T(0)-\Lambda z}{T(0)}\right)^{\frac{\Gamma}{\Lambda}} \tag{11}$$

2.4.

From (11) we obtain

$$T_{\text{parcel}}(z) = T_{\text{parcel}}(0)\left(1-\frac{\Lambda z}{T(0)}\right)^{\frac{\Gamma}{\Lambda}}$$

If $\Lambda z << T(0)$, then by putting $x = \dfrac{-T(0)}{\Lambda z}$, we obtain

$$T_{\text{parcel}}(z) = T_{\text{parcel}}(0)\left(\left(1+\frac{1}{x}\right)^{x}\right)^{-\frac{\Gamma z}{T(0)}}$$

$$\approx T_{\text{parcel}}(0)\mathbf{e}^{-\frac{\Gamma z}{T(0)}} \approx T_{\text{parcel}}(0)\left(1-\frac{\Gamma z}{T(0)}\right) \approx T_{\text{parcel}}(0) - \Gamma z$$

hence,

$$T_{\text{parcel}}(z) \approx T_{\text{parcel}}(0) - \Gamma z \tag{12}$$

3. Atmospheric stability

In order to know the stability of atmosphere, we can study the stability of the equilibrium of an air parcel in this atmosphere.

At the altitude z_0, where $T_{\text{parcel}}(z_0) = T(z_0)$, the air parcel is in equilibrium. Indeed, in this case the specific mass ρ of air in the parcel equals ρ'- that of the surrounding air in the atmosphere. Therefore, the buoyant force of the surrounding air on the parcel equals the weight of the parcel. The resultant of these two forces is zero.

Remember that the temperature of the air parcel $T_{\text{parcel}}(z)$ is given by (7), in which we can assume approximately $G = \Gamma$ at any altitude z near $z = z_0$.

Now, consider the stability of the air parcel equilibrium:

Suppose that the air parcel is lifted into a higher position, at the altitude $z_0 + d$ (with d>0), $T_{\text{parcel}}(z_0 + d) = T_{\text{parcel}}(z_0) - \Gamma d$ and $T(z_0 + d) = T(z_0) - \Lambda d$.

- In the case the atmosphere has temperature lapse rate $\Lambda > \Gamma$, we have $T_{\text{parcel}}(z_0 + d) > T(z_0 + d)$, then $\rho < \rho'$. The buoyant force is then larger than the air parcel weight, their resultant is oriented upward and tends to push the parcel away from the equilibrium position.

Conversely, if the air parcel is lowered to the altitude $z_0 - d$ (d>0), $T_{\text{parcel}}(z_0 - d) < T(z_0 - d)$ and then $\rho > \rho'$.

The buoyant force is then smaller than the air parcel weight; their resultant is oriented downward and tends to push the parcel away from the equilibrium position (see Figure 1)

So the equilibrium of the parcel is unstable, and we found that: *An atmosphere with a temperature lapse rate $\Lambda > \Gamma$ is unstable.*

- In an atmosphere with temperature lapse rate $\Lambda < \Gamma$, if the air parcel is lifted to a higher position, at altitude $z_0 + d$ (with d>0), $T_{\text{parcel}}(z_0 + d) < T(z_0 + d)$, then

$\rho > \rho'$. The buoyant force is then smaller than the air parcel weight, their resultant is oriented downward and tends to push the parcel back to the equilibrium position.

Conversely, if the air parcel is lowered to altitude $z_0 - d$ ($d > 0$), $T_{\text{parcel}}(z_0 - d) > T(z_0 - d)$ and then $\rho < \rho'$. The buoyant force is then larger than the air parcel weight, their resultant is oriented upward and tends to push the parcel also back to the equilibrium position (see Figure 2).

So the equilibrium of the parcel is stable, and we found that: *An atmosphere with a temperature lapse rate* $\Lambda < \Gamma$ *is stable.*

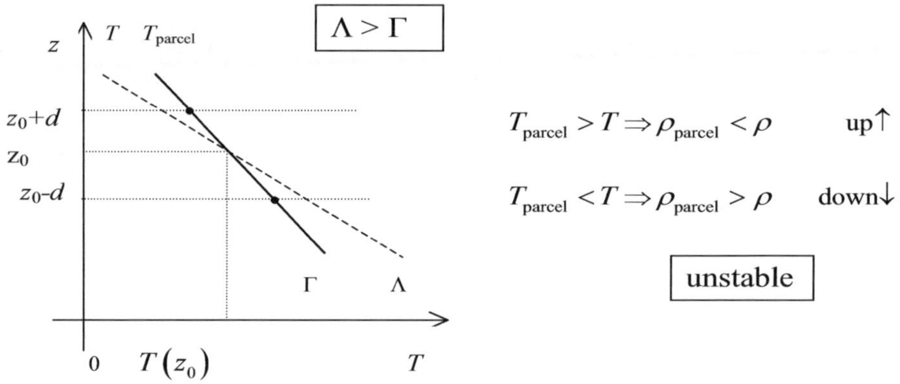

Figure 1

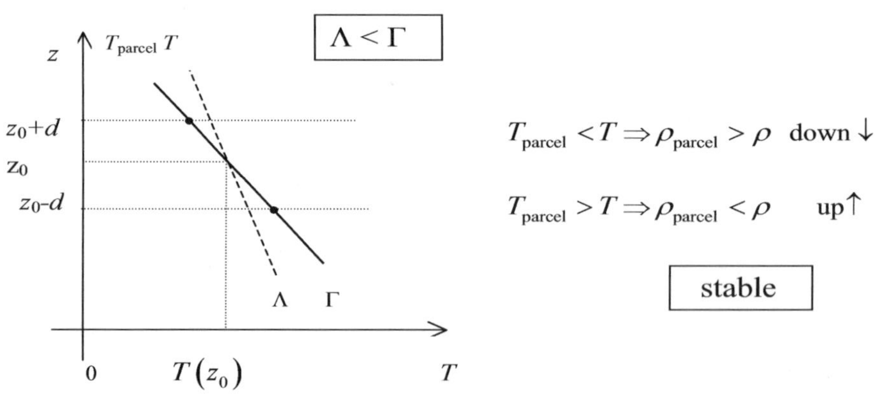

Figure 2

• In an atmosphere with lapse rate $\Lambda = \Gamma$, if the parcel is brought from equilibrium position and put in any other position, it will stay there, the equilibrium is indifferent. *An atmosphere with a temperature lapse rate* $\Lambda = \Gamma$ *is neutral*

3.2. In a stable atmosphere, with $\Lambda < \Gamma$, a parcel, which on ground has temperature $T_{\text{parcel}}(0) > T(0)$ and pressure $p(0)$ equal to that of the atmosphere, can rise and reach a maximal altitude h, where $T_{\text{parcel}}(h) = T(h)$.

In vertical motion from the ground to the altitude h, the air parcel realizes an adiabatic quasi-static process, in which its temperature changes from $T_{\text{parcel}}(0)$ to $T_{\text{parcel}}(h) = T(h)$. Using (11), we can write:

$$\left(1-\frac{\Lambda h}{T(0)}\right)^{-\frac{\Gamma}{\Lambda}} = \frac{T_{\text{parcel}}(0)}{T(h)} = \frac{T_{\text{parcel}}(0)}{T(0)\left(1-\frac{\Lambda h}{T(0)}\right)}$$

$$\left(1-\frac{\Lambda h}{T(0)}\right)^{1-\frac{\Gamma}{\Lambda}} = T_{\text{parcel}}(0)\times T^{-1}(0)$$

$$1-\frac{\Lambda h}{T(0)} = T_{\text{parcel}}^{\frac{\Lambda}{\Lambda\text{-}\Gamma}}(0)\times T^{-\frac{\Lambda}{\Lambda\text{-}\Gamma}}(0)$$

$$h = \frac{1}{\Lambda}T(0)\left[1-T_{\text{parcel}}^{\frac{\Lambda}{\Lambda\text{-}\Gamma}}(0)\times T^{-\frac{\Lambda}{\Lambda\text{-}\Gamma}}(0)\right]$$

$$= \frac{1}{\Lambda}\left[T(0)-T_{\text{parcel}}^{-\frac{\Lambda}{\Lambda-\Gamma}}(0)\,T^{\frac{\Gamma}{\Gamma-\Lambda}}(0)\right]$$

So that the maximal altitude h has the following expression:

$$h = \frac{1}{\Lambda}\left[T(0)-\left(\frac{\left(T(0)\right)^{\Gamma}}{\left(T_{\text{parcel}}(0)\right)^{\Lambda}}\right)^{\frac{1}{\Gamma-\Lambda}}\right] \qquad (13)$$

4.

Using data from the Table, we obtain the plot of z versus T shown in Figure 3.

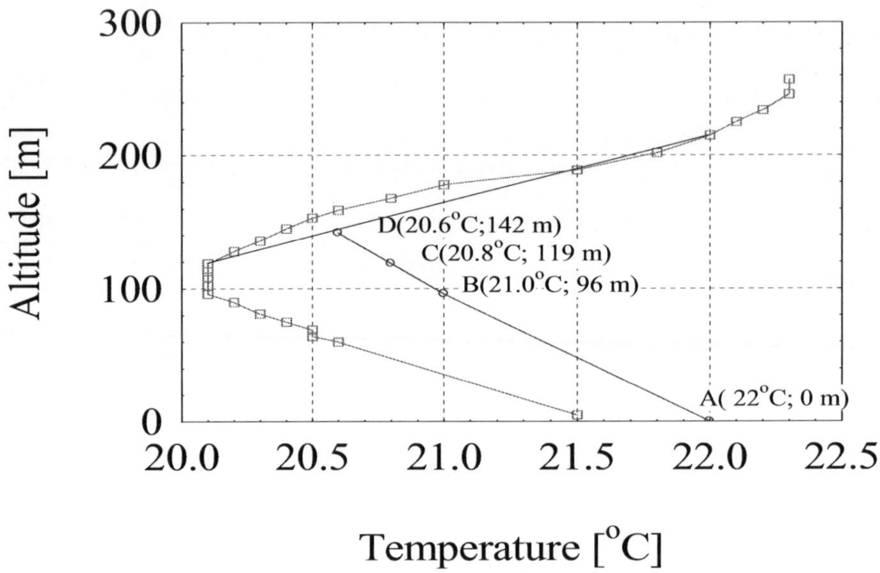

Figure 3

4.1. We can divide the atmosphere under 200m into three layers, corresponding to the following altitudes:

1) $0 < z < 96$ m, $\Lambda_1 = \dfrac{21.5 - 20.1}{91} = 15.4 \times 10^{-3}\,\dfrac{\text{K}}{\text{m}}$.

2) $96\text{ m} < z < 119$ m, $\Lambda_2 = 0$, isothermal layer.

3) $119\text{ m} < z < 215$ m, $\Lambda_3 = -\dfrac{22 - 20.1}{215 - 119} = -0.02\,\dfrac{\text{K}}{\text{m}}$.

In the layer 1), the parcel temperature can be calculated by using (11)

$$T_{\text{parcel}}(96\,\text{m}) = 294.04\text{ K} \approx 294.0\text{ K} \quad \text{that is } 21.0^{\circ}\text{C}$$

In the layer 2), the parcel temperature can be calculated by using its expression in isothermal atmosphere $T_{\text{parcel}}(z) = T_{\text{parcel}}(0)\,\mathbf{exp}\left[-\dfrac{\Gamma z}{T(0)}\right]$.

The altitude 96 m is used as origin, corresponding to 0 m. The altitude 119 m corresponds to 23 m. We obtain the following value for parcel temperature:

$$T_{\text{parcel}}(119\text{ m}) = 293.81\text{ K} \quad \text{that is } 20.8^{\circ}\text{C}$$

4.2. In the layer 3), starting from 119 m, by using (13) we find the maximal elevation $h = 23$ m, and the corresponding temperature 293.6 K (or 20.6 °C).

Finally, the mixing height is

$$H = 119 + 23 = 142\text{ m}.$$

And

$$T_{\text{parcel}}(142\text{ m}) = 293.6\text{ K} \quad \text{that is} \quad 20.6^{\circ}\text{C}$$

From this relation, we can find $T_{\text{parcel}}(119\text{ m}) \approx 293.82\text{ K}$ and $h = 23\text{ m}$.

Note: By using approximate expression (12) we can easily find $T_{\text{parcel}}(z) = 294$ K and 293.8 K at elevations 96 m and 119 m, respectively. At 119 m elevation, the difference between parcel and surrounding air temperatures is 0.7 K (= 293.8 – 293.1), so that the maximal distance the parcel will travel in the third layer is $0.7/(\Gamma - \Lambda_3) = 0.7/0.03 = 23$ m.

5.

Consider a volume of atmosphere of Hanoi metropolitan area being a parallelepiped with height H, base sides L and W. The emission rate of CO gas by motorbikes from 7:00 am to 8:00 am

$$M = 800\,000 \times 5 \times 12 / 3600 = 13\,300 \text{ g/s}$$

The CO concentration in air is uniform at all points in the parallelepiped and denoted by $C(t)$.

5.1. After an elementary interval of time dt, due to the emission of the motorbikes, the mass of CO gas in the box increases by Mdt. The wind blows parallel to the short sides W, bringing away an amount of CO gas with mass $LHC(t)udt$. The remaining part raises the CO concentration by a quantity dC in all over the box. Therefore:

$$Mdt - LHC(t)udt = LWHdC$$

or

$$\frac{dC}{dt}+\frac{u}{W}C(t)=\frac{M}{LWH} \tag{14}$$

5.2. The general solution of (14) is :

$$C(t)=K\,\mathbf{exp}\left(-\frac{ut}{W}\right)+\frac{M}{LHu} \tag{15}$$

From the initial condition $C(0)=0$, we can deduce :

$$C(t)=\frac{M}{LHu}\left[1-\mathbf{exp}\left(-\frac{ut}{W}\right)\right] \tag{16}$$

5.3. Taking as origin of time the moment 7:00 am, then 8:00 am corresponds to t =3600 s. Putting the given data in (15), we obtain :

$$C(3600\text{ s})=6.35\times(1-0.64)=2.3\text{ mg/m}^3$$

DIFFERENTIAL THERMOMETRIC METHOD

In this problem, we use the differential thermometric method to fulfill the two following tasks:

1. Finding the temperature of solidification of a crystalline solid substance.
2. Determining the efficiency of a solar cell.

A. Differential thermometric method

In this experiment forward biased silicon diodes are used as temperature sensors to measure temperature. If the electric current through the diode is constant, then the voltage drop across the diode depends on the temperature according to the relation

$$V(T) = V(T_0) - \alpha(T - T_0) \qquad (1)$$

where $V(T)$ and $V(T_0)$ are respectively the voltage drops across the diode at temperature T and at room temperature T_0 (measured in °C), and the factor

$$\alpha = (2.00 \pm 0.03)\,\mathrm{mV/^{o}C} \qquad (2)$$

The value of $V(T_0)$ may vary slightly from diode to diode.

If two such diodes are placed at different temperatures, the difference between the temperatures can be measured from the difference of the voltage drops across the two diodes. The difference of the voltage drops, called the *differential voltage*, can be measured with high precision; hence the temperature difference can also be measured with high precision. This method is called the *differential thermometric method*. The electric circuit used with the diodes in this experiment is shown in Figure 1. Diodes D_1 and D_2 are forward biased by a 9V battery, through 10 kΩ resistors, R_1 and R_2. This circuit keeps the current in the two diodes approximately constant.

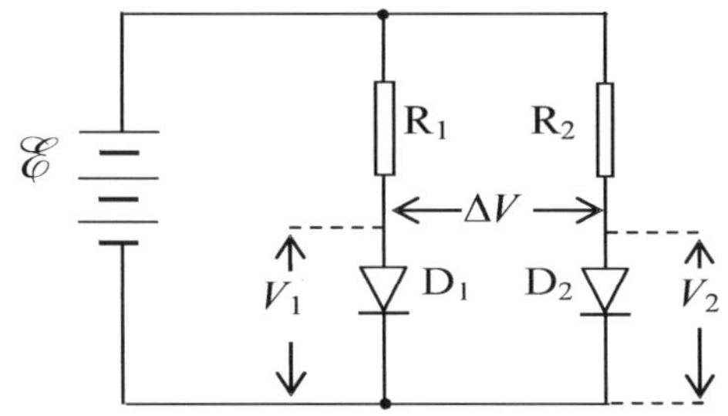

Figure 1. Electric circuit of the diode

If the temperature of diode D_1 is T_1 and that of D_2 is T_2, then according to (1), we have:

$$V_1(T_1) = V_1(T_0) - \alpha(T_1 - T_0)$$

and

$$V_2(T_2) = V_2(T_0) - \alpha(T_2 - T_0)$$

The differential voltage is:

$$\Delta V = V_2(T_2) - V_1(T_1) = V_2(T_0) - V_1(T_0) - \alpha(T_2 - T_1) = \Delta V(T_0) - \alpha(T_2 - T_1)$$

$$\Delta V = \Delta V(T_0) - \alpha \Delta T \qquad (3)$$

in which $\Delta T = T_2 - T_1$. By measuring the differential voltage ΔV, we can determine the temperature difference.

To bias the diodes, we use a circuit box, the diagram of which is shown in Figure 2.

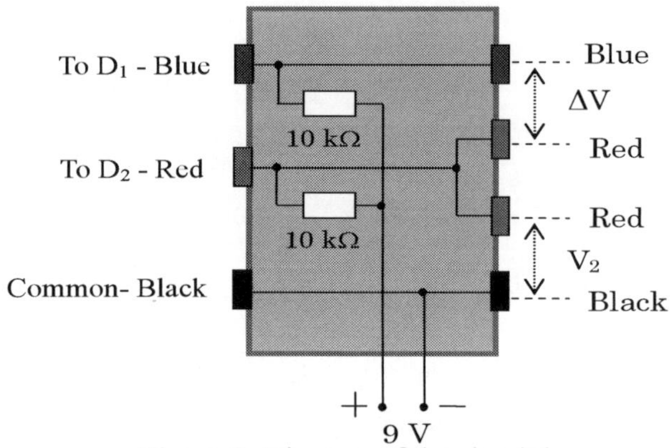

Figure 2. Diagram of the circuit box (top view)

The circuit box contains two biasing resistors of 10 kΩ for the diodes, electrical leads to the 9 V battery, sockets for connecting to the diodes D_1 and D_2, and sockets for connecting to digital multimeters to measure the voltage drop V_2 on diode D_2 and the differential voltage ΔV of the diodes D_1 and D_2.

B. Task 1: Finding the temperature of solidification of a crystalline substance

1. Aim of the experiment

If a crystalline solid substance is heated to the melting state and then cooled down, it solidifies at a fixed temperature T_s, called *temperature of solidification*, also called the *melting point* of the substance. The *traditional method* to determine T_s is to follow the change in temperature with time during the cooling process. Due to the fact that the solidification process is accompanied by the release of the latent heat of the phase transition, the temperature of the substance does not change while the substance is solidifying. If the amount of the substance is large enough, the time interval in which the temperature remains constant is rather long, and one can easily determine this temperature. On the contrary, if the amount of substance is small, this time interval is too short to be observed and hence it is difficult to determine T_s.

In order to determine T_s in case of small amount of substance, we use the *differential thermometric method*, whose principle can be summarized as follows. We use two identical small dishes, one containing a small amount of the substance to be studied, called the *sample dish*, and the other not containing the substance, called the *reference dish*. The two dishes are put on a heat source, whose temperature varies slowly with time. The thermal flows to and from the two dishes are nearly the same. Each dish contains a temperature sensor (a forward biased silicon diode). While there is no phase change in the substance, the temperature T_{samp} of the sample dish and the temperature T_{ref} of the reference dish vary at nearly the same rate, and thus $\Delta T = T_{ref} - T_{samp}$ varies slowly with T_{samp}. If there is a phase change in the substance, and during the phase change T_{samp} does not vary and equals T_s, while T_{ref} steadily varies, then ΔT varies quickly. The plot of ΔT versus T_{samp} shows an abrupt change. The value of T_{samp} corresponding to the abrupt change of ΔT is indeed T_s.

The aim of this experiment is to determine the temperature of solidification T_s of a pure crystalline substance, having T_s in the range from 50°C to 70°C, by using the traditional and differential thermal analysis methods. The amount of substance used in the experiment is about 20 mg.

2. Apparatus and materials

1. The heat source is a 20 W halogen lamp.

2. The dish holder is a bakelite plate with a square hole in it. A steel plate is fixed on the hole. Two small magnets are put on the steel plate.

3. Two small steel dishes, each contains a silicon diode soldered on it. One dish is used as the reference dish, the other - as the sample dish.

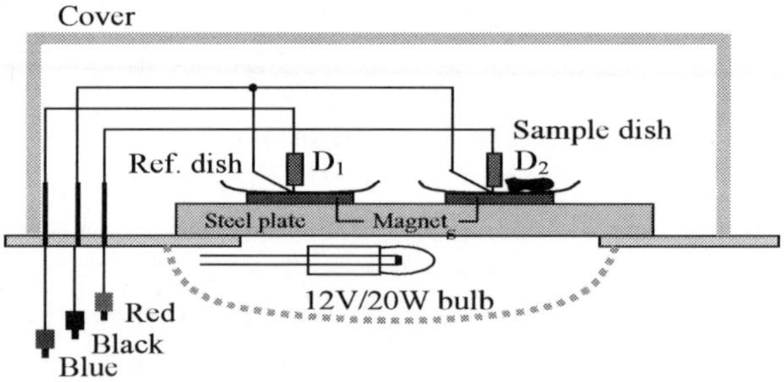

Figure 3. Apparatus for measuring the solidification temperature

Each dish is placed on a magnet. The magnetic force maintains the contact between the dish, the magnet and the steel plate. The magnets also keep a moderate thermal contact between the steel plate and the dishes.

A grey plastic box used as a cover to protect the dishes from the outside influence.

Figure 3 shows the arrangement of the dishes and the magnets on the dish holder and the light bulb.

4. Two digital multimeters are used as voltmeters. They can also measure room temperature by turning the Function selector to the "°C/°F" function. The voltage function of the multimeter has an error of ±2 on the last digit.

Note: to prevent the multimeter (see Figure 9) from going into the "Auto power

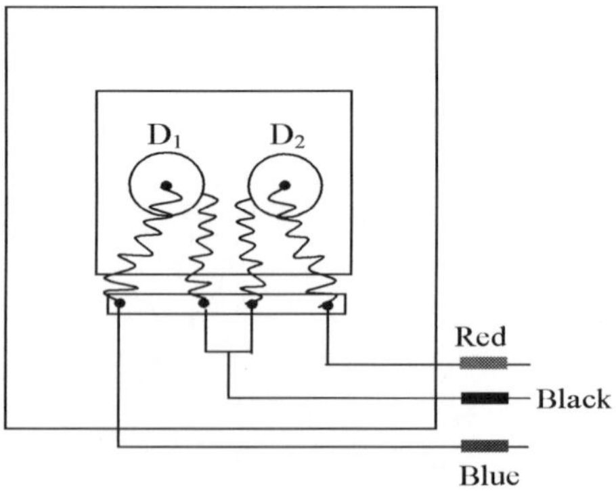

Figure 4. The dishes on the dish holder (top view)

off" function, turn the Function selector from OFF position to the desired function while pressing and holding the SELECT button.

5. A circuit box as shown in Figure 2.
6. A 9 V battery.
7. Electrical leads.
8. A small ampoule containing about 20 mg of the substance to be measured.
9. A stop watch
10. A calculator
11. Graph papers.

3. Experiment

1. The magnets are placed on two equivalent locations on the steel plate. The reference dish and the empty sample dish are put on the magnets as shown in the Figure 4. We use the dish on the left side as the reference dish, with diode D_1 on it (D_1 is called the reference diode), and the dish on the right side as the sample dish, with diode D_2 on it (D_2 is called the measuring diode).

Put the lamp-shade up side down as shown in Figure 5. Do not switch the lamp on. Put the dish holder on the lamp. Connect the apparatuses so that you can measure the voltage drop on the diode D_2, that is $V_{samp} = V_2$, and the differential voltage ΔV.

In order to eliminate errors due to the warming up period of the instruments and devices, it is strongly recommended that the complete measurement circuit be switched on for about 5 minutes before starting real experiments.

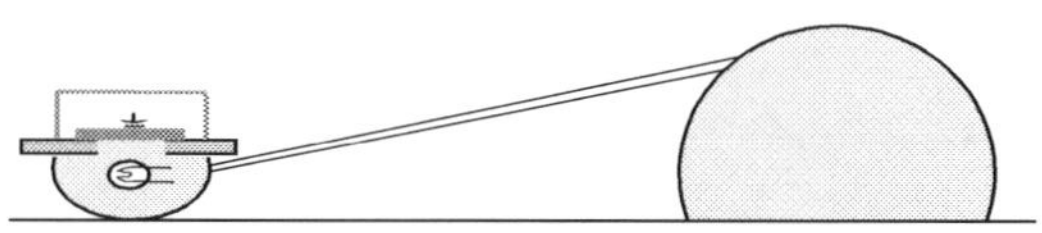

Figure 5.
Using the halogen lamp as a heat source

1.1. Measure the room temperature T_0 and the voltage drop $V_{samp}(T_0)$ across diode D_2 fixed to the sample dish, at room temperature T_0.

1.2. Calculate the voltage drops $V_{samp}(50^\circ C)$, $V_{samp}(70^\circ C)$ and $V_{samp}(80^\circ C)$ on the measuring diode at temperatures 50°C, 70°C and 80°C, respectively.

2. With both dishes still empty, switch the lamp on. Follow V_{sam}. When the temperature of the sample dish reaches $T_{samp} \sim 80^{\circ}C$, switch the lamp off.

2.1. Wait until $T_{samp} \sim 70^{\circ}C$, and then follow the change in V_{samp} and ΔV with time, while the steel plate is cooling down. Note down the values of V_{samp} and ΔV every 10 s to 20 s in the table provided in the answer sheet. If ΔV varies quickly, the time interval between consecutive measurements may be shorter. When the temperature of the sample dish decreases to $T_{samp} \sim 50^{\circ}C$, the measurement is stopped.

2.2. Plot the graph of V_{samp} versus t, called Graph 1, on a graph paper provided.

2.3. Plot the graph of ΔV versus V_{samp}, called Graph 2, on a graph paper provided.

Note: for 2.2 and 2.3 do not forget to write down the correct name of each graph.

3. Pour the substance from the ampoule into the sample dish. Repeat the experiment identically as mentioned in section **2**.

3.1. Write down the data of V_{samp} and ΔV with time t in the table provided in the answer sheet.

3.2. Plot the graph of V_{samp} versus t, called Graph 3, on a graph paper provided.

3.3. Plot the graph of ΔV versus V_{samp}, called Graph 4, on a graph paper provided.

Note: for 3.2 and 3.3 do not forget to write down the correct name of each graph.

4. By comparing the graphs in section **2** and section **3**, determine the temperature of solidification of the substance.

4.1. Using the traditional method to determine T_s: by comparing the graphs of V_{samp} versus t in sections 3 and 2, i.e. Graph 3 and Graph 1, mark the point on Graph 3 where the substance solidifies and determine the value V_s (corresponding to this point) of V_{samp}.

Find out the temperature of solidification T_s of the substance and estimate its error.

4.2. Using the differential thermometric method to determine T_s: by comparing the graphs of ΔV versus V_{samp} in sections 3 and 2, i.e. Graph 4 and Graph 2, mark the point on Graph 4 where the substance solidifies and determine the value V_s of V_{samp}.

Find out the temperature of solidification T_s of the substance.

4.3. From errors of measurement data and instruments, calculate the error of T_s obtained with the differential thermometric method. Write down the error calculations and finally write down the values of T_s together with its error in the answer sheet.

C. Task 2: Determining the efficiency of a solar cell under illumination of an incandescent lamp

1. Aim of the experiment

The aim of the experiment is to determine the *efficiency* of a solar cell under illumination of an incandescent lamp. Efficiency is defined as the ratio of the electrical power that the solar cell can supply to an external circuit, to the total radiant power received by the cell. The efficiency depends on the incident radiation spectrum. In this experiment the radiation incident to the cell is that of an incandescent halogen lamp. In order to determine the efficiency of the solar cell, we have to measure the *irradiance* E at a point situated under the lamp, at a distance d from the lamp along the vertical direction, and the *maximum power* P_{max} of the solar cell when it is placed at this point. In this experiment, d = 12 cm (Figure 6). Irradiance E can be defined by:

$$E = \Phi / S$$

in which Φ is the radiant flux (radiant power), and S is the area of the illuminated surface.

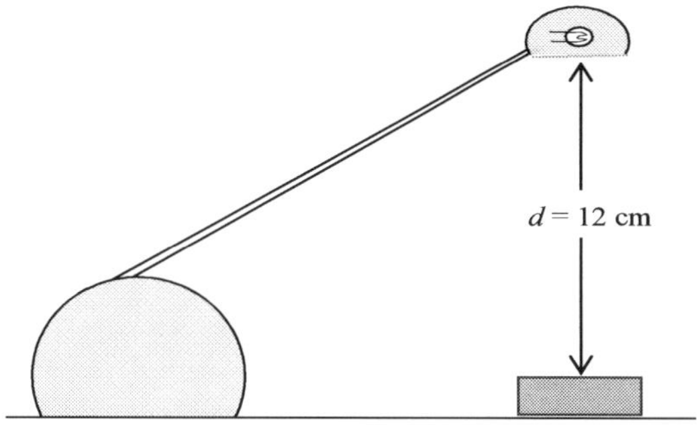

Figure 6. Using the halogen lamp as a light source

2. Apparatus and materials

1. The light source is a 20W halogen lamp.

2. The radiation detector is a hollow cone made of copper, the inner surface of it is blackened with soot (Figure 7). The cone is incompletely thermally isolated from the surrounding. In this experiment, the detector is considered an ideal black body. To measure temperature, we use silicon diodes. The measuring diode is fixed to the radiation detector (D_2 in Figure 1 and Figure 7), so that its temperature equals that of the cone. The reference diode is placed on the inner side of the wall of the box containing the detector; its temperature equals that of the surrounding. The total heat capacity of the detector (the cone and the measuring diode) is $C = (0.69 \pm 0.02)$ J/K. The detector is covered by a very thin polyethylene film; the radiation absorption and reflection of which can be neglected.

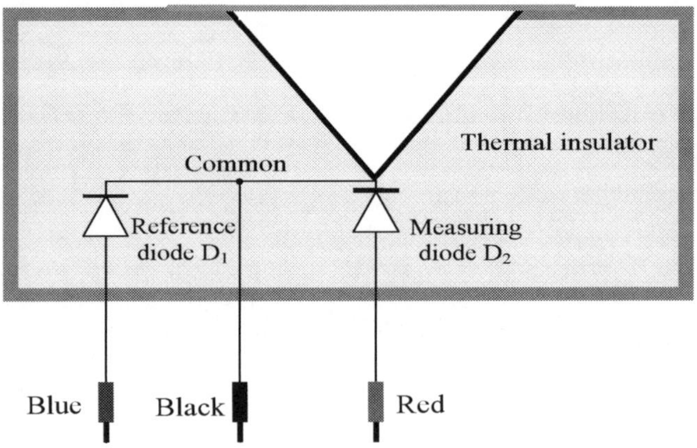

Figure 7. Diagram of the radiation detector

3. A circuit box as shown in Figure 2.

4. A piece of solar cell fixed on a plastic box (Figure 8). The area of the cell includes some metal connection strips. For the efficiency calculation these strips are considered parts of the cell.

5. Two digital multimeters. When used to measure the voltage, they have a very large internal resistance, which can be considered infinitely large. When we use them to measure the current, we cannot neglect their internal resistance. The voltage function of the multimeter has an error of ±2 on the last digit.

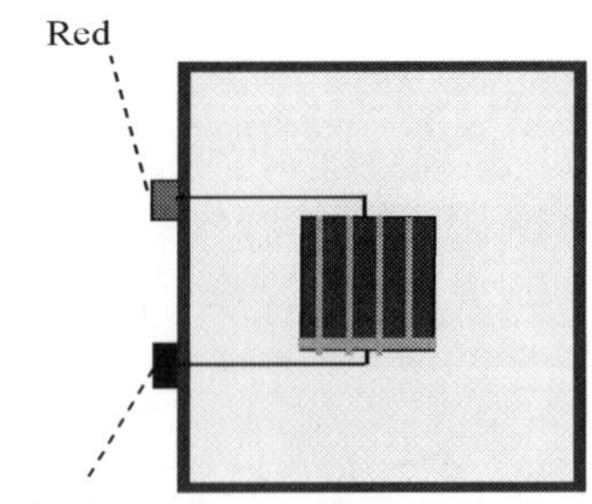

Figure 8.
The solar cell

The multimeters can also measure the room temperature.

Note: to prevent the multimeter (see Figure 9) from going into the "Auto power off" function, turn the Function selector from OFF position to the desired function while pressing and holding the SELECT button.

6. A 9 V battery
7. A variable resistor.
8. A stop watch
9. A ruler with 1mm divisions
10. Electrical leads.
11. Graph papers.

3. Experiment

When the detector receives energy from radiation, it heats up. At the same time, the detector loses its heat by several mechanisms, such as thermal conduction, convection, radiation etc...Thus, the radiant energy received by detector in a time interval dt is equal to the sum of the energy needed to increase the detector temperature and the energy transferred from the detector to the surrounding:

$$\Phi dt = CdT + dQ$$

where C is the heat capacity of the detector and the diode, dT - the temperature increase and dQ - the heat loss.

When the temperature difference between the detector and the surrounding $\Delta T = T - T_0$ is small, we can consider that the heat dQ transferred from the detector to the surrounding in the time interval dt is approximately proportional to ΔT and dt, that is $dQ = k\Delta T dt$, with k being a factor having the dimension of W/K. Hence, assuming that k is constant and ΔT is small, we have:

$$\Phi dt = CdT + k\Delta T dt = Cd(\Delta T) + k\Delta T dt$$

or
$$\frac{d(\Delta T)}{dt} + \frac{k}{C}\Delta T = \frac{\Phi}{C} \qquad (4)$$

The solution of this differential equation determines the variation of the temperature difference ΔT with time t, from the moment the detector begins to receive the light with a constant irradiation, assuming that at t=0, ΔT =0

$$\Delta T(t) = \frac{\Phi}{k}\left(1 - e^{-\frac{k}{C}t}\right) \qquad (5)$$

When the radiation is switched off, the mentioned above differential equation becomes

$$\frac{d(\Delta T)}{dt}+\frac{k}{C}\Delta T=0 \qquad (6)$$

and the temperature difference ΔT varies with the time according to the following formula:

$$\Delta T(t)=\Delta T(0)e^{-\frac{k}{C}t} \qquad (7)$$

where $\Delta T(0)$ is the temperature difference at $t=0$ (the moment when the measurement starts).

1. Determine the room temperature T_0.

2. Compose an electric circuit comprising the diode sensors, the circuit box and the multimeters to measure the temperature of the detector.

In order to eliminate errors due to the warming up period of the instruments and devices, it is strongly recommended that the complete measurement circuit be switched on for about 5 minutes before starting real experiments.

2.1. Place the detector under the light source, at a distance of d = 12 cm to the lamp. The lamp is off. Follow the variation of ΔV for about 2 minutes with sampling intervals of 10 s and determine the value of $\Delta V(T_0)$ in equation (3).

2.2. Switch the lamp on to illuminate the detector. Follow the variation of ΔV. Every 10-15 s, write down a value of ΔV in the table provided in the answer sheet. (Note: columns x and y of the table will be used later in section **4.**). After 2 minutes, switch the lamp off.

2.3. Move the detector away from the lamp. Follow the variation of ΔV for about 2 minutes after that. Every 10-15 s, write down a value of ΔV in the table provided in the answer sheet. (Note: columns x and y of the table will be used later in section **3.**).

Hints: *As the detector has a thermal inertia, it is recommended not to use some data obtained immediately after the moment the detector begins to be illuminated or ceases to be illuminated.*

3. Plot a graph in an x-y system of coordinates, with variables x and y chosen appropriately, in order to prove that after the lamp is switched off, equation (7) is satisfied.

3.1. Write down the expression for variables x and y.

3.2. Plot a graph of y versus x, called Graph 5.

3.3. From the graph, determine the value of k.

4. Plot a graph in an x-y system of coordinates, with variables x and y chosen appropriately, in order to prove that when the detector is illuminated, equation (5) is satisfied.

4.1. Write down the expressions for variables x and y.

4.2. Plot a graph of y versus x, called Graph 6.

4.3. Determine the irradiance E at the orifice of the detector.

5. Put the solar cell to the same place where the radiation detector was. Connect the solar cell to an appropriate electric circuit comprising the multimeters and a variable resistor which is used to change the load of the cell. Measure the current in the circuit and the voltage on the cell at different values of the resistor.

5.1. Draw a diagram of the circuit used in this experiment.

5.2. By rotating the knob of the variable resistor, you change the value of the load. Note the values of current I and voltage V at each position of the knob.

5.3. Plot a graph of the power of the cell, which supplies to the load, as a function of the current through the cell. This is Graph 7.

5.4. From the graph deduce the maximum power P_{max} of the cell and estimate its error.

5.5. Write down the expression for the efficiency of the cell that corresponds to the obtained maximum power. Calculate its value and error.

Contents of the experiment kit (see also Figure 10)

1	Halogen lamp 220 V/ 20 W	9	Stop watch
2	Dish holder	10	Calculator
3	Dish	11	Radiation detector
4	Multimeter	12	Solar cell
5	Circuit box	13	Variable resistor
6	9 V battery	14	Ruler
7	Electrical leads	15	Box used as a cover
8	Ampoule with substance to be measured		

Note: to prevent the multimeter (see Figure 9) from going into the "Auto power off" function, turn the Function selector from OFF position to the desired function while pressing and holding the SELECT button.

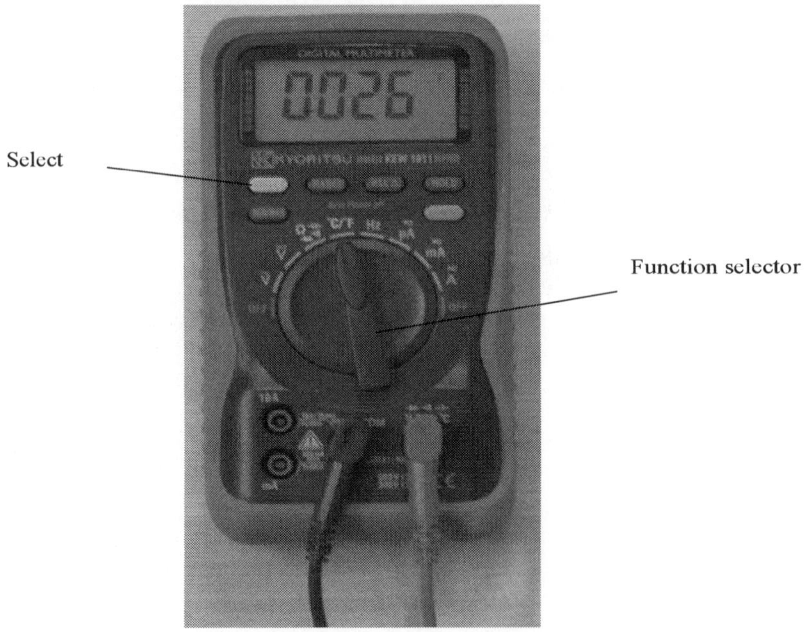

Figure 9. Digital multimeter

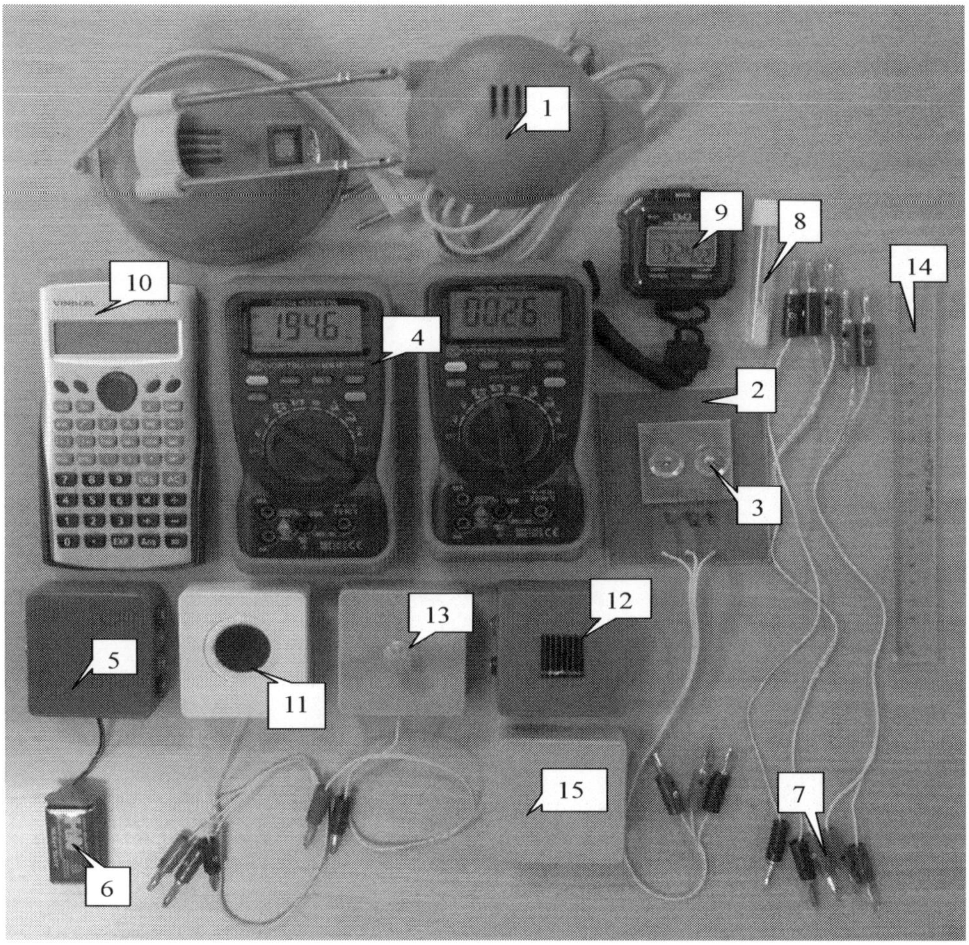

Figure 10. Contents of the experiment kit

Solution

Task 1

1.

1.1. $T_0 =$ 25±1 °C

$$V_{\text{samp}}(T_0) = 573.9 \text{ mV}$$

With different experiment sets, V_{samp} may differ from the above value within ±40 mV.

Note for error estimation:
δV and δV are calculated using the specs of the multimeter: ±0.5% reading digit +2 on the last digit. Example: if V = 500mV, the error $\delta V = 500 \times 0.5\% + 0.2 = 2.7$ mV ≈ 3 mV.

Thus, $V_{samp}(T_0) = 574 \pm 3$ mV.

All values of $V_{samp}(T_0)$ within 505÷585 mV are acceptable.

1.2. Formula for temperature calculation:

From Eq (1): $V_{\text{samp}} = V_{\text{samp}}(T_0) - \alpha(T - T_0)$

$$V_{\text{samp}}(50^{\circ}\text{C}) = 523.9 \text{ mV}$$

$$V_{\text{samp}}(70^{\circ}\text{C}) = 483.9 \text{ mV}$$

$$V_{\text{samp}}(80^{\circ}\text{C}) = 463.9 \text{ mV}$$

Error calculation: $\delta V_{\text{samp}} = \delta V_{\text{samp}}(T_0) + (T - T_0)\delta\alpha$

Example: $V_{\text{samp}} = 495.2$ mV, then $\delta V_{\text{samp}} = 2.7 + 0.03 \times (50 - 25) = 3.45 \text{ mV} \approx 3.5 \text{ mV}$

Thus:

$$V_{\text{samp}}(50^{\circ}\text{C}) = 524 \pm 4 \text{ mV}$$

$$V_{\text{samp}}(70^{\circ}\text{C}) = 484 \pm 4 \text{ mV}$$

$$V_{\text{samp}}\left(80^{\circ}\text{C}\right)=464\pm5\ \text{mV}$$

The same rule for acceptable range of V_{samp} as in 1.1 is applied.

2.

2.1. Data of cooling-down process without sample:

t (s)	V_{samp} (mV) (±3mV)	ΔV (mV) (±0.2mV)
0	492	-0.4
10	493	-0.5
20	493	-0.5
30	494	-0.6
40	495	-0.7
50	496	-0.7
60	497	-0.8
70	497	-0.8
80	498	-0.9
90	499	-1.0
100	500	-1.0
110	500	-1.1
120	501	-1.1
130	502	-1.2
140	503	-1.2
150	503	-1.3
160	504	-1.3
170	504	-1.4
180	505	-1.5
190	506	-1.6
200	507	-1.6
210	507	-1.7
220	508	-1.7
230	508	-1.8
240	509	-1.8
250	509	-1.8
260	510	-1.9
270	511	-1.9

280	512	-1.9
290	512	-2.0
300	513	-2.0
310	514	-2.1
320	515	-2.1
330	515	-2.1
340	516	-2.1
350	516	-2.2
360	517	-2.2
370	518	-2.3
380	518	-2.3
390	519	-2.3
400	520	-2.4
410	520	-2.4
420	521	-2.5
430	521	-2.5
440	522	-2.5
450	523	-2.6
460	523	-2.6

The acceptable range of ΔV is ± 40 mV. There is no fixed rule for the change in ΔV with T (this depends on the positions of the dishes on the plate, etc.)

2.2.

Graph 1

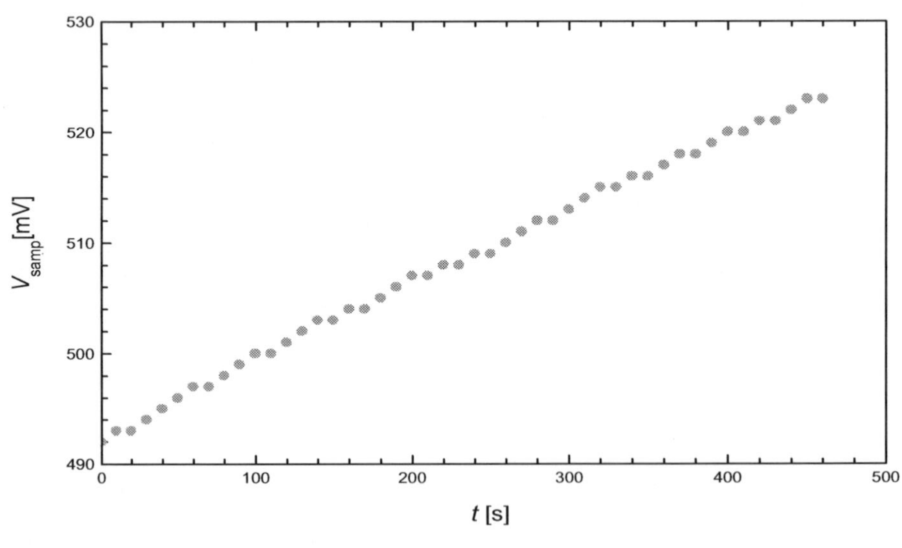

The correct graph should not have any abrupt changes of the slope.

2.3.

Graph 2

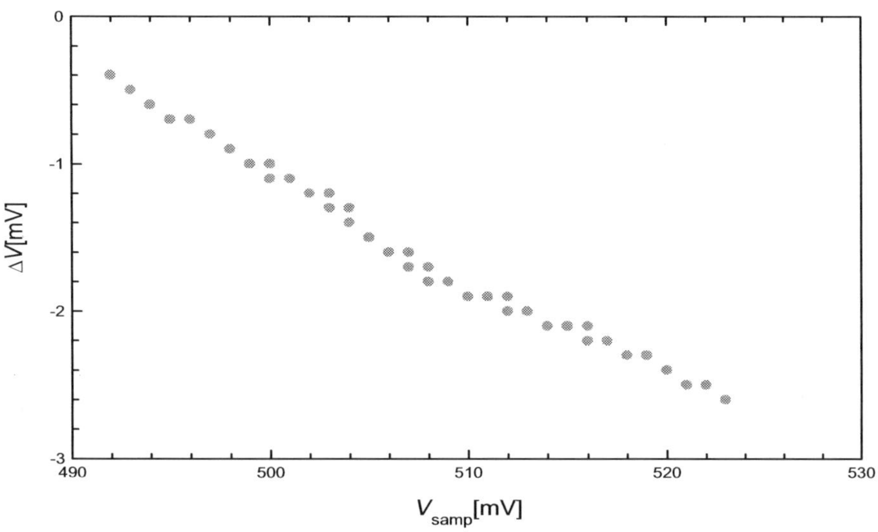

The correct graph should not have any abrupt changes of the slope.

3.

3.1. Dish with substance

t (s)	V_{samp} (mV) (±3mV)	ΔV (mV) (±0.2mV)
0	492	-4.6
10	493	-4.6
20	493	-4.6
30	494	-4.6
40	495	-4.6
50	496	-4.6
60	497	-4.6
70	497	-4.5
80	498	-4.5
90	499	-4.5
100	500	-4.5
110	500	-4.5
120	501	-4.5

130	502	-4.6
140	503	-4.6
150	503	-5.1
160	503	-5.6
170	503	-6.2
180	503	-6.5
190	504	-6.6
200	505	-6.5
210	506	-6.4
220	507	-6.3
230	507	-6.1
240	508	-5.9
250	509	-5.7
260	510	-5.5
270	511	-5.3
280	512	-5.1
290	512	-5.0
300	513	-4.9
310	514	-4.8
320	515	-4.7
330	515	-4.7
340	516	-4.6
350	516	-4.6
360	517	-4.5
370	518	-4.5
380	518	-4.4
390	519	-4.4
400	520	-4.4
410	520	-4.4
420	521	-4.4
430	521	-4.3
440	522	-4.3
450	523	-4.3
460	523	-4.3

3.2.

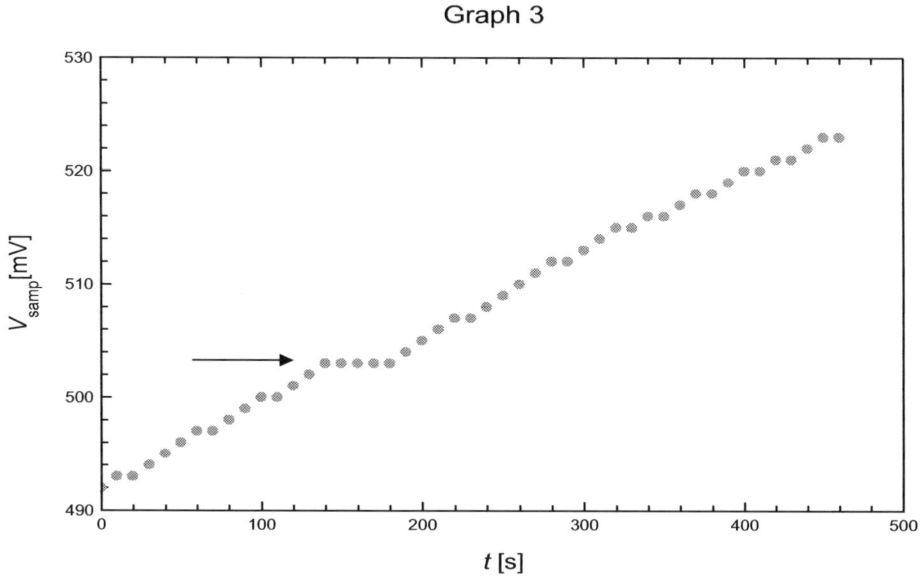

The correct Graph 3 should contain a short plateau as marked by the arrow in the above figure.

3.3.

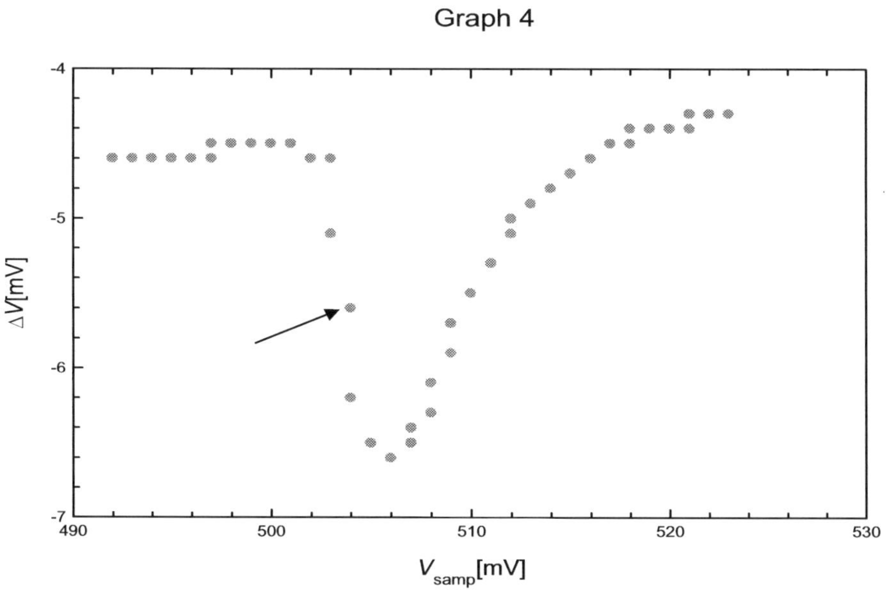

The correct Graph 4 should have an abrupt change in ΔV, as shown by the arrow in the above figure.

Note: when the dish contains the substance, values of ΔV may change compared to those without the substance.

4.

4.1. V_s is shown in Graph 3. Value V_s = (503±3) mV. From that, T_s = 60.5 °C can be deduced.

4.2. V_s is shown in Graph 4. Value V_s = (503±3) mV. From that, T_s = 60.5 °C can be deduced.

4.3. Error calculations, using root mean square method:

Error of T_s: $T_s = T_0 + \dfrac{V(T_0) - V(T_s)}{\alpha} = T_0 + A$, in which A is an intermediate variable.

Therefore error of T_s can be written as $\delta T_s = \sqrt{(\delta T_0)^2 + (\delta A)^2}$, in which δ... is the error.

Error for A is calculated separately:

$$\delta A = \frac{V(T_0) - V(T_s)}{\alpha} \sqrt{\left\{ \frac{\delta[V(T_0) - V(T_s)]}{V(T_0) - V(T_s)} \right\}^2 + \left(\frac{\delta\alpha}{\alpha} \right)^2}$$

in which we have:

$$\delta[V(T_0) - V(T_s)] = \sqrt{[\delta V(T_0)]^2 + [\delta V(T_s)]^2}$$

Errors of other variables in this experiment:

δT_0=1°C

$\delta V(T_0)$= 3 mV, read on the multimeter.

$\delta\alpha$ = 0.03 mV/°C

$\delta V(T_s) \approx$ 3 mV

From the above constituent errors we have:

$$\delta[V(T_0) - V(T_s)] \approx 4.24\,mV$$

$$\delta A \approx 2.1°\text{C}$$

Finally, the error of T_s is: $\delta T_s \approx 2.5°\text{C}$

Hence, the final result is: $T_s = 60 \pm 2.5\ °\text{C}$

Note: if the student uses any other reasonable error calculation method that leads to approximately the same result, it is also accepted.

Task 2

1.

1.1. $T_0 = 26 \pm 1°C$

2.

2.1. Measured data with the lamp off

t (s)	$\Delta V(T_0)$ (mV) ($\pm$0.2mV)
0	19.0
10	19.0
20	19.0
30	19.0
40	19.0
50	18.9
60	18.9
70	18.9
80	18.9
90	18.9
100	19.0
110	19.0
120	19.0

Values of $\Delta V(T_0)$ can be different from one experiment set to another. The acceptable values lie in between -40÷+40 mV.

2.2. Measured data with the lamp on

t (s)	ΔV (mV) ($\pm$0.2mV)
0	19.5
10	21.9
20	23.8
30	25.5
40	26.9
50	28.0
60	29.0
70	29.9
80	30.7
90	31.4

100	32.0
110	32.4
120	32.9

When illuminated (by the lamp) values of ΔV may change 10 ÷ 20 mV compared to the initial situation (lamp off).

2.3. Measured data after turning the lamp off

t (s)	ΔV (mV) (±0.2mV)
0	23.2
10	22.4
20	21.6
30	21.0
40	20.5
50	20.1
60	19.6
70	19.3
80	18.9
90	18.6
100	18.4
110	18.2
120	17.9

3. Plotting graph 5 and calculating k

3.1. $x = t$; $y = \ln\left[\Delta V\left(T_0\right) - \Delta V\left(t\right)\right]$

Note: other reasonable ways of writing expressions for x and y that also leads to a linear relationship using **ln** are also accepted.

3.2. Graph 5

Graph 5

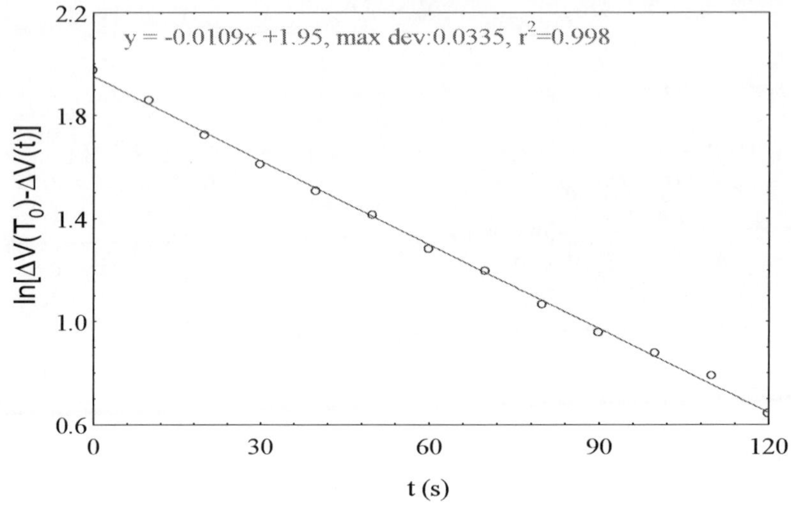

3.3. Calculating k: $\dfrac{k}{C} =$ 0.0109 s^{-1} and $C = 0.69$ J/K, thus: $k = 7.52\times10^{-3}$ W/K

Note: Error of k will be calculated in 5.5. Students are not asked to give error of k in this step. The acceptable value of k lies in between $6\times10^{-3} \div 9\times10^{-3}$ W/K depending on the experiment set.

4. Plotting Graph 6 and calculating E

4.1. $x = \left[1 - \exp\left(\dfrac{-kt}{C}\right)\right]$; $y = \left|\Delta V\left(T_0\right) - \Delta V\left(t\right)\right|$

4.2.

Graph 6

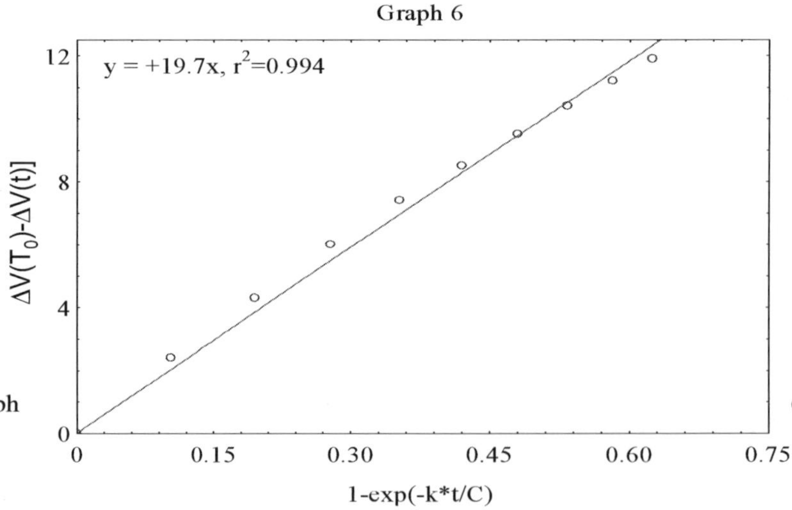

Graph 6 should

be substantially linear, with the slope in between 15÷25 mV, depending on the experiment set.

4.3. From the slope of Graph 6 and the area of the detector orifice we obtain E = 140 W/m^2. The area of the detector orifice is

$$S_{\text{det}} = \pi R_{\text{det}}{}^2 = \pi \times (13 \times 10^{-3})^2 = 5.30 \times 10^{-4}\ \text{m}^2 \ \text{ with error: } \frac{\delta R_{\text{det}}}{R_{\text{det}}} = 5\%$$

Error of E will be calculated in 5.5. Students are not asked to give error of E in this step. The acceptable value of E lies in between 120 ÷ 160 W/m^2, depending on the experiment set.

5.

5.1. Circuit diagram:

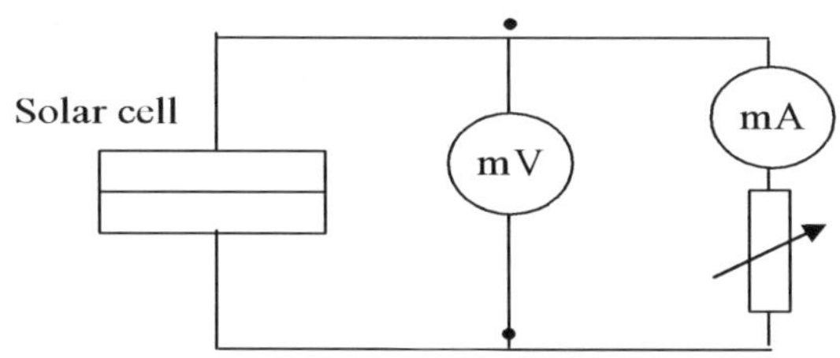

5.2. Measurements of V and I

V (mV) (±0.3÷3mV)	I (mA) (±0.05÷0.1mA)	P (mW)
18.6 ±0.3	11.7	0.21
33.5	11.7	0.39
150	11.5	1.72
157	11.6	1.82
182 ±1	11.4	2.08
267	11.2	3.00
402 ±2	9.23	3.70
448	6.70	3.02
459	5.91	2.74
468	5.07	2.37
473 ±3	4.63	2.20
480	3.81	1.86
485	3.24	1.57

487	3.12	1.54
489	3.13	1.55

5.3.

Graph 7

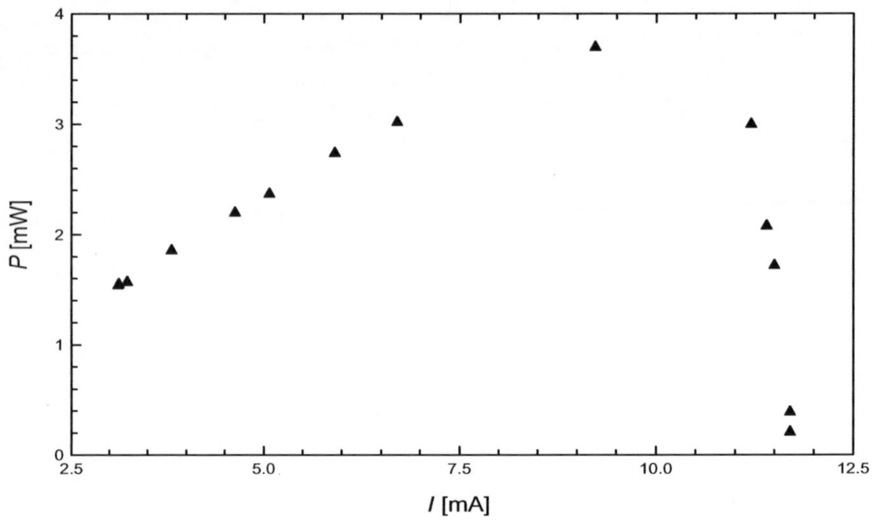

5.4. $P_{\max} = 3.7 \pm 0.2$ mW

The acceptable value of $P_{\max}$ lies in between 3÷4.5 mW, depending on the experiment set.

5.5. Expression for the efficiency

$$S_{\text{cell}} = 19 \times 24\,\text{mm}^2 = 450 \times 10^{-6}\,\text{m}^2$$

Then $\eta_{\max} = \dfrac{P_{\max}}{E \times S_{\text{cell}}} = 0.058$

Error calculation:

$$\delta\eta_{\max} = \eta_{\max}\sqrt{\left(\frac{\delta P_{\max}}{P_{\max}}\right)^2 + \left(\frac{\delta E}{E}\right)^2 + \left(\frac{\delta S_{\text{cell}}}{S_{\text{cell}}}\right)^2},$$ in which S_{cell} is the area of the solar cell.

$\dfrac{\delta P_{\max}}{P_{\max}}$ is estimated from Graph 7, typical value ≈ 6 %

$\frac{\delta S_{\text{cell}}}{S_{\text{cell}}}$: error from the millimeter measurement (with the ruler), typical value ≈ 5 %

E is calculated from averaging the ratio (using Graph 6):

$$B = \frac{\Delta V(T_0) - \Delta V(t)}{1 - \mathbf{exp}\left(-\frac{k}{C}t\right)} = \frac{E\pi R_{\text{det}}^{\;2}\alpha}{k}$$

in which B is an intermediate variable, R_{det} is the radius of the detector orifice.

$$E = \frac{kB}{\pi R_{\text{det}}^{\;2}\alpha}$$

Calculation of error of E:

$$\overline{\left(\frac{\delta E}{E}\right)} = \sqrt{\left(\frac{\delta k}{k}\right)^2 + \left(\frac{\delta B}{B}\right)^2 + 4\left(\frac{\delta R_{\text{det}}}{R_{\text{det}}}\right)^2 + \left(\frac{\delta \alpha}{\alpha}\right)^2}$$

k is calculated from the regression of:

$$\Delta T = \Delta T(0)\mathbf{exp}\left(-\frac{k}{C}t\right), \text{ hence } \mathbf{ln}\,\Delta T = \mathbf{ln}\,\Delta T(0) - \frac{k}{C}t$$

We set $k/C = m$ then $k = mC$

From the regression, we can calculate the error of m:

$$\frac{\delta m}{m} \approx 2(1-r) \approx 0.2\%$$

$$\frac{\delta k}{k} = \sqrt{\left(\frac{\delta m}{m}\right)^2 + \left(\frac{\delta C}{C}\right)^2}$$

We derive the expression for the error of $\eta_{\max}$:

$$\delta\eta_{\max} = \eta_{\max}\sqrt{\left(\frac{\delta P_{\max}}{P_{\max}}\right)^2 + \left(\frac{\delta S_{\text{cell}}}{S_{\text{cell}}}\right)^2 + \left(\frac{\delta B}{B}\right)^2 + 4\left(\frac{\delta R_{\text{det}}}{R_{\text{det}}}\right)^2 + \left(\frac{\delta m}{m}\right)^2 + \left(\frac{\delta C}{C}\right)^2 + \left(\frac{\delta \alpha}{\alpha}\right)^2}$$

Typical values for $\eta_{\max}$ and other constituent errors:

$\eta_{\max} \approx 0.058$

$\frac{\delta P_{\max}}{P_{\max}} = 5\%$; $\frac{\delta B}{B} \approx 0.6\%$; $\frac{\delta m}{m} \approx 0.2\%$; $\frac{\delta S_{\text{cell}}}{S_{\text{cell}}} \approx 5\%$; $\frac{\delta R_{\text{det}}}{R_{\text{det}}} \approx 5\%$;

$$\frac{\delta C}{C} \approx 3\%\,;\ \frac{\delta k}{k} \approx 3\%\,;\ \frac{\delta E}{E} \approx 10.5\%\,;\ \frac{\delta \alpha}{\alpha} \approx 1.5\%$$

Finally:

$$\frac{\delta \eta_{\max}}{\eta_{\max}} = 12.7\%;\ \ \delta \eta_{\max} \approx 0.0074$$

and

$$\eta_{\max} = (5.8 \pm 0.8)\%$$

Note: if the student uses any other reasonable error method that leads to approximately the same result, it is also accepted.

40TH

INTERNATIONAL PHYSICS OLYMPIAD

MEXICO

2009

THEORETICAL PROBLEM No. 1

EVOLUTION OF THE EARTH-MOON SYSTEM

Scientists can determine the distance Earth-Moon with great precision. They achieve this by bouncing a laser beam on special mirrors deposited on the Moon's surface by astronauts in 1969, and measuring the round travel time of the light (see Figure 1).

Figure 1. A laser beam sent from an observatory is used to measure accurately the distance between the Earth and the Moon.

With these observations, they have directly measured that the Moon is slowly receding from the Earth. That is, the Earth-Moon distance is increasing with time. This is happening because due to tidal torques the Earth is transferring angular momentum to the Moon, see Figure 2. In this problem you will derive the basic parameters of the phenomenon.

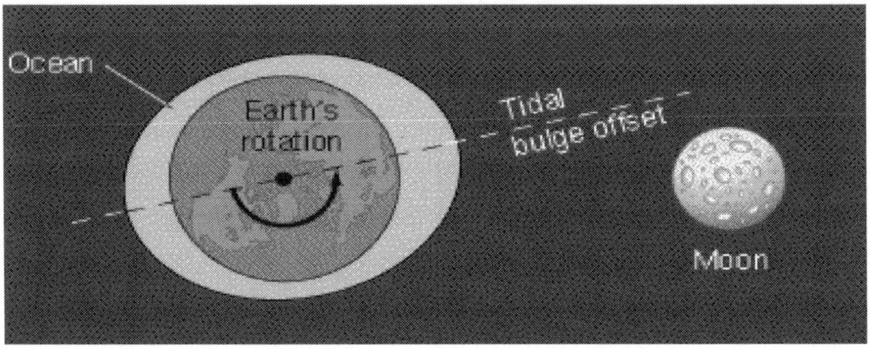

Figure 2. The Moon's gravity produces tidal deformations or "bulges" in the Earth. Because of the Earth's rotation, the line that goes through the bulges is not aligned with the line between the Earth and the Moon. This misalignment produces a torque that transfers angular momentum from the Earth's rotation to the Moon's translation. The drawing is not to scale.

1. Conservation of Angular Momentum.

Let L_1 be the present total angular momentum of the Earth-Moon system. Now, make the following assumptions: i) L_1 is the sum of the rotation of the Earth around its axis and the translation of the Moon in its orbit around the Earth only. ii) The Moon's orbit is circular and the Moon can be taken as a point. iii) The Earth's axis of rotation and the Moon's axis of revolution are parallel. iv) To simplify the calculations, we take the motion to be around the center of the Earth and not the center of mass. Throughout the problem, all moments of inertia, torques and angular momenta are defined around the axis of the Earth. v) Ignore the influence of the Sun.

1a	Write down the equation for the present total angular momentum of the Earth-Moon system. Set this equation in terms of I_E, the moment of inertia of the Earth; ω_{E1}, the present angular frequency of the Earth's rotation; I_{M1}, the present moment of inertia of the Moon with respect to the Earth's axis; and ω_{M1}, the present angular frequency of the Moon's orbit.	0.2

This process of transfer of angular momentum will end when the period of rotation of the Earth and the period of revolution of the Moon around the Earth have the same duration. At this point the tidal bulges produced by the Moon on the Earth will be aligned with the line between the Moon and the Earth and the torque will disappear.

1b	Write down the equation for the final total angular momentum L_2 of the Earth-Moon system. Make the same assumptions as in Question 1a. Set this equation in terms of I_E, the moment of inertia of the Earth; ω_2, the final angular frequency of the Earth's rotation and Moon's translation; and I_{M2}, the final moment of inertia of the Moon.	0.2

1c	Neglecting the contribution of the Earth's rotation to the final total angular momentum, write down the equation that expresses the angular momentum conservation for this problem.	0.3

2. Final Separation and Final Angular Frequency of the Earth-Moon System.

Assume that the gravitational equation for a circular orbit (of the Moon around the Earth) is always valid. Neglect the contribution of the Earth's rotation to the final total angular momentum.

2a	Write down the gravitational equation for the circular orbit of the Moon around the Earth, at the final state, in terms of M_E, ω_2, G and the final separation D_2 between the Earth and the Moon. M_E is the mass of the Earth and G is the gravitational constant.	0.2

2b	Write down the equation for the final separation D_2 between the Earth and the Moon in terms of the known parameters, L_1, the total angular momentum of the system, M_E and M_M, the masses of the Earth and Moon, respectively, and G.	0.5

2c	Write down the equation for the final angular frequency ω_2 of the Earth-Moon system in terms of the known parameters L_1, M_E, M_M and G.	0.5

Below you will be asked to find the numerical values of D_2 and ω_2. For this you need to know the moment of inertia of the Earth.

2d	Write down the equation for the moment of inertia of the Earth I_E assuming it is a sphere with inner density ρ_i from the center to a radius r_i, and with outer density ρ_o from the radius r_i to the surface at a radius r_o (see Figure 3).	0.5

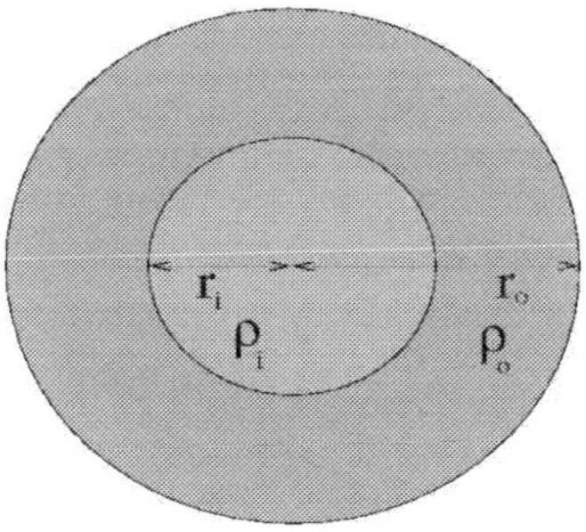

Figure 3. The Earth as a sphere with two densities, ρ_i and ρ_o.

Determine the numerical values requested in this problem always to *two significant digits*.

2e	Evaluate the moment of inertia of the Earth I_E, using $\rho_i = 1.3 \times 10^4$ kg m^{-3}, $r_i = 3.5 \times 10^6$ m, $\rho_o = 4.0 \times 10^3$ kg m^{-3}, and $r_o = 6.4 \times 10^6$ m.	0.2

The masses of the Earth and Moon are $M_E = 6.0 \times 10^{24}$ kg and $M_M = 7.3 \times 10^{22}$ kg, respectively. The present separation between the Earth and the Moon is $D_1 = 3.8 \times 10^8$ m. The present angular frequency of the Earth's rotation is $\omega_{E1} = 7.3 \times 10^{-5}$ s^{-1}. The present angular frequency of the Moon's translation around the Earth is $\omega_{M1} = 2.7 \times 10^{-6}$ s^{-1}, and the gravitational constant is $G = 6.7 \times 10^{-11}$ m^3 kg^{-1} s^{-2}.

2f	Evaluate the numerical value of the total angular momentum of the system, L_1.	0.2

2g	Find the final separation D_2 in meters and in units of the present separation D_1.	0.3

2h	Find the final angular frequency ω_2 in s^{-1}, as well as the final duration of the day in units of present days.	0.3

Verify that the assumption of neglecting the contribution of the Earth's rotation to the final total angular momentum is justified by finding the ratio of the final angular momentum of the Earth to that of the Moon. This should be a small quantity.

2i	Find the ratio of the final angular momentum of the Earth to that of the Moon.	0.2

3. How much is the Moon receding per year?

Now, you will find how much the Moon is receding from the Earth each year. For this, you will need to know the equation for the torque acting at present on the Moon. Assume that the tidal bulges can be approximated by two point masses, each of mass m, located on the surface of the Earth, see Fig. 4. Let θ be the angle between the line that goes through the bulges and the line that joins the centers of the Earth and the Moon.

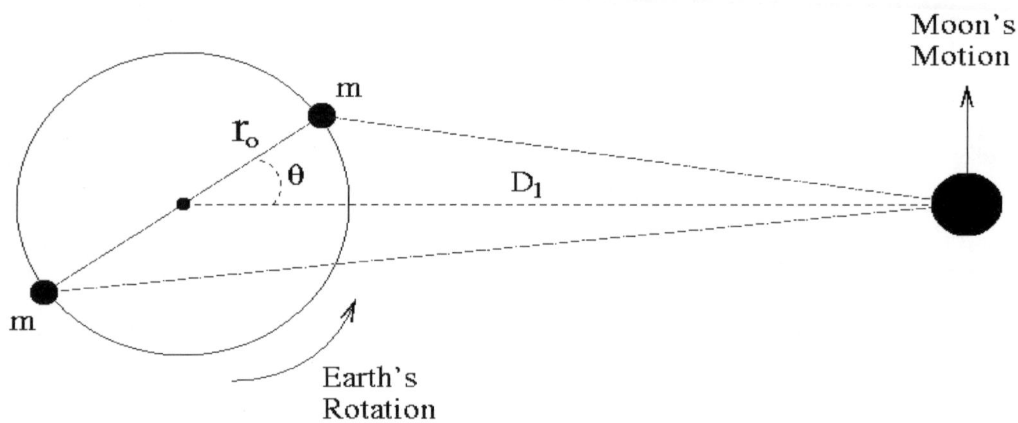

Figure 4. Schematic diagram to estimate the torque produced on the Moon by the bulges on the Earth. The drawing is not to scale.

3a	Find F_c, the magnitude of the force produced on the Moon by the closest point mass.	0.4

3b	Find F_f, the magnitude of the force produced on the Moon by the farthest point mass.	0.4

You may now evaluate the torques produced by the point masses.

3c	Find the magnitude of τ_c, the torque produced by the closest point mass.	0.4

3d	Find the magnitude of τ_f, the torque produced by the farthest point mass.	0.4

3e	Find the magnitude of the total torque τ produced by the two masses. Since $r_o << D_1$ you should approximate your expression to lowest significant order in r_o / D_1. You may use that $(1+x)^a$ ..$1+ax$, if $x<<1$.	1.0

3f	Calculate the numerical value of the total torque τ, taking into account that $\theta = 3^\circ$ and that $m = 3.6 \propto 10^{16}$ kg (note that this mass is of the order of 10^{-8} times the mass of the Earth).	0.5

Since the torque is the rate of change of angular momentum with time, find the increase in the distance Earth-Moon at present, per year. For this step, express the angular momentum of the Moon in terms of M_M, M_E, D_1 and G only.

3g	Find the increase in the distance Earth-Moon at present, per year.	1.0

Finally, estimate how much the length of the day is increasing each year.

3h	Find the decrease of ω_{E1} per year and how much is the length of the day at present increasing each year.	1.0

4. Where is the energy going?

In contrast to the angular momentum, that is conserved, the total (rotational plus gravitational) energy of the system is not. We will look into this in this last section.

4a	Write down an equation for the total (rotational plus gravitational) energy of the Earth-Moon system at present, E. Put this equation in terms of I_E, ω_{E1}, M_M, M_E, D_1 and G only.	0.4

4b	Write down an equation for the change in E, ΔE, as a function of the changes in D_1 and in ω_{E1}. Evaluate the numerical value of ΔE for a year, using the values of changes in D_1 and in ω_{E1} found in questions 3g and 3h.	0.4

Verify that this loss of energy is consistent with an estimate for the energy dissipated as heat in the tides produced by the Moon on the Earth. Assume that the tides rise, on the average by 0.5 m, a layer of water $h =$ 0.5 m deep that covers the surface of the Earth (for simplicity assume that all the surface of the Earth is covered with water). This happens twice a day. Further assume that 10% of this gravitational energy is dissipated as heat due to viscosity when the water descends. Take the density of water to be $\rho_{water} = 10^3$ kg m^{-3}, and the gravitational acceleration on the surface of the Earth to be $g = 9.8$ m s^{-2}.

4c	What is the mass of this surface layer of water?	0.2

4d	Calculate how much energy is dissipated in a year? How does this compare with the energy lost per year by the Earth-Moon system at present?	0.3

THEORETICAL PROBLEM No. 1

EVOLUTION OF THE EARTH-MOON SYSTEM

SOLUTIONS

1. Conservation of Angular Momentum

1a	$L_1 = I_E \omega_{E1} + I_{M1} \omega_{M1}$	0.2

1b	$L_2 = I_E \omega_2 + I_{M2} \omega_2$	0.2

1c	$I_E \omega_{E1} + I_{M1} \omega_{M1} = I_{M2} \omega_2 = L_1$	0.3

2. Final Separation and Angular Frequency of the Earth-Moon System.

2a	$\omega_2^2 D_2^3 = GM_E$	0.2

2b	$D_2 = \dfrac{L_1^2}{GM_E M_M^2}$	0.5

2c	$\omega_2 = \dfrac{G^2 M_E^2 M_M^3}{L_1^3}$	0.5

2d	The moment of inertia of the Earth will be the addition of the moment of inertia of a sphere with radius r_o and density ρ_o and of a sphere with radius r_i and density $\rho_i - \rho_o$: $I_E = \dfrac{2}{5}\dfrac{4\pi}{3}[r_o^5 \rho_o + r_i^5 (\rho_i - \rho_o)]$.	0.5

2e	$I_E = \dfrac{2}{5}\dfrac{4\pi}{3}[r_o^5 \rho_o + r_i^5 (\rho_i - \rho_o)] = 8.0 \times 10^{37}$ kg m^2	0.2

2f	$L_1 = I_E \omega_{E1} + I_{M1} \omega_{M1} = 3.4 \times 10^{34}$ kg m^2 s^{-1}	0.2

2g	$D_2 = 5.4\times10^8$ m, that is $D_2 = 1.4 D_1$	0.3

2h	$\omega_2 = 1.6\times10^{-6}$ s^{-1}, that is, a period of 46 days.	0.3

2i	Since $I_E\,\omega_2 = 1.3\times10^{32}$ kg m^2 s^{-1} and $I_{M2}\,\omega_2 = 3.4\times10^{34}$ kg m^2 s^{-1}, the approximation is justified since the final angular momentum of the Earth is 1/260 of that of the Moon.	0.2

3. How much is the Moon receding per year?

3a	Using the law of cosines, the magnitude of the force produced by the mass m closest to the Moon will be: $F_c = \dfrac{G\,m\,M_M}{D_1^2 + r_o^2 - 2D_1 r_o \cos(\theta)}$	0.4

3b	Using the law of cosines, the magnitude of the force produced by the mass m farthest to the Moon will be: $F_f = \dfrac{G\,m\,M_M}{D_1^2 + r_o^2 + 2D_1 r_o \cos(\theta)}$	0.4

3c	Using the law of sines, the torque will be $\tau_c = F_c \dfrac{\sin(\theta)\, r_0\, D_1}{[D_1^2 + r_o^2 - 2\,D_1\, r_o \cos(\theta)]^{1/2}} = \dfrac{G\,m\,M_M \sin(\theta)\, r_0\, D_1}{[D_1^2 + r_o^2 - 2D_1 r_o \cos(\theta)]^{3/2}}$	0.4

3d	Using the law of sines, the torque will be $\tau_f = F_f \dfrac{\sin(\theta)\, r_0\, D_1}{[D_1^2 + r_o^2 + 2\,D_1\, r_o \cos(\theta)]^{1/2}} = \dfrac{G\,m\,M_M \sin(\theta)\, r_0\, D_1}{[D_1^2 + r_o^2 + 2D_1 r_o \cos(\theta)]^{3/2}}$	0.4

3e	$\tau_c - \tau_f = G\,m\,M_M \sin(\theta) r_0\, D_1^{-2}\,(1 - \dfrac{3r_o^2}{2D_1^2} + \dfrac{3r_o\cos(\theta)}{D_1} - 1 + \dfrac{3r_o^2}{2D_1^2} + \dfrac{3r_o\cos(\theta)}{D_1})$ $= \dfrac{6G\,m\,M_M\, r_o^2 \sin(\theta)\cos(\theta)}{D_1^3}$	1.0

3f	$\tau = \frac{6 G m M_M r_o^2 \sin(\theta)\cos(\theta)}{D_1^3} = 4.1\times10^{16}\,\text{N m}$	0.5

3g	Since $\omega_{M1}^2 D_1^3 = G M_E$, we have that the angular momentum of the Moon is $I_{M1}\,\omega_{M1} = M_M\,D_1^2\left[\frac{G\,M_E}{D_1^3}\right]^{1/2} = M_M\,[D_1\,G\,M_E]^{1/2}$ The torque will be: $\tau = \frac{M_M\,[G M_E]^{1/2}\,\Delta(D_1^{1/2})}{\Delta t} = \frac{M_M\,[G M_E]^{1/2}\,\Delta D_1}{2[D_1]^{1/2}\,\Delta t}$ So, we have that $\Delta D_1 = \frac{2\,\tau\,\Delta t}{M_M}\left[\frac{D_1}{G M_E}\right]^{1/2}$ That for $\Delta t = 3.1\times10^7$ s = 1 year, gives $\Delta D_1 = 0.034$ m. This is the yearly increase in the Earth-Moon distance.	1.0

3h	We now use that $\tau = -\frac{I_E\,\Delta\omega_{E1}}{\Delta t}$ from where we get $\Delta\omega_{E1} = -\frac{\tau\,\Delta t}{I_E}$ that for $\Delta t = 3.1\times10^7$ s = 1 year gives $\Delta\omega_{E1} = -1.6\times10^{-14}\,\text{s}^{-1}$. If P_E is the period of time considered, we have that: $\frac{\Delta P_E}{P_E} = -\frac{\Delta\omega_{E1}}{\omega_E}$ since $P_E = 1\,day = 8.64\times10^4$ s, we get $\Delta P_E = 1.9\times10^{-5}$ s. This is the amount of time that the day lengthens in a year.	1.0

4. Where is the energy going?

4a	The present total (rotational plus gravitational) energy of the system is: $E = \frac{1}{2} I_E\,\omega_{E1}^2 + \frac{1}{2} I_M\,\omega_{M1}^2 - \frac{G M_E M_M}{D_1}$. Using that $\omega_{M1}^2\,D_1^3 = G\,M_E$, we get	0.4

	$E=\frac{1}{2}I_E\,\omega_{E1}^2-\frac{1}{2}\frac{GM_EM_M}{D_1}$	
4b	$\Delta E=I_E\,\omega_{E1}\,\Delta\omega_{E1}+\frac{1}{2}\frac{GM_EM_M}{D_1^2}\Delta D_1$, that gives $\Delta E=-9.0\times10^{19}$ J	0.4
4c	$M_{water}=4\pi r_o^2\times h\times\rho_{water}$ kg $=2.6\times10^{17}$ kg.	0.2
4d	$\Delta E_{water}=-g\,M_{water}\times0.5\,m\times2\,day^{-1}\times365\,days\times0.1=-9.3\times10^{19}$ J. Then, the two energy estimates are comparable.	0.3

THEORETICAL PROBLEM 2

DOPPLER LASER COOLING AND OPTICAL MOLASSES

The purpose of this problem is to develop a simple theory to understand the so-called "laser cooling" and "optical molasses" phenomena. This refers to the cooling of a beam of neutral atoms, typically alkaline, by counterpropagating laser beams with the same frequency. This is part of the Physics Nobel Prize awarded to S. Chu, P. Phillips and C. Cohen-Tannoudji in 1997.

The image above shows sodium atoms (the bright spot in the center) trapped at the intersection of three orthogonal pairs of opposing laser beams. The trapping region is called "optical molasses" because the dissipative optical force resembles the viscous drag on a body moving through molasses.

In this problem you will analyze the basic phenomenon of the interaction between a photon incident on an atom and the basis of the dissipative mechanism in one dimension.

PART I: BASICS OF LASER COOLING

Consider an atom of mass m moving in the $+x$ direction with velocity v. For simplicity, we shall consider the problem to be one-dimensional, namely, we shall ignore the y and z directions (see figure 1). The atom has two internal energy levels. The energy of the lowest state is considered to be zero and the energy of the excited state to be $\hbar\omega_0$, where $\hbar = h/2\pi$. The atom is initially in the lowest state. A laser beam with frequency ω_L in the laboratory is directed in the $-x$ direction and it is incident on the atom. Quantum mechanically the laser is composed of a large number of photons, each with energy $\hbar\omega_L$ and momentum $-\hbar q$. A photon can be absorbed by the atom and later spontaneously emitted; this emission can occur with equal probabilities along the $+x$ and $-x$ directions. Since the atom moves at non-relativistic speeds, $v/c \ll 1$ (with c the speed of light) keep terms up to first order in this quantity only. Consider also $\hbar q/mv \ll 1$, namely, that the momentum of the atom is much larger than the momentum of a single photon. In writing your answers, keep only corrections linear in either of the above quantities.

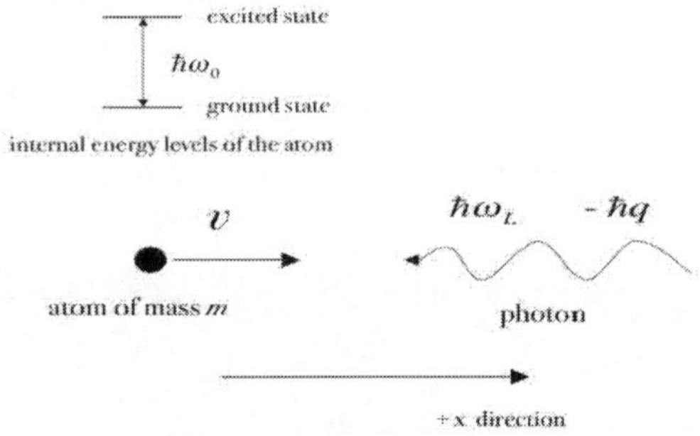

Fig.1 Sketch of an atom of mass m with velocity v in the $+x$ direction, colliding with a photon with energy $\hbar\omega_L$ and momentum $-\hbar q$. The atom has two internal states with energy difference $\hbar\omega_0$.

Assume that the laser frequency ω_L is tuned such that, as seen by the moving atom, it is in resonance with the internal transition of the atom. Answer the following questions:

1. Absorption.

1a	Write down the resonance condition for the absorption of the photon.	0.2

1b	Write down the momentum p_{at} of the atom after absorption, as seen in the laboratory.	0.2

1c	Write down the total energy ε_{at} of the atom after absorption, as seen in the laboratory.	0.2

2. Spontaneous emission of a photon in the $-x$ direction.

At some time after the absorption of the incident photon, the atom may emit a photon in the $-x$ direction.

2a	Write down the energy of the emitted photon, ε_{ph}, after the emission process in the $-x$ direction, as seen in the laboratory.	0.2

2b	Write down the momentum of the emitted photon p_{ph}, after the emission process in the $-x$ direction, as seen in the laboratory.	0.2

2c	Write down the momentum of the atom p_{at}, after the emission process in the $-x$ direction, as seen in the laboratory.	0.2

2d	Write down the total energy of the atom ε_{at}, after the emission process in the $-x$ direction, as seen in the laboratory.	0.2

3. Spontaneous emission of a photon in the $+x$ direction.

At some time after the absorption of the incident photon, the atom may instead emit a photon in the $+x$ direction.

3a	Write down the energy of the emitted photon, ε_{ph}, after the emission process in the $+x$ direction, as seen in the laboratory.	0.2

3b	Write down the momentum of the emitted photon p_{ph}, after the emission process in the $+x$ direction, as seen in the laboratory.	0.2

3c	Write down the momentum of the atom p_{at}, after the emission process in the $+x$ direction, as seen in the laboratory.	0.2

3d	Write down the total energy of the atom ε_{at}, after the emission process in the $+x$ direction, as seen in the laboratory.	0.2

4. Average emission after the absorption.

The spontaneous emission of a photon in the $-x$ or in the $+x$ directions occurs with the same probability. Taking this into account, answer the following questions.

4a	Write down the average energy of an emitted photon, ε_{ph}, after the emission process.	0.2

4b	Write down the average momentum of an emitted photon p_{ph}, after the emission process.	0.2

4c	Write down the average total energy of the atom ε_{at}, after the emission process.	0.2

4d	Write down the average momentum of the atom p_{at}, after the emission process.	0.2

5. Energy and momentum transfer.

Assuming a complete one-photon absorption-emission process only, as described above, there is a net average momentum and energy transfer between the laser radiation and the atom.

5a	Write down the average energy change $\Delta\varepsilon$ of the atom after a complete one-photon absorption-emission process.	0.2

5b	Write down the average momentum change Δp of the atom after a complete one-photon absorption-emission process.	0.2

6. Energy and momentum transfer by a laser beam along the $+x$ direction.

Consider now that a laser beam of frequency ω_L' is incident on the atom along the $+x$ direction, while the atom moves also in the $+x$ direction with velocity v. Assuming a resonance condition between the internal transition of the atom and the laser beam, as seen by the atom, answer the following questions:

6a	Write down the average energy change $\Delta\varepsilon$ of the atom after a complete one-photon absorption-emission process.	0.3

6b	Write down the average momentum change Δp of the atom after a complete one-photon absorption-emission process.	0.3

PART II: DISSIPATION AND THE FUNDAMENTALS OF OPTICAL MOLASSES

Nature, however, imposes an inherent uncertainty in quantum processes. Thus, the fact that the atom can spontaneously emit a photon in a *finite* time after absorption, gives as a result that the resonance condition does not have to be obeyed *exactly* as in the discussion above. That is, the frequency of the laser beams ω_L and ω_L' may have any value and the absorption-emission process can still occur. These will happen with different (quantum) probabilities and, as one should expect, the maximum probability is found at the exact resonance condition. On the average, the time elapsed between a single process of absorption and emission is called the lifetime of the excited energy level of the atom and it is denoted by Γ^{-1}.

Consider a collection of N atoms at *rest* in the laboratory frame of reference, and a laser beam of frequency ω_L incident on them. The atoms absorb and emit continuously such that there is, on average, N_{exc} atoms in the excited state (and therefore, $N - N_{exc}$ atoms in the ground state). A quantum mechanical calculation yields the following result:

$$N_{exc} = N\frac{\Omega_R^2}{(\omega_0 - \omega_L)^2 + \frac{\Gamma^2}{4} + 2\Omega_R^2}$$

where ω_0 is the resonance frequency of the atomic transition and Ω_R is the so-called Rabi frequency; Ω_R^2 is proportional to the *intensity* of the laser beam. As mentioned above, you can see that this number is different from zero even if the resonance frequency ω_0 is different from the frequency of the laser beam ω_L. An alternative way of expressing the previous result is that the number of absorption-emission processes per unit of time is $N_{exc}\Gamma$.

Consider the physical situation depicted in Figure 2, in which two counter propagating laser beams with the *same* but *arbitrary* frequency ω_L are incident on a gas of N atoms that move in the $+x$ direction with velocity v.

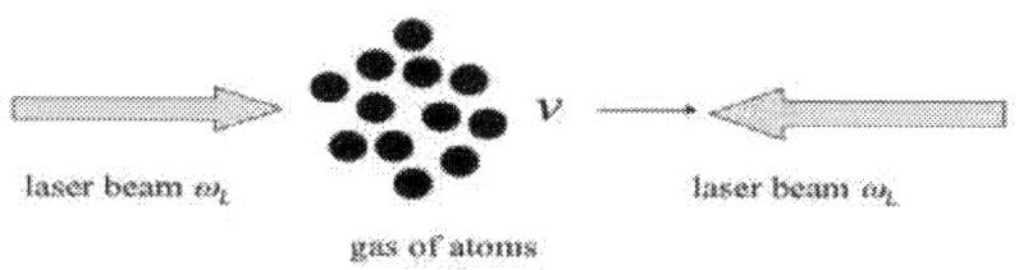

Figure 2. Two counter propagating laser beams with the *same* but *arbitrary* frequency ω_L are incident on a gas of N atoms that move in the $+x$ direction with velocity v.

7. Force on the atomic beam by the lasers.

7a	With the information found so far, find the force that the lasers exert on the atomic beam. You should assume that $mv >> \hbar q$.	1.5

8. Low velocity limit.

Assume now that the velocity of the atoms is small enough, such that you can expand the force up to first order in v.

8a	Find an expression for the force found in Question (7a), in this limit.	1.5

Using this result, you can find the conditions for speeding up, slowing down, or no effect at all on the atoms by the laser radiation.

8b	Write down the condition to obtain a positive force (speeding up the atoms).	0.25

8c	Write down the condition to obtain a zero force.	0.25

8d	Write down the condition to obtain a negative force (slowing down the atoms).	0.25

8e	Consider now that the atoms are moving with a velocity $-v$ (in the $-x$ direction). Write down the condition to obtain a slowing down force on the atoms.	0.25

9. Optical molasses.

In the case of a negative force, one obtains a frictional dissipative force. Assume that initially, at $t=0$, the gas of atoms has velocity v_0.

9a	In the limit of low velocities, find the velocity of the atoms after the laser beams have been on for a time τ.	1.5

9b	Assume now that the gas of atoms is in thermal equilibrium at a temperature T_0. Find the temperature T after the laser beams have been on for a time τ.	0.5

This model does not allow you to go to arbitrarily low temperatures.

THEORETICAL PROBLEM 2

SOLUTION

DOPPLER LASER COOLING AND OPTICAL MOLASSES

The key to this problem is the Doppler effect (to be precise, the longitudinal Doppler effect): The frequency of a monochromatic beam of light detected by an observer depends on its state of motion relative to the emitter, i.e. the observed frequency is

$$\omega' = \omega\sqrt{\frac{1 \pm v/c}{1 \mp v/c}} \approx \omega\left(1 \pm \frac{v}{c}\right)$$

where v is the relative speed of emitter and observer and ω the frequency of the emitter. The upper-lower signs correspond, respectively, when source and observer move towards or away from each other. The second equality holds in the limit of low velocities (non-relativistic limit).

The frequency of the laser in the lab is ω_L; ω_0 is the transition frequency of the atom; the atom moves with speed v towards the incident direction of the laser:

It is important to point out that the results must be given to first significant order in v/c or $\hbar q/mv$.

PART I: BASICS OF LASER COOLING

1. Absorption.

1a	Write down the resonance condition for the absorption of the photon. $\omega_0 \approx \omega_L\left(1 + \frac{v}{c}\right)$	0.2

1b	Write down the momentum p_{at} of the atom after absorption, as seen in the laboratory $p_{at} = p - \hbar q \approx mv - \frac{\hbar\omega_L}{c}$	0.2

1c	Write down the energy ε_{at} of the atom after absorption, as seen in the laboratory $\varepsilon_{at} = \frac{p_{at}^2}{2m} + \hbar\omega_0 \approx \frac{mv^2}{2} + \hbar\omega_L$	0.2

2. Spontaneous emission in the $-x$ direction.

First, one calculates the energy of the emitted photon, as seen in the lab reference frame. One must be careful to keep the correct order; this is because the velocity of the atom changes after the absorption, however, this is second order correction for the emitted frequency:

$$\omega_{ph} \approx \omega_0\left(1-\frac{v'}{c}\right) \quad \text{with} \quad v' \approx v-\frac{\hbar q}{m}$$

thus,

$$\begin{aligned}\omega_{ph} &\approx \omega_0\left(1-\frac{v}{c}+\frac{\hbar q}{mc}\right)\\ &\approx \omega_L\left(1+\frac{v}{c}\right)\left(1-\frac{v}{c}+\frac{\hbar q}{mc}\right)\\ &\approx \omega_L\left(1+\frac{\hbar q}{mc}\right)\\ &\approx \omega_L\left(1+\left(\frac{\hbar q}{mv}\right)\left(\frac{v}{c}\right)\right)\\ &\approx \omega_L\end{aligned}$$

2a	Write down the energy of the emitted photon, ε_{ph}, after the emission process in the $-x$ direction, as seen in the laboratory. $\varepsilon_{ph} \approx \hbar\omega_L$	0.2

2b	Write down the momentum of the emitted photon p_{ph}, after the emission process in the $-x$ direction, as seen in the laboratory. $p_{ph} \approx -\hbar\omega_L / c$	0.2

Use conservation of momentum (see 1b):

$$p_{at}+p_{ph} \approx p-\hbar q$$

2c	Write down the momentum of the atom p_{at}, after the emission process in the $-x$ direction, as seen in the laboratory. $p_{at} \approx p = mv$	0.2

2d	Write down the energy of the atom ε_{at}, after the emission process in the $-x$ direction, as seen in the laboratory. $\varepsilon_{at} \approx \frac{p^2}{2m}=\frac{mv^2}{2}$	0.2

3. Spontaneous emission in the $+x$ direction.

The same as in the previous questions, keeping the right order

3a	Write down the energy of the emitted photon, ε_{ph}, after the emission process in the $+x$ direction, as seen in the laboratory. $\varepsilon_{ph} \approx \hbar\omega_0\left(1+\frac{v}{c}\right) \approx \hbar\omega_L\left(1+\frac{v}{c}\right)\left(1+\frac{v}{c}\right) \approx \hbar\omega_L\left(1+2\frac{v}{c}\right)$	0.2

3b	Write down the momentum of the emitted photon p_{ph}, after the emission process in the $+x$ direction, as seen in the laboratory. $p_{ph} \approx \frac{\hbar\omega_L}{c}\left(1+2\frac{v}{c}\right)$	0.2

3c	Write down the momentum of the atom p_{at}, after the emission process in the $+x$ direction, as seen in the laboratory. $p_{at} = p - \hbar q - p_{ph} \approx p - \hbar q - \frac{\hbar\omega_L}{c}\left(1+2\frac{v}{c}\right) \approx mv - 2\frac{\hbar\omega_L}{c}$	0.2

3d	Write down the energy of the atom ε_{at}, after the emission process in the $+x$ direction, as seen in the laboratory. $\varepsilon_{at} = \frac{p_{at}^2}{2m} \approx \frac{mv^2}{2}\left(1-2\frac{\hbar q}{mv}\right)$	0.2

4. Average emission after absorption.

The spontaneous emission processes occur with equal probabilities in both directions.

4a	Write down the average energy of an emitted photon, ε_{ph}, after the emission process. $\varepsilon_{ph} = \frac{1}{2}\varepsilon_{ph}^{+} + \frac{1}{2}\varepsilon_{ph}^{-} \approx \hbar\omega_L\left(1+\frac{v}{c}\right)$	0.2

4b	Write down the average momentum of an emitted photon p_{ph}, after the emission process. $p_{ph} = \frac{1}{2}p_{ph}^{+} + \frac{1}{2}p_{ph}^{-} \approx \frac{\hbar\omega_L}{c}\frac{v}{c} = mv\left(\frac{\hbar q}{mv}\frac{v}{c}\right) \approx 0$ second order	0.2

4c	Write down the average energy of the atom ε_{at}, after the emission process. $\varepsilon_{at} = \frac{1}{2}\varepsilon_{at}^{+} + \frac{1}{2}\varepsilon_{at}^{-} \approx \frac{mv^2}{2}\left(1-\frac{\hbar q}{mv}\right)$	0.2

4d	Write down the average momentum of the atom p_{at}, after the emission process. $p_{at} = \frac{1}{2}p_{at}^{+} + \frac{1}{2}p_{at}^{-} \approx p - \frac{\hbar\omega_L}{c}$	0.2

5. Energy and momentum transfer.

Assuming a complete one-photon absorption-emission process only, as described above, there is a net average momentum and energy transfer between the laser and the atom.

5a	Write down the average energy change $\Delta\varepsilon$ of the atom after a complete one-photon absorption-emission process. $\Delta\varepsilon = \varepsilon_{at}^{after} - \varepsilon_{at}^{before} \approx -\frac{1}{2}\hbar qv = -\frac{1}{2}\hbar\omega_L\frac{v}{c}$	0.2

5b	Write down the average momentum change Δp of the atom after a complete one-photon absorption-emission process. $\Delta p = p_{at}^{after} - p_{at}^{before} \approx -\hbar q = -\frac{\hbar\omega_L}{c}$	0.2

6. Energy and momentum transfer by a laser beam along the $+x$ direction.

6a	Write down the average energy change $\Delta\varepsilon$ of the atom after a complete one-photon absorption-emission process. $\Delta\varepsilon = \varepsilon_{at}^{after} - \varepsilon_{at}^{before} \approx +\frac{1}{2}\hbar qv = +\frac{1}{2}\hbar\omega_L'\frac{v}{c}$	0.3

6b	Write down the average momentum change Δp of the atom after a complete one-photon absorption-emission process. $\Delta p = p_{at}^{after} - p_{at}^{before} \approx +\hbar q = +\frac{\hbar\omega_L'}{c}$	0.3

PART II: DISSIPATION AND THE FUNDAMENTALS OF OPTICAL MOLASSES

Two counterpropagating laser beams with the *same* but *arbitrary* frequency ω_L are incident on a beam of N atoms that move in the $+x$ direction with (average) velocity v.

7. Force on the atomic beam by the lasers.

On the average, the fraction of atoms found in the excited state is given by,

$$P_{exc}=\frac{N_{exc}}{N}=\frac{\Omega_R^2}{(\omega_0-\omega_L)^2+\frac{\Gamma^2}{4}+2\Omega_R^2}$$

where ω_0 is the resonance frequency of the atoms and Ω_R is the so-called Rabi frequency; Ω_R^2 is proportional to the *intensity* of the laser beam. The lifetime of the excited energy level of the atom is Γ^{-1}.

The force is calculated as the number of absorption-emission cycles, times the momentum exchange in each event, divided by the time of each event. CAREFUL! One must take into account the Doppler shift of each laser, as seen by the atoms:

7a	With the information found so far, find the force that the lasers exert on the atomic beam. You must assume that $mv >> \hbar q$. $F=N\Delta p^- P_{exc}^-\Gamma+N\Delta p^+ P_{exc}^+\Gamma$ $=\left(\frac{\Omega_R^2}{\left(\omega_0-\omega_L+\omega_L\frac{v}{c}\right)^2+\frac{\Gamma^2}{4}+2\Omega_R^2}-\frac{\Omega_R^2}{\left(\omega_0-\omega_L-\omega_L\frac{v}{c}\right)^2+\frac{\Gamma^2}{4}+2\Omega_R^2}\right)N\Gamma\hbar q$	1.5

8. Low velocity limit.

Assume now the velocity to be small enough in order to expand the force to first order in v.

8a	Find an expression for the force found in Question (7a), in this limit. $F\approx-\frac{4N\hbar q^2\Omega_R^2\Gamma}{\left((\omega_0-\omega_L)^2+\frac{\Gamma^2}{4}+2\Omega_R^2\right)^2}(\omega_0-\omega_L)v$	1.5

8b	Write down the condition to obtain a positive force (speeding up the atom). $\omega_0<\omega_L$	0.25

8c	Write down the condition to obtain a zero force. $\omega_0=\omega_L$	0.25

8d	Write down the condition to obtain a negative force (slowing down the atom). $\omega_0 > \omega_L$ ... this is the famous rule "tune below resonance for cooling down"	0.25

8e	Consider now that the atoms are moving with a velocity $-v$ (in the $-x$ direction). Write down the condition to obtain a slowing down force on the atoms. $\omega_0 > \omega_L$ ... i.e. independent of the direction motion of the atom.	0.25

9. Optical molasses

In the case of a negative force, one obtains a frictional dissipative force. Assume that initially, $t = 0$, the gas of atoms has velocity v_0.

9a	In the limit of low velocities, find the velocity of the atoms after the laser beams have been on for a time τ. $F = -\beta v \Rightarrow m\frac{dv}{dt} \approx -\beta v$ β can be read from (8a) $\Rightarrow v = v_0 e^{-\beta t/m}$	1.5

9b	Assume now that the gas of atoms is in thermal equilibrium at a temperature T_0. Find the temperature T after the laser beams have been on for a time τ. Recalling that $\frac{1}{2}mv^2 = \frac{1}{2}kT$ in 1 dimension, and using v as the average thermal velocity in the equation of (9a), we can write down $T = T_0 e^{-2\beta t/m}$	0.5

THEORETICAL PROBLEM No. 3

WHY ARE STARS SO LARGE?

The stars are spheres of hot gas. Most of them shine because they are fusing hydrogen into helium in their central parts. In this problem we use concepts of both classical and quantum mechanics, as well as of electrostatics and thermodynamics, to understand why stars have to be big enough to achieve this fusion process and also derive what would be the mass and radius of the smallest star that can fuse hydrogen.

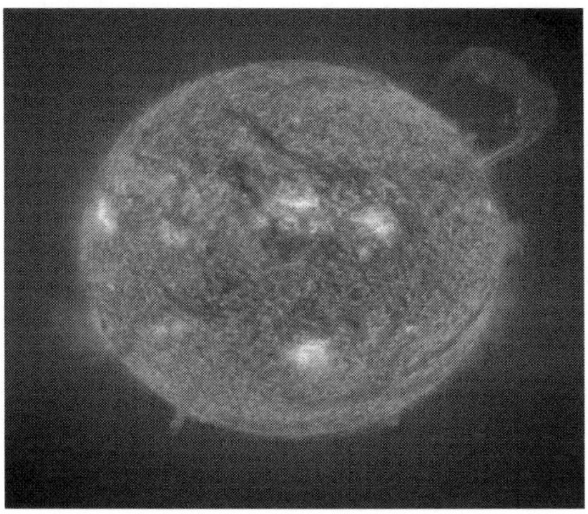

Figure 1. Our Sun, as most stars, shines as a result of thermonuclear fusion of hydrogen into helium in its central parts.

USEFUL CONSTANTS

Gravitational constant = $G = 6.7 \propto 10^{-11}$ m^3 kg^{-1} s^2

Boltzmann's constant = $k = 1.4 \propto 10^{-23}$ J K^{-1}

Planck's constant = $h = 6.6 \propto 10^{-34}$ m^2 kg s^{-1}

Mass of the proton = $m_p = 1.7 \propto 10^{-27}$ kg

Mass of the electron = $m_e = 9.1 \propto 10^{-31}$ kg

Unit of electric charge = $q = 1.6 \propto 10^{-19}$ C

Electric constant (vacuum permittivity) = $\varepsilon_0 = 8.9 \propto 10^{-12}$ C^2 N^{-1} m^{-2}

Radius of the Sun = $R_S = 7.0 \propto 10^8$ m

Mass of the Sun = $M_S = 2.0 \propto 10^{30}$ kg

1. A classical estimate of the temperature at the center of the stars.

Assume that the gas that forms the star is pure ionized hydrogen (electrons and protons in equal amounts), and that it behaves like an ideal gas. From the classical point of view, to fuse two protons, they need to get as close as 10^{-15} m for the short range strong nuclear force, which is attractive, to become dominant. However, to bring them together they have to overcome first the repulsive action of Coulomb's force. Assume classically that the two protons (taken to be point sources) are moving in an antiparallel way, each with velocity v_{rms}, the root-mean-square (rms) velocity of the protons, in a one-dimensional frontal collision.

1a	What has to be the temperature of the gas, T_c, so that the distance of closest approach of the protons, d_c, equals 10^{-15} m? Give this and all numerical values in this problem up to two significant figures.	1.5

2. Finding that the previous temperature estimate is wrong.

To check if the previous temperature estimate is reasonable, one needs an independent way of estimating the central temperature of a star. The structure of the stars is very complicated, but we can gain significant understanding making some assumptions. Stars are in equilibrium, that is, they do not expand or contract because the inward force of gravity is balanced by the outward force of pressure (see Figure 2). For a slab of gas the equation of hydrostatic equilibrium at a given distance r from the center of the star, is given by

$$\frac{\Delta P}{\Delta r} = -\frac{GM_r \rho_r}{r^2},$$

where P is the pressure of the gas, G the gravitational constant, M_r the mass of the star within a sphere of radius r, and ρ_r is the density of the gas in the slab.

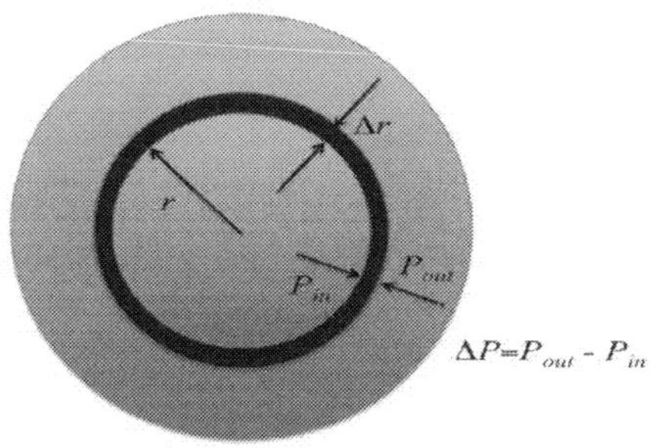

Figure 2. The stars are in hydrostatic equilibrium, with the pressure difference balancing gravity.

An order of magnitude estimate of the central temperature of the star can be obtained with values of the parameters at the center and at the surface of the star, making the following approximations:

$\Delta P \;..\; P_o - P_c$,

where P_c and P_o are the pressures at the center and surface of the star, respectively. Since $P_c >> P_o$, we can assume that

$\Delta P \;..\; -P_c$.

Within the same approximation, we can write

$\Delta r \;..\; R$,

where R is the total radius of the star, and

$M_r \;..\; M_R = M$,

with M the total mass of the star.

The density may be approximated by its value at the center,

$\rho_r \;..\; \rho_c$.

You can assume that the pressure is that of an ideal gas.

2a	Find an equation for the temperature at the center of the star, T_c, in terms of the radius and mass of the star and of physical constants only.	0.5

We can use now the following prediction of this model as a criterion for its validity:

2b	Using the equation found in (2a) write down the ratio M/R expected for a star in terms of physical constants and T_c only.	0.5

2c	Use the value of T_c derived in section (1a) and find the numerical value of the ratio M/R expected for a star.	0.5

2d	Now, calculate the ratio $M(Sun)/R(Sun)$, and verify that this value is much smaller than the one found in (2c).	0.5

3. A quantum mechanical estimate of the temperature at the center of the stars

The large discrepancy found in (2d) suggests that the classical estimate for T_c obtained in (1a) is not correct. The solution to this discrepancy is found when we consider quantum mechanical effects, that tell us that the protons behave as waves and that a single proton is smeared on a size of the order of λ_p, the de Broglie wavelength. This implies that if d_c, the distance of closest approach of the protons is of the order of λ_p, the protons in a quantum mechanical sense overlap and can fuse.

3a	Assuming that $d_c = \frac{\lambda_p}{2^{1/2}}$ is the condition that allows fusion, for a proton with velocity v_{rms}, find an equation for T_c in terms of physical constants only.	1.0

3b	Evaluate numerically the value of T_c obtained in (3a).	0.5

3c	Use the value of T_c derived in (3b) to find the numerical value of the ratio M/R expected for a star, using the formula derived in (2b). Verify that this value is quite similar to the ratio $M(Sun)/R(Sun)$ observed.	0.5

Indeed, stars in the so-called *main sequence* (fusing hydrogen) approximately do follow this ratio for a large range of masses.

4. The mass/radius ratio of the stars.

The previous agreement suggests that the quantum mechanical approach for estimating the temperature at the center of the Sun is correct.

4a	Use the previous results to demonstrate that for any star fusing hydrogen, the ratio of mass M to radius R is the same and depends only on physical constants. Find the equation for the ratio M/R for stars fusing hydrogen.	0.5

5. The mass and radius of the smallest star.

The result found in (4a) suggests that there could be stars of any mass as long as such a relationship is fulfilled; however, this is not true.

The gas inside normal stars fusing hydrogen is known to behave approximately as an ideal gas. This means that d_e, the typical separation *between electrons* is on the average larger that λ_e, their typical de Broglie wavelength. If closer, the electrons would be in a so-called degenerate state and the stars would behave differently. Note the distinction in the ways we treat protons and electrons inside the star. For protons, their de Broglie waves should overlap closely as they collide in order to fuse, whereas for electrons their de Broglie waves should not overlap in order to remain as an ideal gas.

The density in the stars increases with decreasing radius. Nevertheless, for this order-of-magnitude estimate assume they are of uniform density. You may further use that $m_p >> m_e$.

5a	Find an equation for n_e, the average electron number density inside the star.	0.5

5b	Find an equation for d_e, the typical separation between electrons inside the star.	0.5

5c	Use the $d_e \times \frac{\lambda_e}{2^{1/2}}$ condition to write down an equation for the radius of the smallest normal star possible. Take the temperature at the center of the star as typical for all the stellar interior.	1.5

5d	Find the numerical value of the radius of the smallest normal star possible, both in meters and in units of solar radius.	0.5

5e	Find the numerical value of the mass of the smallest normal star possible, both in kg and in units of solar masses.	0.5

6. Fusing helium nuclei in older stars.

As stars get older they will have fused most of the hydrogen in their cores into helium (He), so they are forced to start fusing helium into heavier elements in order to continue shining. A helium nucleus has two protons and two neutrons, so it has twice the charge and approximately four times the mass of a proton. We saw before that $d_c = \frac{\lambda_p}{2^{1/2}}$ is the condition for the protons to fuse.

6a	Set the equivalent condition for helium nuclei and find $v_{rms}(He)$, the rms velocity of the helium nuclei and $T(He)$, the temperature needed for helium fusion.	0.5

Answers

Theoretical Problem No. 3

Why are stars so large?

1) A first, classic estimate of the temperature at the center of the stars.

1a	We equate the initial kinetic energy of the two protons to the electric potential energy at the distance of closest approach: $2(\frac{1}{2}m_p\, v_{rms}^2) = \frac{q^2}{4\pi\,\varepsilon_0\, d_c}$; and since $\frac{3}{2}kT_c = \frac{1}{2}m_p\, v_{rms}^2$, we obtain $T_c = \frac{q^2}{12\pi\,\varepsilon_0\, d_c\, k} = 5.5\times10^9$ K	1.5

2) Finding that the previous temperature estimate is wrong.

2a	Since we have that $\frac{\Delta P}{\Delta r} = -\frac{GM_r\,\rho_r}{r^2}$, making the assumptions given above, we obtain that: $P_c = \frac{GM\,\rho_c}{R}$. Now, the pressure of an ideal gas is $P_c = \frac{2\rho_c\, kT_c}{m_p}$, where k is Boltzmann´s constant, T_c is the central temperature of the star, and m_p is the proton mass. The factor of 2 in the previous equation appears because we have two particles (one proton and one electron) per proton mass and that both contribute equally to the pressure. Equating the two previous equations, we finally obtain that: $T_c = \frac{GM\, m_p}{2kR}$	0.5

2b	From section (2a) we have that: $\frac{M}{R} = \frac{2kT_c}{Gm_p}$	0.5

2c	From section (2b) we have that, for $T_c = 5.5\times10^9$ K: $$\frac{M}{R} = \frac{2kT_c}{Gm_p} = 1.4\times10^{24}\ \text{kg m}^{-1}.$$	0.5

2d	For the Sun we have that: $\frac{M(Sun)}{R(Sun)} = 2.9\times10^{21}$ kg m^{-1} , that is, three orders of magnitude smaller.	0.5

3) A quantum mechanical estimate of the temperature at the center of the stars

3a	We have that $\lambda_p = \frac{h}{m_p\, v_{rms}}$, and since $\frac{3}{2}kT_c = \frac{1}{2}m_p\, v_{rms}^2$, and $T_c = \frac{q^2}{12\pi\,\varepsilon_0\, d_c\, k}$, we obtain: $$T_c = \frac{q^4\, m_p}{24\pi^2\,\varepsilon_0^2\, k\, h^2}\ .$$	1.0

3b	$$T_c = \frac{q^4\, m_p}{24\pi^2\,\varepsilon_0^2\, k\, h^2} = 9.7\times10^6\ \text{K}.$$	0.5

3c	From section (2b) we have that, for $T_c = 9.7\times10^6$ K: $\frac{M}{R} = \frac{2kT_c}{Gm_p} = 2.4\times10^{21}$ kg m^{-1} ; while for the Sun we have that: $\frac{M(Sun)}{R(Sun)} = 2.9\times10^{21}$ kg m^{-1}.	0.5

4) The mass/radius ratio of the stars.

4a	Taking into account that	0.5

	$\frac{M}{R}=\frac{2kT_c}{Gm_p}$, and that $T_c=\frac{q^4 m_p}{24\pi^2\varepsilon_0^2 k h^2}$, we obtain: $\frac{M}{R}=\frac{q^4}{12\pi^2\varepsilon_0^2 G h^2}$.	

5) The mass and radius of the smallest star.

5a	$n_e=\frac{M}{(4/3)\pi R^3 m_p}$	0.5

5b	$d_e=n_e^{-1/3}=\left(\frac{M}{(4/3)\pi R^3 m_p}\right)^{-1/3}$	0.5

5c	We assume that $d_e\geq\frac{\lambda_e}{2^{1/2}}$. Since $\lambda_e=\frac{h}{m_e\, v_{rms}(electron)}$, $\frac{3}{2}kT_c=\frac{1}{2}m_e\, v_{rms}^2(electron)$, $T_c=\frac{q^4 m_p}{24\pi^2\varepsilon_0^2 k h^2}$, $\frac{M}{R}=\frac{q^4}{12\pi^2\varepsilon_0^2 G h^2}$, and $d_e=\left(\frac{M}{(4/3)\pi R^3 m_p}\right)^{-1/3}$, we get that $R\geq\frac{\varepsilon_o^{1/2} h^2}{4^{1/4} q m_e^{3/4} m_p^{5/4} G^{1/2}}$	1.5

5d	$R\geq\frac{\varepsilon_o^{1/2} h^2}{4^{1/4} q m_e^{3/4} m_p^{5/4} G^{1/2}}=6.9\times10^7\,\text{m}=0.10\ R(Sun)$	0.5

5e	The mass to radius ratio is: $\frac{M}{R}=\frac{q^4}{12\pi^2\varepsilon_0^2 Gh^2}=2.4\times10^{21}$ kg m^{-1}, from where we derive that $M\geq 1.7\times10^{29}$ kg $=0.09\ M(Sun)$	0.5

6) Fusing helium nuclei in older stars.

6a	For helium we have that $\frac{4q^2}{4\pi\varepsilon_0 m_{He} v_{rms}^2(He)}=\frac{h}{2^{1/2} m_{He} v_{rms}(He)}$; from where we get $v_{rms}(He)=\frac{2^{1/2}q^2}{\pi\varepsilon_0 h}=2.0\times10^6$ m s^{-1}. We now use: $T(He)=\frac{v_{rms}^2(He)m_{He}}{3k}=6.5\times10^8$ K. This value is of the order of magnitude of the estimates of stellar models.	0.5

IPhO2009

Experimental Competition
Wednesday, July 15, 2009

The experimental part of this Olympiad consists of two problems. In Problem 1 the aim is to measure the wavelength of a diode laser, and in Problem 2 the goal is to measure the birefringence of a material called mica.

Please read this first:

1. The total time available is 5 hours for the experimental competition.
2. Use only the pencils provided.
3. Use only the front side of the paper sheets.
4. Each problem is presented in the ***question form***, marked with a **Q** in the upper left corner.
5. You *must* summarize the answers you have obtained in the ***answer form,*** marked with an **A** in the upper left corner.
6. In addition, there is a set of ***working sheets,*** marked with a **W** in the upper left corner, where you may write your calculations.
7. In addition, write down the ***Problem Number*** (1 or 2) on the top of the ***answer forms*** and ***working sheets.***
8. Write on the working sheets of paper whatever you consider is required for the solution of the problem. Please use *as little text as possible*; express yourself primarily in equations, numbers, figures, and plots.
9. For each problem and each of the forms (question form, answer form and working sheets), fill in the boxes at the top of each sheet of paper used by writing your student number (***Student Number***), the progressive page number ***(Page No.)*** and the total number of pages used ***(Total No. of Pages).*** If you use some working sheets of paper for notes that you do not wish to be marked, do **not** destroy it. Instead, mark it with a large X across the entire working sheet and do **not** include it in your numbering.
10. At the end of the exam, arrange all sheets for each problem *in the following order:*
    - answer form (including graph paper for your plots).
    - used working sheets in order
    - the working sheets that you do not wish to be considered (marked with the large X)
    - unused working sheets
    - printed question form.

    Place the papers of each problem set inside the folder and leave everything on your desk. You are not allowed to take *any* sheets of paper out of the room nor any device of the experimental kit.
11. The devices and materials for the experiments are contained in **two** separate packing layers within the box. The photographs of the sets are in the next page. Some devices are LABELED. For each experiment check that all the material is in the box. **If during the experiments you find that any of the devices is not working properly, please ask for a replacement.**

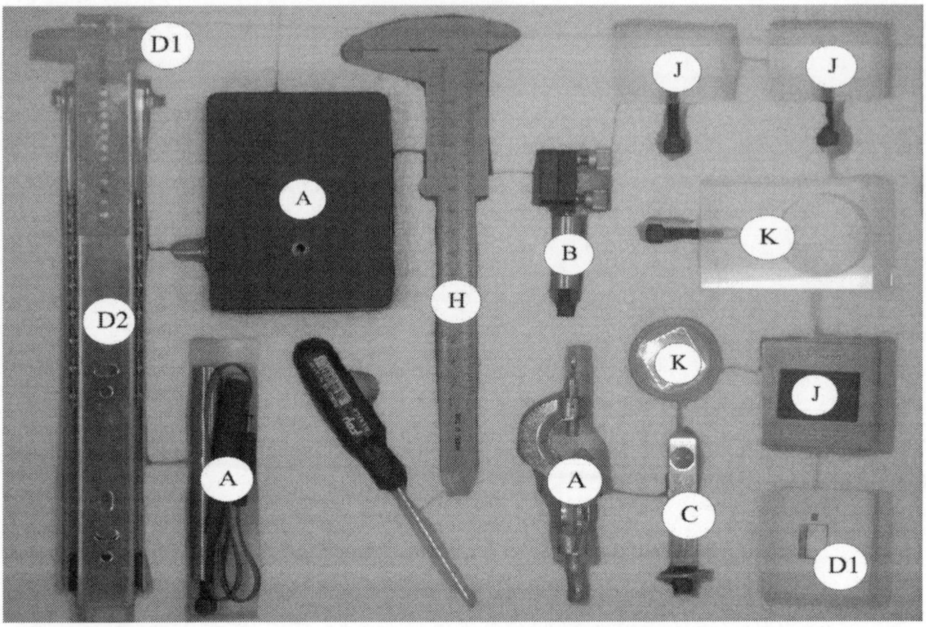
D1
J
J
A
B
K
H
D2
K
J
A
A
C
D1

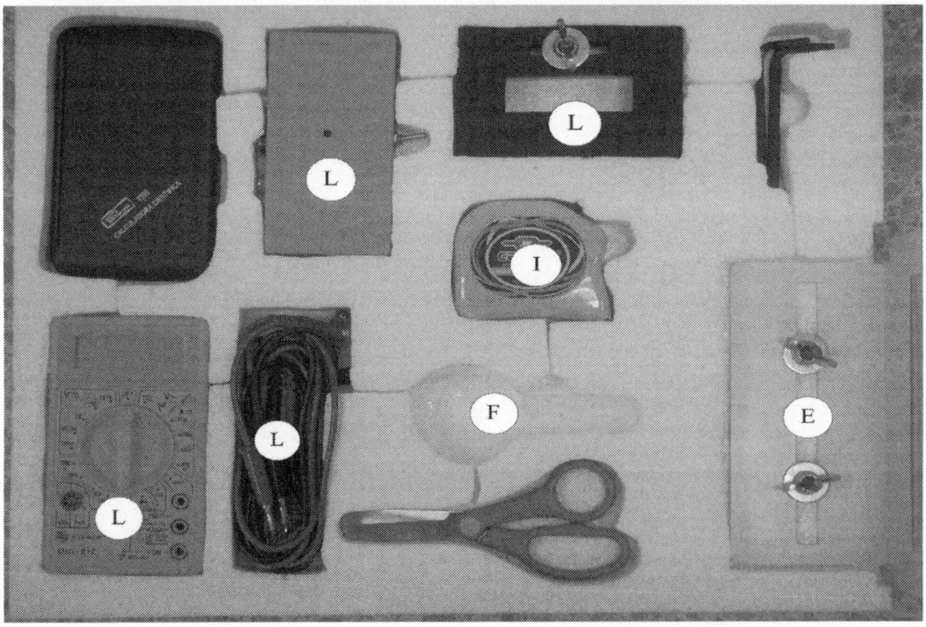
L
L
I
F
E
L
L

DIODE LASER EQUIPMENT AND MOVABLE MIRROR.

In both experimental setups you should need a diode laser, with its holder and power supply, and a mirror on a mechanical movable mount.

Before you decide on which problem to work first, we suggest that you mount the laser and the mirror, as indicated in Figure 0. Use the following material:

1) Wooden optical table.

2) Diode laser equipment. Includes the diode laser, support post, "S" clamp and power supply box (LABEL A). See photograph for mounting instructions. **DO NOT LOOK DIRECTLY INTO THE LASER BEAM.**

3) A mirror on a movable mount with two adjusting knobs and support post (LABEL B). See photograph for mounting instructions. **CAUTION: mount the support post to the optical table without touching the mirror. Take off the paper cover after you have mounted the mirror.**

Mount the above devices as indicated in Figure 0. The alignment of the laser beam will be done later on. NOTE: Although we have provided you with optional Allen wrenches, everything can be left fingertight.

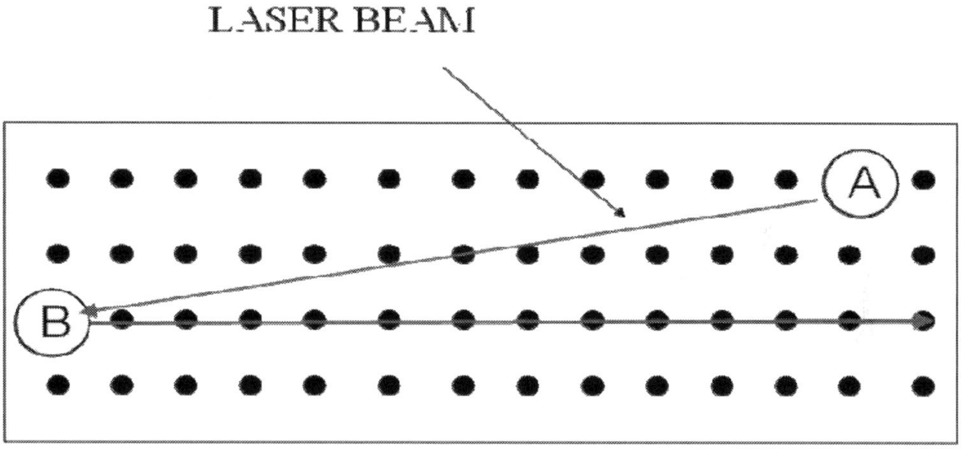

Figure 0. Mounting the laser and the mirror.

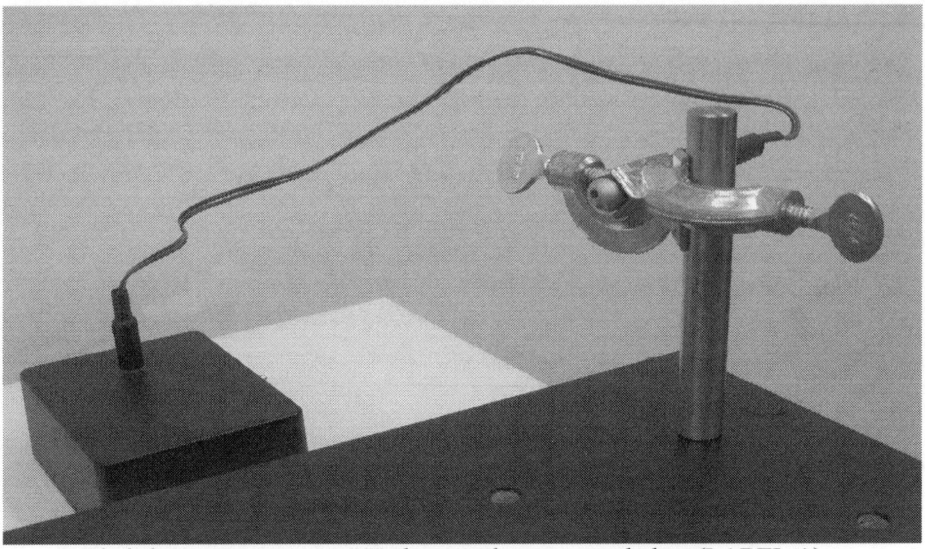

Diode laser, support post, "S" clamp and power supply box (LABEL A).

Mirror on a movable movable mount with two adjusting knobs and support post (LABEL B).

EXPERIMENTAL PROBLEM 1

DETERMINATION OF THE WAVELENGTH OF A DIODE LASER

MATERIAL

In addition to items 1), 2) and 3), you should use:

4) A lens mounted on a square post (LABEL C).
5) A razor blade in a slide holder to be placed in acrylic support, (LABEL D1) and mounted on sliding rail (LABEL D2). Use the screwdriver to tighten the support if necessary. See photograph for mounting instructions.
6) An observation screen with a caliper scale (1/20 mm) (LABEL E).
7) A magnifying glass (LABEL F).
8) 30 cm ruler (LABEL G).
9) Caliper (LABEL H).
10) Measuring tape (LABEL I).
11) Calculator.
12) White index cards, masking tape, stickers, scissors, triangle squares set.
13) Pencils, paper, graph paper.

Razor blade in a slide holder to be placed in acrylic support (LABEL D1) and mounted on sliding rail (LABEL D2).

EXPERIMENT DESCRIPTION

You are asked to determine a diode laser wavelength. The particular feature of this measurement is that no exact micrometer scales (such as prefabricated diffraction gratings) are used. The smallest lengths measured are in the millimetric range. The wavelength is determined using light diffraction on a sharp edge of a razor blade.

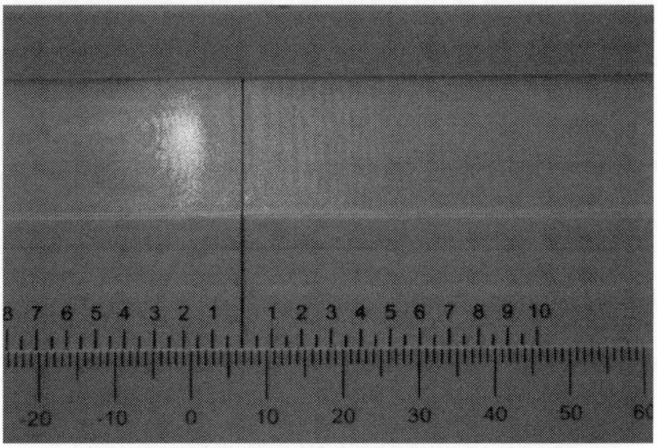

Figure 1.1 Typical interference fringe pattern.

Once the laser beam (A) is reflected on the mirror (B), it must be made to pass through a lens (C), which has a focal length of *a few centimeters*. It can now be assumed that the focus is a light point source from which a spherical wave is emitted. After the lens, and along its path, the laser beam hits a sharp razor blade edge as an obstacle. This can be considered to be a light source from which a cylindrical wave is emitted. These two waves interfere with each other, in the forward direction, creating a diffractive pattern that can be observed on a screen. See Figure 1.1 with a photograph of a typical pattern.

There are two important cases, see Figures 1.2 and 1.3.

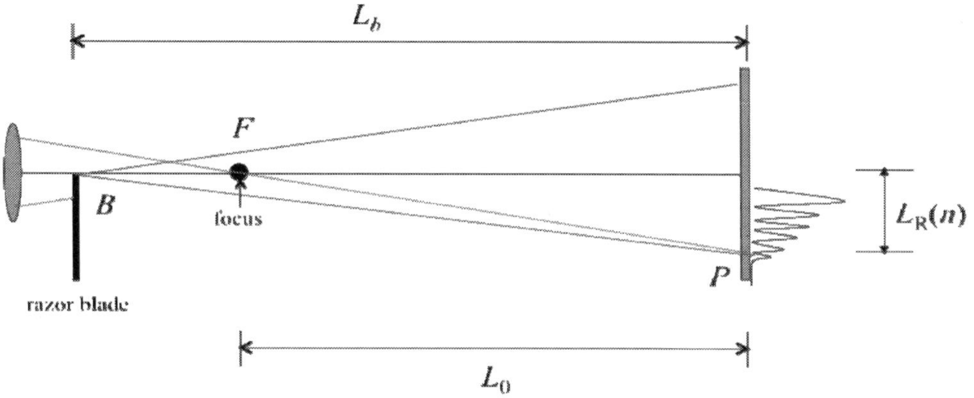

Figure 1.2. Case (I). The razor blade is *before* the focus of the lens. Figure is not at scale. B in this diagram is the edge of the blade and F is the focal point.

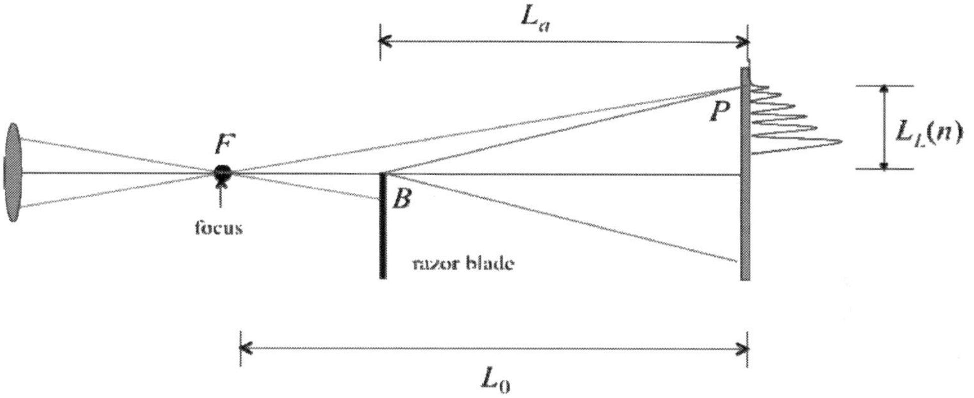

Figure 1.3. Case (II). The razor blade is *after* the focus of the lens. Figure is not at scale. B in this diagram is the edge of the blade and F is the focal point.

EXPERIMENTAL SETUP

Task 1.1 Experimental setup (1.0 points). Design an experimental setup to obtain the above described interference patterns. The distance L_0 from the focus to the screen should be much larger than the focal length.

- Make a sketch of your experimental setup in the drawing of the optical table provided. Do this by writing the LABELS of the different devices on the drawing of the optical table. You can make additional simple drawings to help clarify your design.
- You may align the laser beam by using one of the white index cards to follow the path.
- Make a sketch of the laser beam path on the drawing of the optical table and write down the height h of the beam as measured from the optical table.

WARNING: Ignore the larger circular pattern that may appear. This is an effect due to the laser diode itself.

Spend some time familiarizing yourself with the setup. You should be able to see of the order of 10 or more vertical linear fringes on the screen. The readings are made using the positions of the **dark** fringes. You may use the magnifying glass to see more clearly the position of the fringes. **The best way to observe the fringes is to look at the back side of the illuminated screen (E).** Thus, the scale of the screen should face out of the optical table. If the alignment of the optical devices is correct, you should see both patterns (of Cases I and II) by simply sliding the blade (D1) through the rail (D2).

THEORETICAL CONSIDERATIONS

Refer to Figure 1.2 and 1.3 above. There are five basic lengths:

L_0 : distance from the focus to the screen.
L_b : distance from the razor blade to the screen, Case I.
L_a : distance from the razor blade to the screen, Case II.
$L_R(n)$: position of the n-th **dark** fringe for Case I.
$L_L(n)$: position of the n-th **dark** fringe for Case II.

The first dark fringe, for both Cases I and II, is the widest one and corresponds to $n = 0$.

Your experimental setup must be such that $L_R(n) << L_0, L_b$ for Case I and $L_L(n) << L_0, L_a$ for Case II.

The phenomenon of wave interference is due to the difference in optical paths of a wave starting at the same point. Depending on their phase difference, the waves may cancel each other (destructive interference) giving rise to dark fringes; or the waves may add (constructive interference) yielding bright fringes.

A detailed analysis of the interference of these waves gives rise to the following condition to obtain a **dark** fringe, for Case I:

$$\Delta_{\mathrm{I}}(n) = \left(n + \frac{5}{8}\right)\lambda \qquad \text{with} \qquad n = 0, 1, 2, \ldots \tag{1.1}$$

and for Case II:

$$\Delta_{\mathrm{II}}(n) = \left(n + \frac{7}{8}\right)\lambda \qquad \text{with} \qquad n = 0, 1, 2, \ldots \tag{1.2}$$

where λ is the wavelength of the laser beam, and Δ_{I} and Δ_{II} are the optical path differences for each case.

The difference in optical paths for Case I is,

$$\Delta_{\mathrm{I}}(n) = (BF + FP) - BP \qquad \text{for each} \qquad n = 0, 1, 2, \ldots \tag{1.3}$$

while for Case II is,

$$\Delta_{\mathrm{II}}(n) = (FB + BP) - FP \qquad \text{for each} \qquad n = 0, 1, 2, \ldots \tag{1.4}$$

Task 1.2 Expressions for optical paths differences (0.5 points). Assuming $L_R(n) << L_0, L_b$ for Case I and $L_L(n) << L_0, L_a$ for Case II in equations (1.3) and (1.4) (make sure your setup satisfies these conditions), find approximated expressions for $\Delta_{\mathrm{I}}(n)$ and $\Delta_{\mathrm{II}}(n)$ in terms of L_0, L_b, L_a, $L_R(n)$ and $L_L(n)$. You may find useful the approximation $(1+x)^r \,..1 + rx$ if $x << 1$.

The experimental difficulty with the above equations is that L_0, $L_R(n)$ and $L_L(n)$ cannot be accurately measured. The first one because it is not easy to find the position of the focus of the lens, and the two last ones because the origin from which they are defined may be very hard to find due to misalignments of your optical devices.

To solve the difficulties with $L_R(n)$ and $L_L(n)$, first choose the zero (0) of the scale of the screen (LABEL E) as the origin for all your measurements of the fringes. Let l_{0R} and l_{0L} be the (unknown) positions from which $L_R(n)$ and $L_L(n)$ are defined. Let $l_R(n)$ and $l_L(n)$ be the positions of the fringes as measured from the origin (0) you chose. Therefore

$$L_R(n) = l_R(n) - l_{0R} \qquad \text{and} \qquad L_L(n) = l_L(n) - l_{0L} \tag{1.5}$$

PERFORMING THE EXPERIMENT. DATA ANALYSIS.

Task 1.3 Measuring the dark fringe positions and locations of the blade (3.25 points).

- For both Case I and Case II, measure the positions of the dark fringes $l_R(n)$ and $l_L(n)$ as a function of the number fringe n. Write down your measurements in Table I; you should report no less than 8 measurements for each case.
- Report also the positions of the blade L_b and L_a, and indicate with its LABEL the intrument you used.
- **IMPORTANT SUGGESTION:** For purposes of both simplification of analysis and better accuracy, measure *directly* the distance $d = L_b - L_a$ with a better accuracy than that of L_b and L_a; that is, do not calculate it from the measurements of L_b and L_a. Indicate with its LABEL the instrument you used.

Make sure that you include the uncertainty of your measurements.

Task 1.4 Data analysis. (3.25 points). With all the previous information you should be able to find out the values of l_{0R} and l_{0L}, and, of course, of the wavelength λ.

- Devise a procedure to obtain those values. Write down the expressions and/or equations needed.
- Include the analysis of the errors. You may use Table I or you can use another one to report your findings; make sure that you label clearly the contents of the columns of your tables.
- Plot the variables analyzed. Use the graph paper provided.
- Write down the values for l_{0R} and l_{0L}, with uncertainties.

Task 1.5 Calculating λ. Write down the calculated value for λ. Include its uncertainty and the analysis to obtain it. **SUGGESTION:** In your formula for λ, wherever you find $(L_b - L_a)$ replace it by d and use its measured value. **(2 points).**

Answer Form

Experimental Problem No. 1

Diode laser wavelength

Task 1.1 Experimental setup.

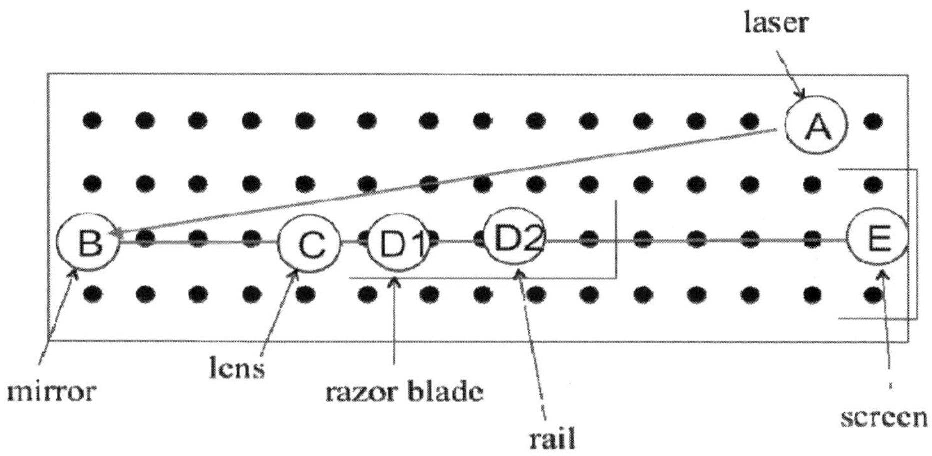

(0.75)

1.1	Sketch the laser path in drawing of Task 1.1 and Write down the height h of the beam as measured from the table $h''\ \Delta h = (5.0''\ 0.05) \propto 10^{-2}$ m (0.25)	1.0

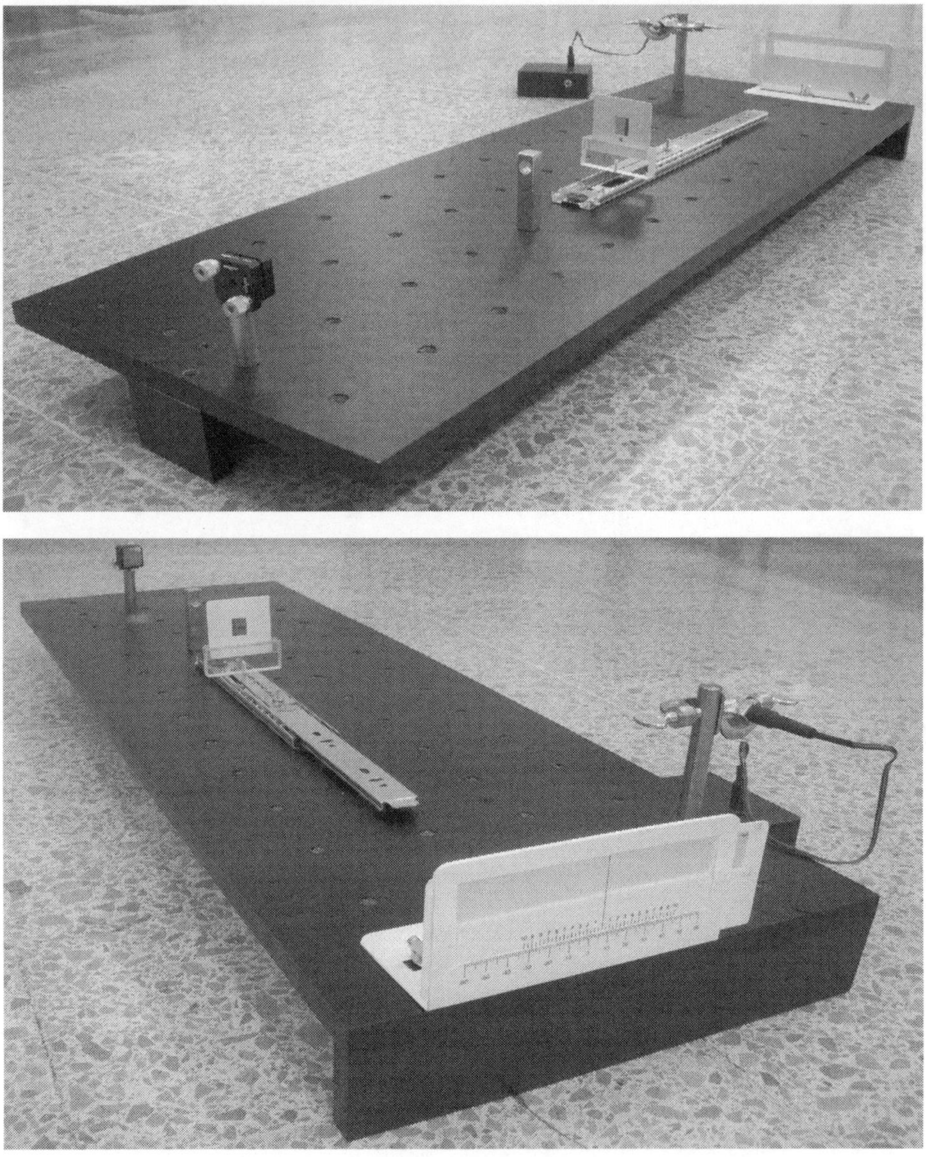

Experimental setup for measurement of diode laser wavelength

Task 1.2 Expressions for optical path differences.

1.2	The path differences are Case I: (0.25) $\Delta_{\mathrm{I}}(n)=(BF+FP)-BP=(L_b-L_0)+\sqrt{L_0^2+L_R^2(n)}-\sqrt{L_b^2+L_R^2(n)}$ $=(L_b-L_0)+L_0\sqrt{1+\frac{L_R^2(n)}{L_0^2}}-L_b\sqrt{1+\frac{L_R^2(n)}{L_b^2}}$ using $\sqrt{1+x}\approx 1+\frac{1}{2}x$ $\approx(L_b-L_0)+L_0\left(1+\frac{1}{2}\frac{L_R^2(n)}{L_0^2}\right)-L_b\left(1+\frac{1}{2}\frac{L_R^2(n)}{L_b^2}\right)$ $\Rightarrow \quad \Delta_{\mathrm{I}}(n)\approx\frac{1}{2}L_R^2(n)\left(\frac{1}{L_0}-\frac{1}{L_b}\right)$ Case II: (0.25) $\Delta_{\mathrm{II}}(n)=(FB+BP)-FP=(L_0-L_a)+\sqrt{L_a^2+L_L^2(n)}-\sqrt{L_0^2+L_L^2(n)}$ $\approx(L_0-L_a)+L_a\sqrt{1+\frac{L_L^2(n)}{L_a^2}}-L_0\sqrt{1+\frac{L_L^2(n)}{L_0^2}}$ using $\sqrt{1+x}\approx 1+\frac{1}{2}x$ $\approx(L_0-L_a)+L_a\left(1+\frac{1}{2}\frac{L_L^2(n)}{L_a^2}\right)-L_0\left(1+\frac{1}{2}\frac{L_L^2(n)}{L_0^2}\right)$ $\Rightarrow \quad \Delta_{\mathrm{II}}(n)\approx\frac{1}{2}L_L^2(n)\left(\frac{1}{L_a}-\frac{1}{L_0}\right)$	0.5

Task 1.3 Measuring the dark fringe positions and locations of the blade. Use additional sheets if necessary.

TABLE I

n	$(l_R(n)''\ 0.1)\propto 10^{-3}$ m	$(l_L(n)''\ 0.1)\propto 10^{-3}$ m	x_R	x_L
0	-7.5	1.1	0.791	0.935
1	-10.1	3.7	1.275	1.369
2	-12.4	6.4	1.620	1.696
3	-14.0	8.2	1.903	1.968
4	-15.6	10.0	2.151	2.208
5	-17.2	11.4	2.372	2.424
6	-18.4	12.2	2.574	2.622
7	-19.7		2.761	
8	-20.7		2.937	
9	-22.0		3.102	
10	-23.0		3.260	
11	-24.1		3.410	

1.3	Report positions of the blade and their difference with higher precision: $L_b \,''\, \Delta L_b = (653 \,''\, 1) \propto 10^{-3}$ m (0.25) LABEL (I) (measuring tape) $L_a \,''\, \Delta L_a = (628 \,''\, 1) \propto 10^{-3}$ m (0.25) LABEL (I) (measuring tape) $d = L_b - L_a = (24.6 \,''\, 0.1) \propto 10^{-3}$ m (0.25) LABEL (H) (caliper)	3.25

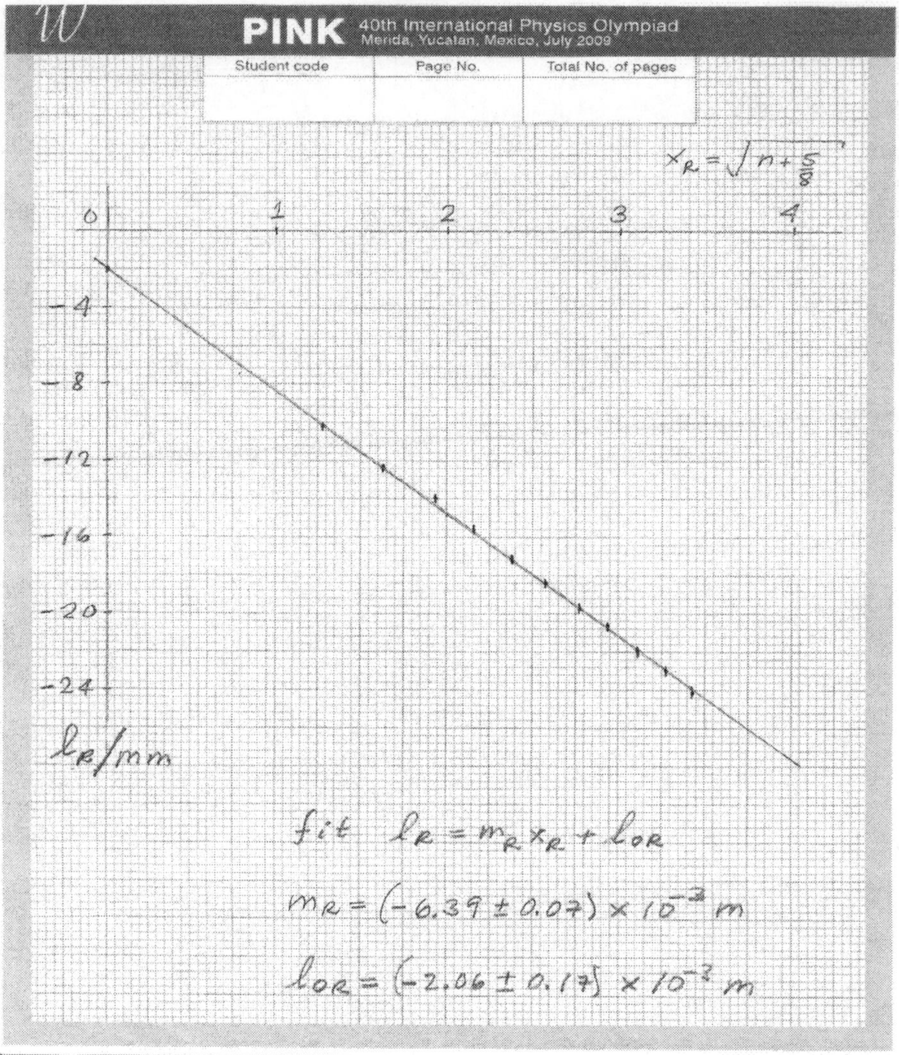
PINK 40th International Physics Olympiad
Merida, Yucatan, Mexico, July 2009
Student code
Page No.
Total No. of pages
$x_R = \sqrt{n + \frac{5}{8}}$
0
1
2
3
4
-4
-8
-12
-16
-20
-24
l_R/mm
fit $l_R = m_R x_R + l_{0R}$
$m_R = (-6.39 \pm 0.07) \times 10^{-3}$ m
$l_{0R} = (-2.06 \pm 0.17) \times 10^{-3}$ m

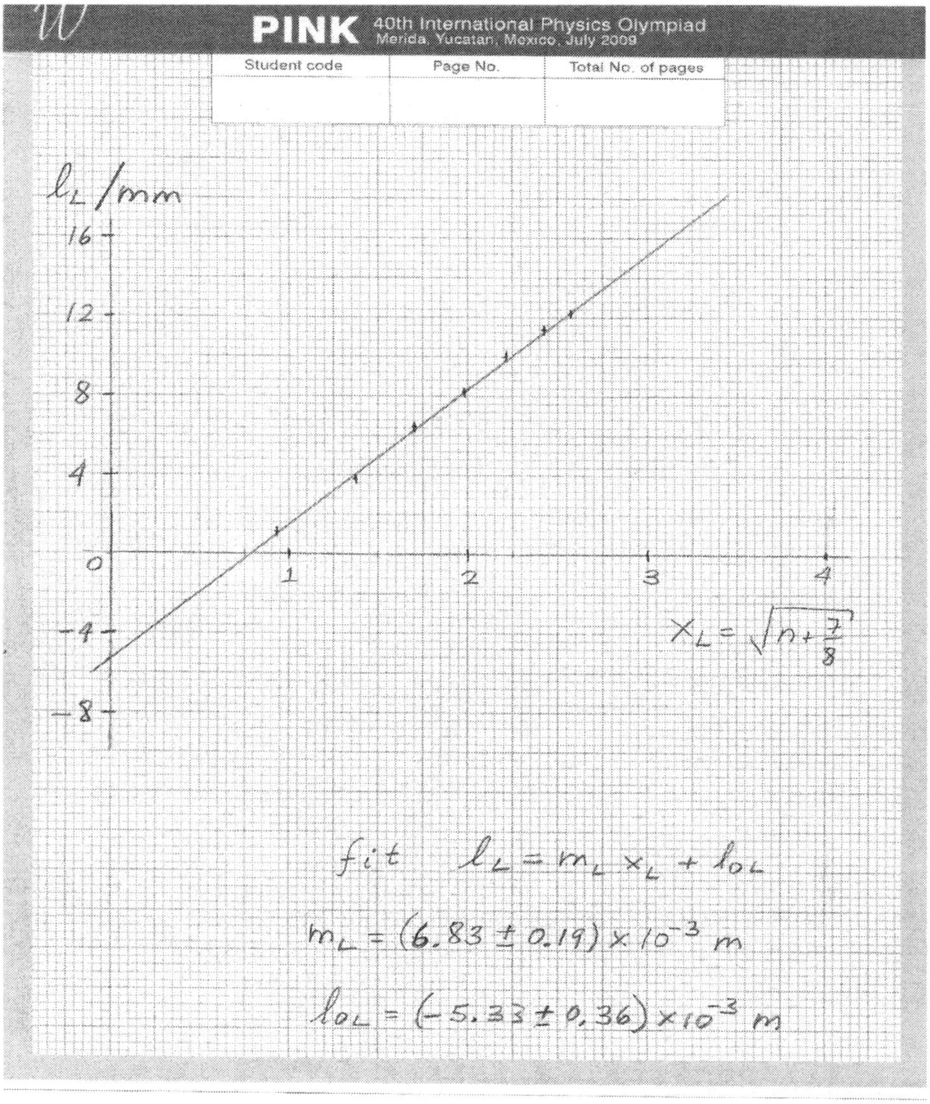

Task 1.4 Performing a statistical and graphical analysis.

1.4 A procedure: 3.25

From the condition of dark fringes and Task 1.2, we have

$$\frac{1}{2}L_R^2(n)\left(\frac{1}{L_0}-\frac{1}{L_b}\right)=\left(n+\frac{5}{8}\right)\lambda$$

and

$$\frac{1}{2}L_L^2(n)\left(\frac{1}{L_a}-\frac{1}{L_0}\right)=\left(n+\frac{7}{8}\right)\lambda$$

Using (1.5), $L_R(n)=l_R(n)-l_{0R}$ and $L_L(n)=l_L(n)-l_{0L}$ we can rewrite

$$\frac{1}{2}\left(l_R(n)-l_{0R}\right)^2\left(\frac{1}{L_0}-\frac{1}{L_b}\right)=\left(n+\frac{5}{8}\right)\lambda$$

$$\Rightarrow \quad l_R(n)=\sqrt{\frac{2L_bL_0}{L_b-L_0}\lambda}\sqrt{n+\frac{5}{8}}+l_{0R}$$

and

$$\frac{1}{2}\left(l_L(n)-l_{0L}\right)^2\left(\frac{1}{L_a}-\frac{1}{L_0}\right)=\left(n+\frac{7}{8}\right)\lambda$$

$$\Rightarrow \quad l_L(n)=\sqrt{\frac{2L_aL_0}{L_0-L_a}\lambda}\sqrt{n+\frac{7}{8}}+l_{0L}$$

These can be cast as equations of a straight line, $y=mx+b$.

Case I:

$y_R=l_R \qquad x_R=\sqrt{n+\frac{5}{8}} \qquad m_R=\sqrt{\frac{2L_bL_0}{L_b-L_0}\lambda} \qquad b_R=l_{0R}$

Case II:

$y_L=l_L \qquad x_L=\sqrt{n+\frac{7}{8}} \qquad m_L=\sqrt{\frac{2L_aL_0}{L_0-L_a}\lambda} \qquad b_L=l_{0L}$

Perform least squares analysis of above equations. In Table I, we write down the values x_R and x_L.

One finds:

$m_R \pm \Delta m_R=(-6.39 \pm 0.07)\times 10^{-3}$ m

$m_L \, '' \, \Delta m_L = (6.83'' \, 0.19) \propto 10^{-3}$ m

and (values of l_{0R} and l_{0L})

$l_{0R} \, '' \, \Delta l_{0R} = b_R \, '' \, \Delta b_R = (-2.06'' \, 0.17) \propto 10^{-3}$ m

$l_{0L} \, '' \, \Delta l_{0L} = b_L \, '' \, \Delta b_L = (-5.33'' \, 0.36) \propto 10^{-3}$ m

The equations used in the least squares analysis:

$$m = \frac{N\sum_{n=1}^{N} x_n y_n - \sum_{n=1}^{N} x_n \sum_{n\leq 1}^{N} y_{n\leq}}{\Delta}$$

$$b = \frac{\sum_{n=1}^{N} x_n^2 \sum_{n\leq 1}^{N} y_{n\leq} - \sum_{n=1}^{N} x_n \sum_{n\leq 1}^{N} x_{n\leq} y_{n\leq}}{\Delta}$$

where

$$\Delta = N\sum_{n=1}^{N} x_n^2 - \left(\sum_{n=1}^{N} x_n\right)^2$$

with N the number of data points.
The uncertainty is calculated as

$$(\Delta m)^2 = N\frac{\sigma^2}{\Delta} \quad , \quad (\Delta b)^2 = \frac{\sigma^2}{\Delta}\sum_{n=1}^{N} x_n^2 \qquad \text{with,}$$

$$\sigma^2 = \frac{1}{N-2}\sum_{n=1}^{N} (y_n - b - mx_n)^2$$

REFERENCE: P.R. Bevington, *Data Reduction and Error Analysis for the Physical Sciences*, McGraw-Hill, 1969.

Task 1.5 Calculating λ.

1.5 From any slope and the value of L_0 one finds, 2.0

$$\lambda = \frac{L_b - L_a}{2L_a L_b} \frac{m_R^2 m_L^2}{m_R^2 + m_L^2}$$

Using the suggestion to replace $d = L_b - L_a$, we can write

$$\lambda = \frac{d}{2L_a L_b} \frac{m_R^2 m_L^2}{m_R^2 + m_L^2}$$

$$\lambda'' \ \Delta\lambda = (663'' \ 25) \propto 10^{-9} \text{ m}$$

The uncertainty may range from 15 to 30 nanometers.

A precise measurement of the wavelength is $\lambda'' \ \Delta\lambda = (655'' \ 1) \propto 10^{-9}$ m.

The formula for the uncertainty,

$$\Delta\lambda = \sqrt{\left(\frac{\bullet\lambda}{\bullet d}\right)^2 \Delta d^2 + \left(\frac{\bullet\lambda}{\bullet L_a}\right)^2 \Delta L_a^2 + \left(\frac{\bullet\lambda}{\bullet L_b}\right)^2 \Delta L_b^2 + \left(\frac{\bullet\lambda}{\bullet m_R}\right)^2 \Delta m_R^2 + \left(\frac{\bullet\lambda}{\bullet m_L}\right)^2 \Delta m_L^2}$$

one finds,

$$\frac{\bullet\lambda}{\bullet d} = \frac{\lambda}{d}, \quad \frac{\bullet\lambda}{\bullet L_b} = \frac{\lambda}{L_b}, \quad \frac{\bullet\lambda}{\bullet L_a} = \frac{\lambda}{L_a} \quad \text{and} \quad \frac{\bullet\lambda}{\bullet m_R} = \frac{2m_L^2}{m_R} \frac{\lambda}{m_L^2 + m_R^2}$$

and analogously for the other slope.

One can calculate directly these quantities. However, one may note that the errors due to L_a, L_b and d are negligible. Moreover, $m_R^2 \ldots m_L^2$ and $L_a \ldots L_b$. This implies,

$$\frac{\bullet\lambda}{\bullet m_R} \ldots \frac{\lambda}{m_R} \ldots \frac{\bullet\lambda}{\bullet m_L}. \text{ Thus,}$$

$$\Delta\lambda \ldots \sqrt{2} \frac{\lambda}{m_L} \Delta m_L \ldots \left(25 \propto 10^{-9}\right) \text{ m}$$

EXPERIMENTAL PROBLEM 2
BIREFRINGENCE OF MICA

In this experiment you will measure the birefringence of mica (a crystal widely used in polarizing optical components).

MATERIAL

In addition to items 1), 2) and 3), you should use,

14) Two polarizing films mounted in slide holders, each with an additional acrylic support (LABEL J). See photograph for mounting instructions.
15) A thin mica plate mounted in a plastic cylinder with a scale with no numbers; acrylic support for the cylinder (LABEL K). See photograph for mounting instructions.
16) Photodetector equipment. A photodetector in a plastic box, connectors and foam support. A multimeter to measure the voltage of the photodetector (LABEL L). See photograph for mounting and connecting instructions.
17) Calculator.
18) White index cards, masking tape, stickers, scissors, triangle squares set.
19) Pencils, paper, graph paper.

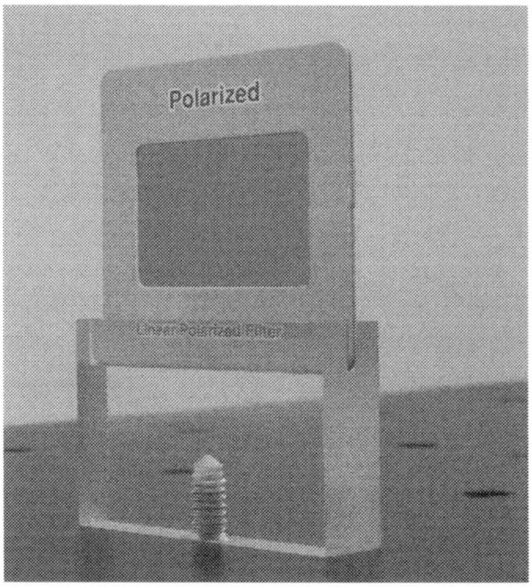

Polarizer mounted in slide holder with acrylic support (LABEL J).

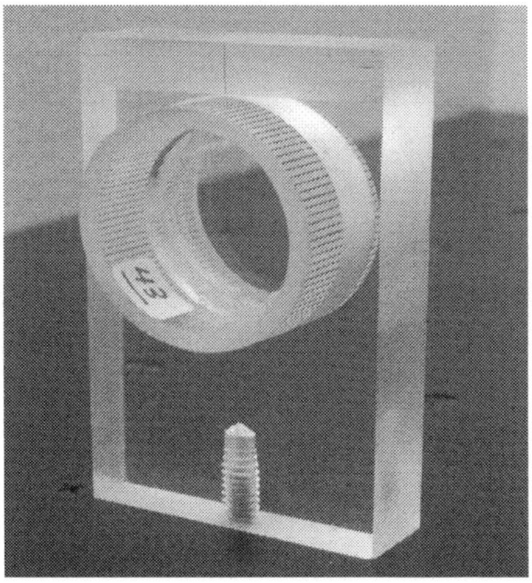

Thin mica plate mounted in cylinder with a scale with no numbers, and acrylic support (LABEL K).

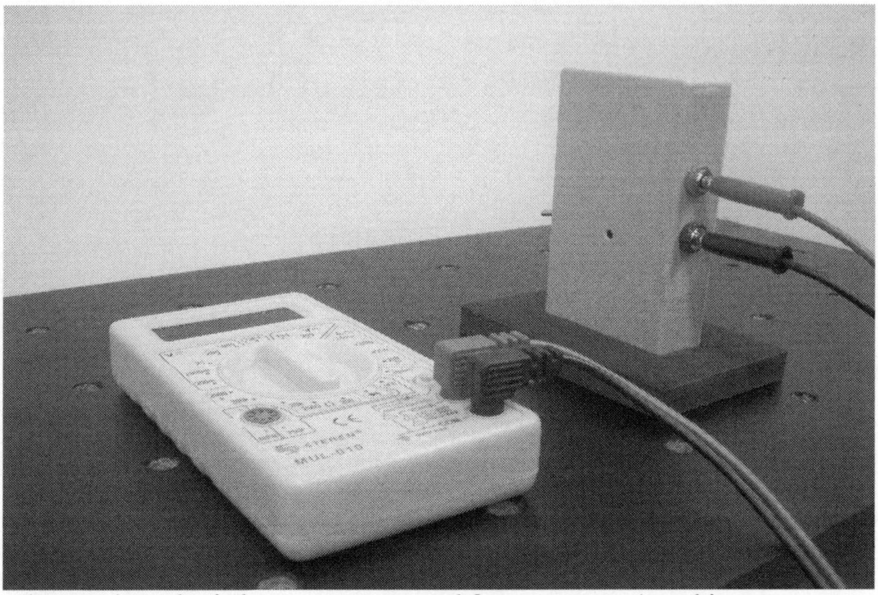

A photodetector in a plastic box, connectors and foam support. A multimeter to measure the voltage of the photodetector (LABEL L). Set the connections as indicated.

DESCRIPTION OF THE PHENOMENON

Light is a transverse electromagnetic wave, with its electric field lying on a plane perpendicular to the propagation direction and oscillating in time as the light wave travels.

If the direction of the electric field remains in time oscillating *along a single line*, the wave is said to be linearly polarized, or simply, polarized. See Figure 2.1.

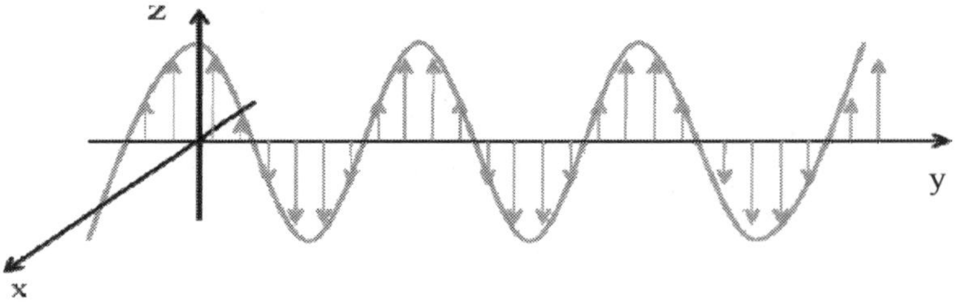

Figure 2.1 A wave travelling in the y-direction and polarized in the z-direction.

A polarizing film (or simply, a polarizer) is a material with a privileged axis parallel to its surface, such that, transmitted light emerges polarized along the axis of the polarizer. Call (+) the privileged axis and (-) the perpendicular one.

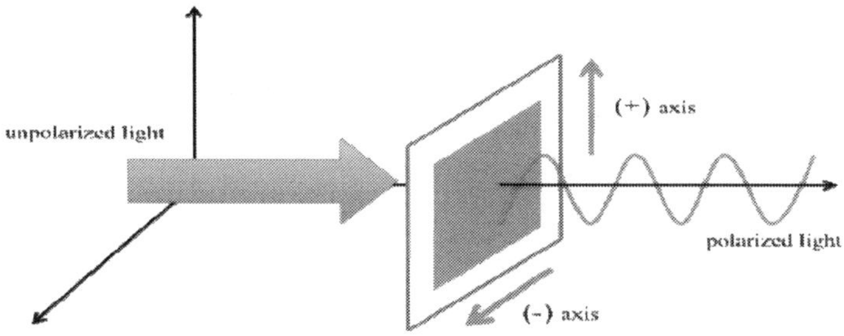

Figure 2.2 Unpolarized light normally incident on a polarizer. Transmitted light is polarized in the (+) direction of the polarizer.

Common transparent materials (such as window glass), transmit light with the same polarization as the incident one, because its index of refraction does not depend on the direction and/or polarization of the incident wave. Many crystals, including mica, however, are sensitive to the direction of the electric field of the wave. For propagation perpendicular to its surface, the mica sheet has two characteristic orthogonal axes, which we will call Axis 1 and Axis 2. This leads to the phenomenon called birefringence.

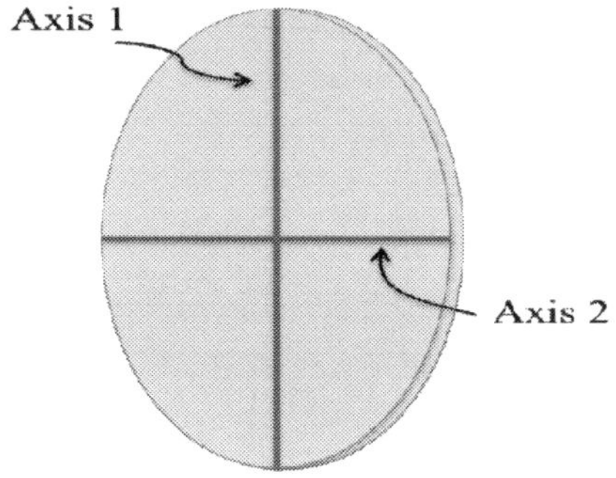

Figure 2.3 Thin slab of mica with its two axes, Axis 1 (red) and Axis 2 (green).

Let us analyze two simple cases to exemplify the birefringence. Assume that a wave **polarized in the vertical direction** is normally incident on one of the surfaces of the thin slab of mica.

Case 1) Axis 1 or Axis 2 is parallel to the polarization of the incident wave. The trasmitted wave passes without changing its polarization state, but the propagation is characterized as if the material had either an index of refraction n_1 or n_2. See Figs. 2.4 and 2.5.

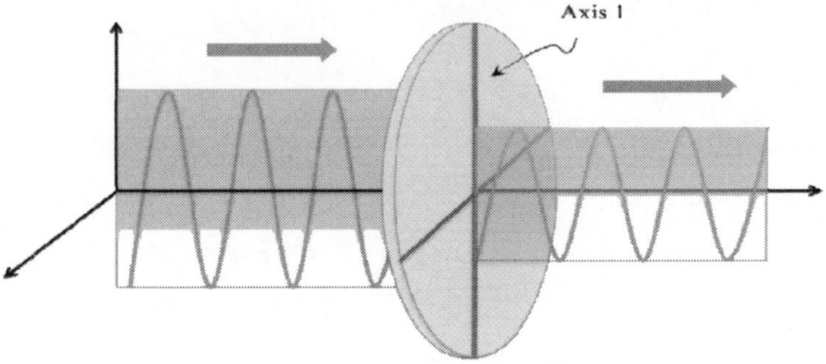

Figure 2.4 Axis 1 is parallel to polarization of incident wave. Index of refraction is n_1.

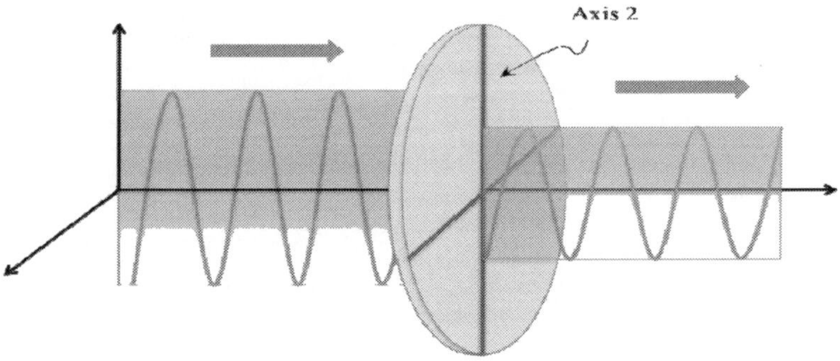

Figure 2.5 Axis 2 is parallel to polarization of incident wave. Index of refraction is n_2.

Case 2) Axis 1 makes an angle θ with the direction of polarization of the incident wave. The transmitted light has a more complicated polarization state. This wave, however, can be seen as the *superposition* of two waves with different phases, one that has polarization **parallel** to the polarization of the incident wave (i.e. "vertical") and another that has polarization **perpendicular** to the polarization of the incident wave (i.e. "horizontal").

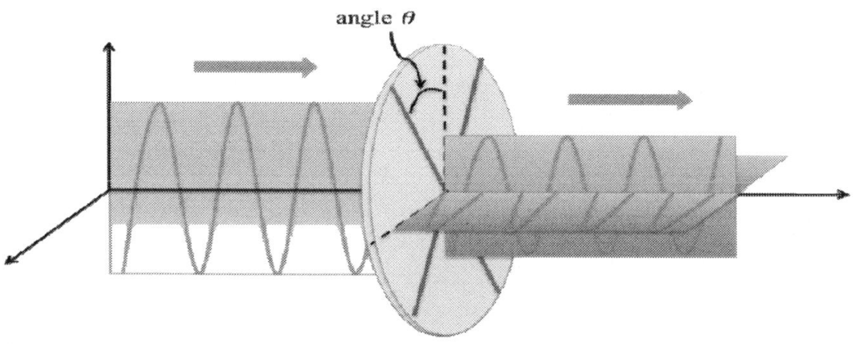

Figure 2.6 Axis 1 makes and angle θ with polarization of incident wave

Call I_P the *intensity* of the wave transmitted *parallel* to the polarization of the incident wave, and I_O the *intensity* of the wave transmitted *perpendicular* to polarization of the incident wave. These intensities depend on the angle θ, on the wavelength λ of the light source, on the thickness L of the thin plate, and on the absolute value of the difference of the refractive indices, $|n_1 - n_2|$. This last quantity is called the *birefringence* of the material. The measurement of this quantity is the goal of this problem. Together with polarizers, birefringent materials are useful for the control of light polarization states.

We point out here that the photodetector measures the intensity of the light incident on it, independent of its polarization.

The dependence of $I_P(\theta)$ and $I_O(\theta)$ on the angle θ is complicated due to other effects not considered, such as the absorption of the incident radiation by the mica. One can obtain, however, approximated but very simple expressions for the normalized intensities $\bar{I}_P(\theta)$ and $\bar{I}_O(\theta)$, defined as,

$$\bar{I}_P(\theta) = \frac{I_P(\theta)}{I_P(\theta) + I_O(\theta)} \tag{2.1}$$

and

$$\bar{I}_O(\theta) = \frac{I_O(\theta)}{I_P(\theta) + I_O(\theta)} \tag{2.2}$$

It can be shown that the normalized intensities are (approximately) given by,

$$\bar{I}_P(\theta) = 1 - \frac{1}{2}(1 - \cos\Delta\phi)\sin^2(2\theta) \tag{2.3}$$

and

$$\bar{I}_O(\theta) = \frac{1}{2}(1 - \cos\Delta\phi)\sin^2(2\theta) \qquad (2.4)$$

where $\Delta\phi$ is the difference of phases of the parallel and perpendicular transmitted waves. This quantity is given by,

$$\Delta\phi = \frac{2\pi L}{\lambda}|n_1 - n_2| \qquad (2.5)$$

where L is the thickness of the thin plate of mica, λ the wavelength of the incident radiation and $|n_1 - n_2|$ the birefringence.

EXPERIMENTAL SETUP

Task 2.1 Experimental setup for measuring intensities. Design an experimental setup for measuring the intensities I_P and I_O of the transmitted wave, as a function of the angle θ of any of the optical axes, as shown in Fig. 2.6. *Do this by writing the LABELS of the different devices on the drawing of the optical table*. Use the convention (+) and (-) for the direction of the polarizers. You can make additional simple drawings to help clarify your design.

Task 2.1 a) Setup for I_P **(0.5 points).**
Task 2.1 b) Setup for I_O **(0.5 points).**

Laser beam alignment. Align the laser beam in such a way that it is parallel to the table and is incident on the center of the cylinder holding the mica. You may align by using one the white index cards to follow the path. Small adjustments can be made with the movable mirror.

Photodetector and the multimeter. The photodetector produces a voltage as light impinges on it. Measure this voltage with the multimeter provided. The voltage produced is linearly proportional to the intensity of the light. Thus, report the intensities as the voltage produced by the photodetector. Without any laser beam incident on the photodetector, you can measure the background light intensity of the detector. This should be less than 1 mV. *Do not correct* for this background when you perform the intensity measurements.

WARNING: The laser beam is partially polarized but it is not known in which direction. Thus, to obtain polarized light with good intensity readings, place a polarizer with either its (+) or (–) axes vertically in such a way that you obtain the maximum transmitted intensity in the absence of any other optical device.

MEASURING INTENSITIES

Task 2.2 The scale for angle settings. The cylinder holding the mica has a regular graduation for settings of the angles. Write down the value in degrees of the smallest interval (*i.e.* between two black consecutive lines). **(0.25 points).**

Finding (approximately) the zero of θ and/or the location of the mica axes. To facilitate the analysis, it is very important that you find the appropriate zero of the angles. We suggest that, first, you identify the location of one of the mica axes, and call it Axis 1. It is almost sure that this position will not coincide with a graduation line on the cylinder. Thus, consider the nearest graduation line in the mica cylinder as the provisional origin for the angles. Call $\overline{\theta}$ the angles measured from such an origin. Below you will be asked to provide a more accurate location of the zero of θ.

Task 2.3 Measuring I_P and I_O. Measure the intensities I_P and I_O for as many angles $\overline{\theta}$ as you consider necessary. Report your measurements in Table I. Try to make the measurements for I_P and I_O for the *same* setting of the cylinder with the mica, that is, for a *fixed* angle $\overline{\theta}$. **(3.0 points).**

Task 2.4 Finding an appropriate zero for θ. The location of Axis 1 defines the zero of the angle θ. As mentioned above, it is mostly sure that the location of Axis 1 does not coincide with a graduation line on the mica cylinder. To find the zero of the angles, you may proceed either graphically or numerically. Recognize that the relationship near a maximum or a minimum may be approximated by a parabola where:

$$I(\overline{\theta})\ ..\ a\overline{\theta}^2 + b\overline{\theta} + c$$

and the minimum or maximum of the parabola is given by,

$$\overline{\theta}_m = -\frac{b}{2a}.$$

Either of the above choices gives rise to a shift $\delta\overline{\theta}$ of all your values of $\overline{\theta}$ given in Table I of Task 2.3, such that they can now be written as angles θ from the appropriate zero, $\theta = \overline{\theta} + \delta\overline{\theta}$. Write down the value of the shift $\delta\overline{\theta}$ in degrees. **(1.0 points).**

DATA ANALYSIS.

Task 2.5 Choosing the appropriate variables. Choose $\overline{I}_P(\theta)$ or $\overline{I}_O(\theta)$ to make an analysis to find the difference of phases $\Delta\phi$. Identify the variables that you will use. (**0.5 point).**

Task 2.6 Data analysis and the phase difference.

÷ Use Table II to write down the values of the variables needed for their analysis. Make sure that you use the corrected values for the angles θ. Include uncertainties. Use graph paper to plot your variables. **(1.0 points).**

÷ Perform an analysis of the data needed to obtain the phase difference $\Delta\phi$. Report your results including uncertainties. Write down any equations or formulas used in the analysis. Plot your results. **(1.75 points).**

÷ Calculate the value of the phase difference $\Delta\phi$ in radians, including its uncertainty. Find the value of the phase difference in the interval $[0,\pi]$. **(0.5 points).**

Task 2.7 Calculating the birefringence $|n_1 - n_2|$. You may note that if you add $2N\pi$ to the phase difference $\Delta\phi$, with N any integer, or if you change the sign of the phase, the values of the intensities are unchanged. However, the value of the birefringence $|n_1 - n_2|$ would change. Thus, to use the value $\Delta\phi$ found in Task 2.6 to correctly calculate the birefringence, you must consider the following:

$$\Delta\phi = \frac{2\pi L}{\lambda}|n_1 - n_2| \qquad \text{if} \qquad L < 82 \propto 10^{-6} \text{ m}$$

or

$$2\pi - \Delta\phi = \frac{2\pi L}{\lambda}|n_1 - n_2| \qquad \text{if} \qquad L > 82 \propto 10^{-6} \text{ m}$$

where the value L of the thickness of the slab of mica you used is written on the cylinder holding it. This number is given in micrometers (1 micrometer = 10^{-6} m). Assign $1 \propto 10^{-6}$m as the uncertainty for L. For the laser wavelength, you may use the value you found in Problem 1 or the average value between $620 \propto 10^{-9}$ m and $750 \propto 10^{-9}$ m, the reported range for red in the visible spectrum. Write down the values of L and λ as well as the birefringence $|n_1 - n_2|$ with its uncertainty. Include the formulas that you used to calculate the uncertainties. **(1.0 points).**

Answer Form

Experimental Problem No. 2

Birefringence of mica

Task 2.1 a) Experimental setup for I_P. (0.5 points)

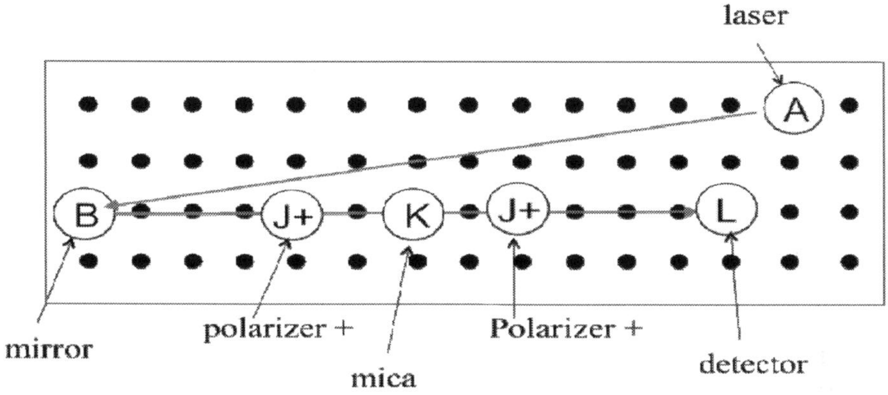

Task 2.1 b) Experimental setup for I_O. (0.5 points)

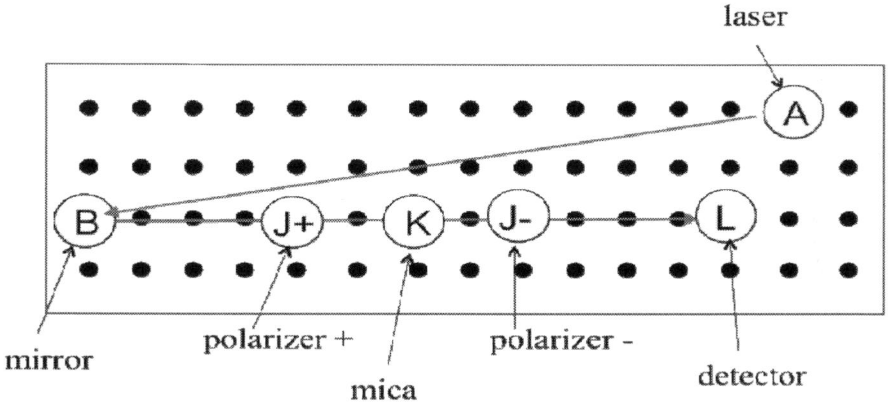

2.1		1.0

Experimental setup for measurement of mica birefringence

Task 2.2 The scale for angles.

2.2	The angle between two consecutive black lines is $\theta_{\text{int}} = 3.6$ degrees because there are 100 lines.	0.25

Tasks 2.3 Measuring I_P and I_O .Use additional sheets if necessary.

TABLE I (3 points)

$\overline{\theta}$ (degrees)	$(I_P\,''\,1)\propto 10^{-3}$ V	$(I_O\,''\,1)\propto 10^{-3}$ V
-3.6	46.4	1.1
0	48.1	0.2
3.6	47.0	0.6
7.2	46.0	2.0
10.8	42.3	4.9
14.4	38.2	9.0
18.0	33.9	12.5

21.6	27.7	17.9
25.2	23.4	22.0
28.8	17.8	27.0
32.4	12.5	31.7
36.0	8.8	34.8
39.6	5.2	38.0
43.2	3.6	39.4
46.8	3.2	39.6
50.4	4.5	38.7
54.0	6.9	36.6
57.6	10.3	33.6
61.2	14.7	29.4
64.8	20.1	24.7
68.4	25.4	19.7
72.0	30.5	14.7
75.6	36.6	10.2
79.2	40.7	6.1
82.8	44.3	3.2
86.4	46.9	1.0
90.0	47.8	0.2
93.6	47.0	0.4
97.2	45.7	2.0

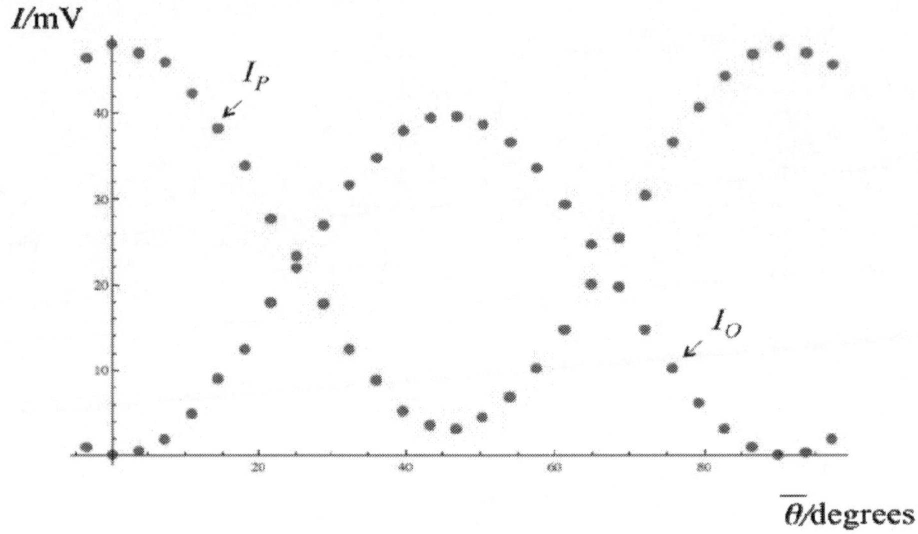

Parallel I_P and perpendicular I_O intensities vs angle $\overline{\theta}$.

GRAPH NOT REQUIRED!

Task 2.4 Finding an appropriate zero for θ.

2.4	*a) Graphical analysis* The value for the shift is $\delta\bar{\theta} = -1.0$ degrees. Add the graph paper with the analysis of this Task. *b) Numerical analysis* From Table I choose the first three points of $\bar{\theta}$ and $I_O(\bar{\theta})$: (intensities in millivolts) $(x_1, y_1) = (-3.6, 1.1) \qquad (x_2, y_2) = (0, 0.2) \qquad (x_3, y_3) = (3.6, 0.6)$ We want to fit $y = ax^2 + bx + c$. This gives three equations: $1.1 = a(3.6)^2 - b(3.6) + c$ $0.2 = c$ $0.6 = a(3.6)^2 + b(3.6) + c$ second in first $\Rightarrow \quad b = \dfrac{-0.9 + a(3.6)^2}{3.6}$ in third $\Rightarrow \quad 0.6 = a\left((3.6)^2 + (3.6)^2\right) - 0.9 + 0.2$ $\Rightarrow \quad a = 0.050 \qquad b = -0.069$ The minimum of the parabola is at: $\bar{\theta}_{\min} = -\dfrac{b}{2a}$ ..0.7 degrees Therefore, $\delta\bar{\theta} = -0.7$ degrees.	1.0

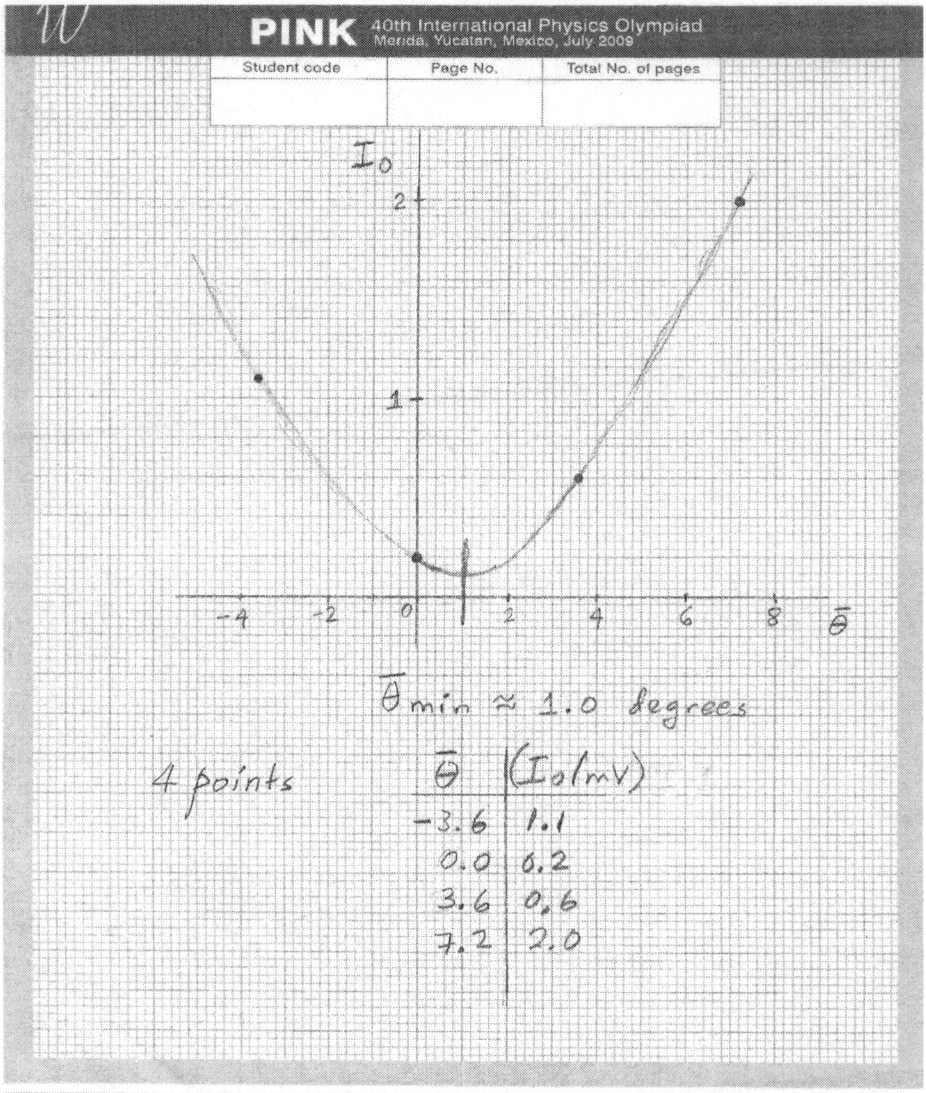

Task 2.5 Choosing the appropriate variables.

2.5	Equation (2.4) for the perpendicular intensity is $\bar{I}_O(\theta)=\frac{1}{2}(1-\cos\Delta\phi)\sin^2(2\theta)$ This can be cast as a straight line $y=mx+b$, with $y=\bar{I}_O(\theta)$, $x=\sin^2(2\theta)$ and $m=\frac{1}{2}(1-\cos\Delta\phi)$ from which the phase may be obtained. **NOTE:** This is not the only way to obtain the phase difference. One may, for instance, analyze the 4 maxima of either $\bar{I}_P(\theta)$ or $\bar{I}_O(\theta)$.	0.5

Task 2.6 Statistical analysis and the phase difference.

2.6	To perform the statistical analysis, we shall then use $y=\bar{I}_O(\theta)$ and $x=\sin^2(2\theta)$.	1.0

Since for $\theta : 0 \downarrow \frac{\pi}{4}$, $x : 0 \downarrow 1$, we use only 12 pairs of data points to cover this range, as given in Table II.

x may be left without uncertainty since it is a setting. The uncertainty in y may be calculated as

$$\Delta \bar{I}_O = \sqrt{\left(\frac{\bullet \bar{I}_O}{\bullet I_O}\right)^2 \Delta I_O^2 + \left(\frac{\bullet \bar{I}_P}{\bullet I_P}\right)^2 \Delta I_P^2}$$ and one gets

$$\Delta \bar{I}_O = \frac{\sqrt{I_O^2 + I_P^2}}{(I_O + I_P)^2} \Delta I_O \;..0.018$$, approximately the same for all values.

TABLE II

$\bar{\theta}$ (degrees)	$x = \sin^2(2\theta)$	$y = \bar{I}_O$ ″ 0.018
2.9	0.010	0.013
6.5	0.051	0.042
10.1	0.119	0.104
13.7	0.212	0.191
17.3	0.322	0.269
20.9	0.444	0.392
24.5	0.569	0.484
28.1	0.690	0.603
31.7	0.799	0.717
35.3	0.890	0.798
38.9	0.955	0.880
42.5	0.992	0.916

2.6 We now perform a least square analysis for the variables y vs x in Table II. The slope and y-intercept are: 1.75

m ″ $\Delta m = 0.913$ ″ 0.012

b ″ $\Delta b = -0.010$ ″ 0.008

The formulas for this analysis are:

$$m = \frac{N\sum_{n=1}^{N} x_n y_n - \sum_{n=1}^{N} x_n \sum_{n\leq 1}^{N} y_{n\leq}}{\Delta}$$

$$b = \frac{\sum_{n=1}^{N} x_n^2 \sum_{n\leq 1}^{N} y_{n\leq} - \sum_{n=1}^{N} x_n \sum_{n\leq 1}^{N} x_{n\leq} y_{n\leq}}{\Delta}$$

where

$$\Delta = N\sum_{n=1}^{N} x_n^2 - \left(\sum_{n=1}^{N} x_n\right)^2$$

with N the number of data points.
The uncertainty is calculated as

$$(\Delta m)^2 = N\frac{\sigma^2}{\Delta} \quad , \quad (\Delta b)^2 = \frac{\sigma^2}{\Delta}\sum_{n=1}^{N} x_n^2 \qquad \text{with,}$$

$$\sigma^2 = \frac{1}{N-2}\sum_{n=1}^{N}(y_n - b - mx_n)^2$$

with $N = 12$ in this example.

Include the accompanying plot or plots.

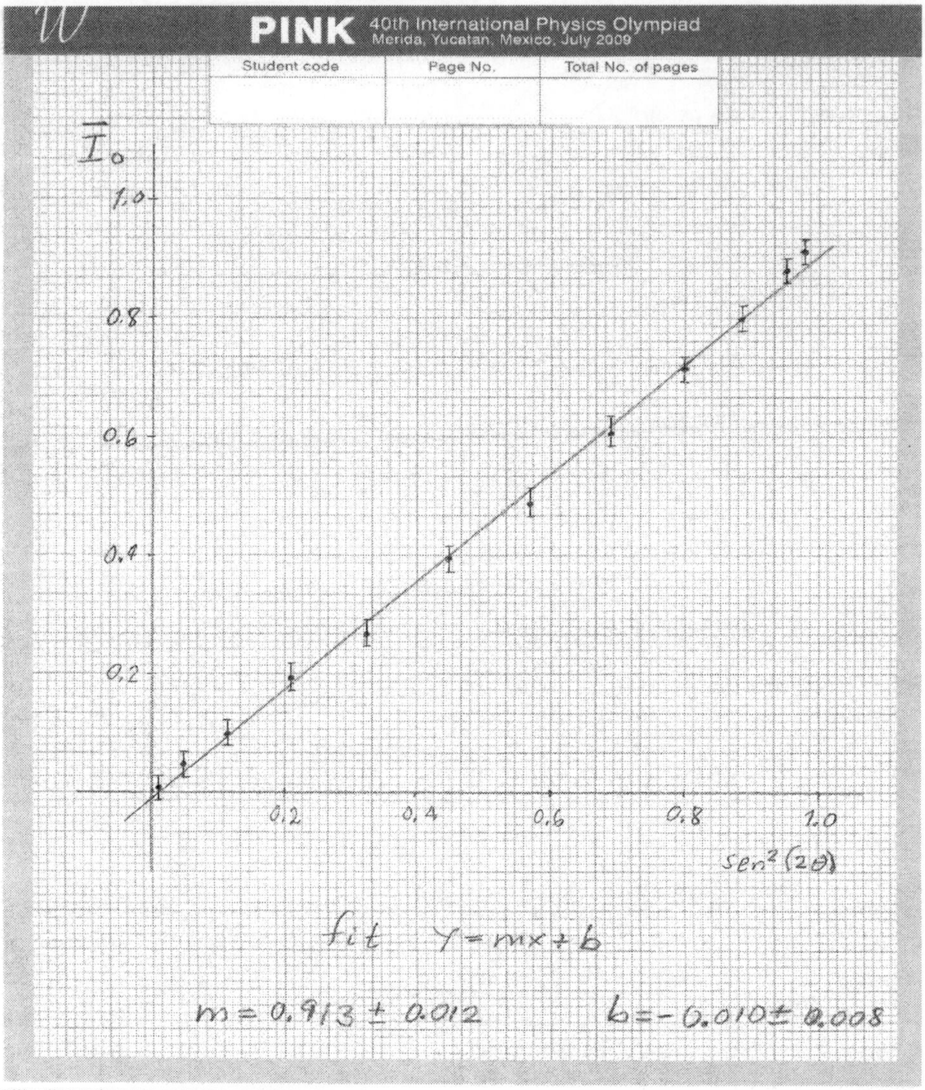

2.6	Calculate the value of the phase $\Delta\phi$ in radians in the interval $[0,\pi]$. From the slope $m = \frac{1}{2}(1-\cos\Delta\phi)$, one finds $\Delta\phi$″ $\Delta(\Delta\phi) = 2.54$″ 0.04 Write down the formulas for the calculation of the uncertainty. We see that,	0.5

<table>
<tr><td></td><td>$\Delta m = \left|\frac{\bullet m}{\bullet \Delta\phi}\right| \Delta(\Delta\phi) = \frac{1}{2}\sin(\Delta\phi)\Delta(\Delta\phi)$, therefore, $\Delta(\Delta\phi) = \frac{2\Delta m}{\sin(\Delta\phi)}$.</td><td></td></tr>
</table>

Task 2.7 Calculating the birefringence $|n_1 - n_2|$.

<table>
<tr><td>2.7</td><td>
Write down the width of the slab of mica you used,

$L'' \ \Delta L = (100'' \ 1) \propto 10^{-6}$ m

Write down the wavelength you use,

$\lambda'' \ \Delta\lambda = (663'' \ 25) \propto 10^{-9}$ m (from Problem 1)

Calculate the birefringence

$|n_1 - n_2|'' \ \Delta|n_1 - n_2| = (3.94'' \ 0.16) \propto 10^{-3}$

The birefringence is between 0.003 and 0.005. Nominal value 0.004

Write down the formulas you used for the calculation of the uncertainty of the birefringence.

Since the width $L > 82$ micrometers, we use

$2\pi - \Delta\phi = \frac{2\pi L}{\lambda}|n_1 - n_2|$

The error is

$\Delta|n_1 - n_2| = \sqrt{\left(\frac{\bullet|n_1 - n_2|}{\bullet\lambda}\right)^2 \Delta\lambda^2 + \left(\frac{\bullet|n_1 - n_2|}{\bullet L}\right)^2 \Delta L^2 + \left(\frac{\bullet|n_1 - n_2|}{\bullet\Delta\phi}\right)^2 \Delta(\Delta\phi)^2}$

$\Delta|n_1 - n_2| = \sqrt{\left(\frac{|n_1 - n_2|}{\lambda}\right)^2 \Delta\lambda^2 + \left(\frac{|n_1 - n_2|}{L}\right)^2 \Delta L^2 + \left(\frac{\lambda}{2\pi L}\right)^2 \Delta(\Delta\phi)^2}$
</td><td>1.0</td></tr>
</table>

Since the data may appear somewhat disperse and/or the errors in the intensities may be large, a graphical analysis may be performed.

In the accompanying plot, it is exemplified a simple graphical analysis: first the main slope is found, then, using the largest deviations one can find two extreme slopes.

The final result is,

$m = 0.91 \pm 0.08$ and $b = -0.01 \pm 0.04$

The calculation of the birefringence and its uncertainty follows as before. One now finds,

$$|n_1 - n_2| \pm \Delta|n_1 - n_2| = (3.94 \pm 0.45) \times 10^{-3}.$$

A larger (more realistic) error.

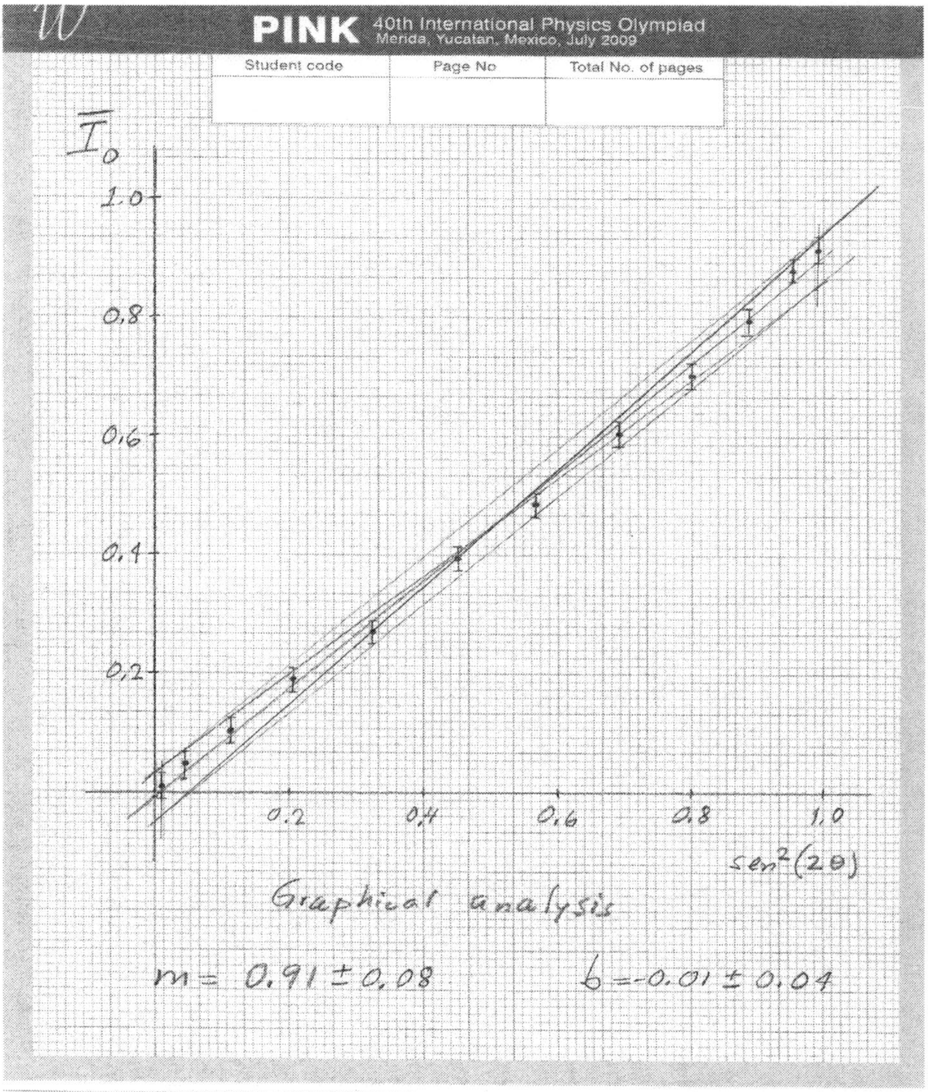
PINK 40th International Physics Olympiad
Merida, Yucatan, Mexico, July 2009
Student code
Page No
Total No. of pages
$\bar{I}_0$
1.0
0.8
0.6
0.4
0.2
0.2
0.4
0.6
0.8
1.0
sen²(2θ)
Graphical analysis
m = 0.91 ± 0.08
b = -0.01 ± 0.04

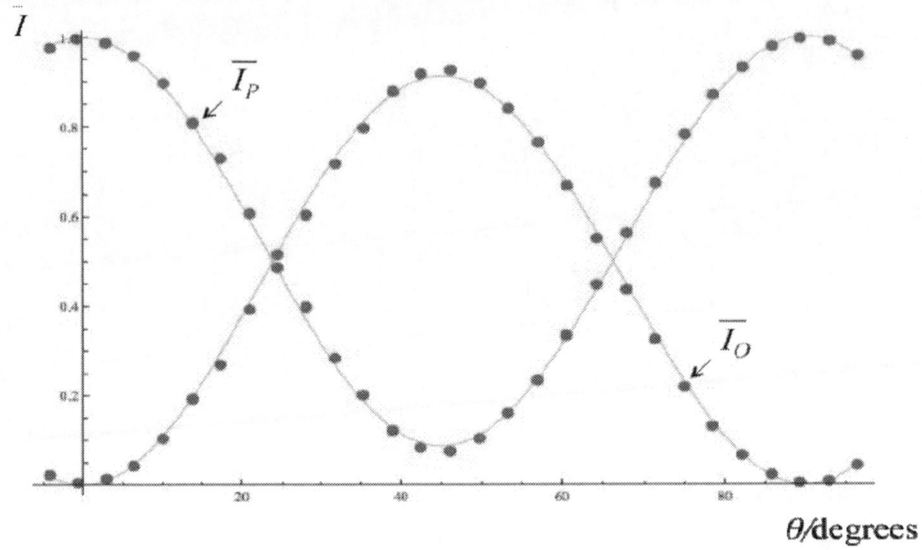

Comparison of experimental data (normalized intensities $\bar{I}_P$ and $\bar{I}_O$) with fitting (equations (2.3) and (2.4)) using the calculated value of the phase difference $\Delta\phi$.

GRAPH NOT REQUIRED!

4676168R00203

Printed in Great Britain
by Amazon.co.uk, Ltd.,
Marston Gate.